T0321787

Quantum Computation

This book presents the mathematics of quantum computation. The purpose is to *introduce* the topic of quantum computing to students in computer science, physics and mathematics, who have no prior knowledge of this field.

The book is written in two parts. The primary mathematical topics required for an initial understanding of quantum computation are dealt with in Part I: sets, functions, complex numbers and other relevant mathematical structures from linear and abstract algebra. Topics are illustrated with examples focussing on the quantum computational aspects which will follow in more detail in Part II.

Part II discusses quantum information, quantum measurement and quantum algorithms. These topics provide foundations upon which more advanced topics may be approached with confidence.

Features:

- A more accessible approach than most competitor texts, which move into advanced, research-level topics too quickly for today's students.
- Part I is comprehensive in providing all necessary mathematical underpinning, particularly for those who need more opportunity to develop their mathematical competence.
- More confident students may move directly to Part II and dip back into Part I as a reference.
- Ideal for use as an introductory text for courses in quantum computing.
- Fully worked examples illustrate the application of mathematical techniques.
- Exercises throughout develop concepts and enhance understanding.
- End-of-chapter exercises offer more practice in developing a secure foundation.

Helmut Bez holds a doctorate in quantum mechanics from Oxford University. He is a visiting fellow in quantum computation in the Department of Computer Science at Loughborough University, England. He has authored around 50 refereed papers in international journals and a further 50 papers in refereed conference proceedings. He has 35 years' teaching experience in computer science, latterly as a reader in geometric computation, at Loughborough University. He has supervised/co-supervised 18 doctoral students.

Tony Croft was the founding director of the Mathematics Education Centre at Loughborough University, one of the largest groups of mathematics education researchers in the UK, with an international reputation for the research into and practice of the learning and teaching of mathematics. He is co-author of several university-level textbooks, has co-authored numerous academic papers and edited academic volumes. He jointly won the IMA Gold Medal 2016 for outstanding contribution to the improvement of the teaching of mathematics and is a UK National Teaching Fellow. He is currently emeritus professor of mathematics education at Loughborough University (https://www.lboro.ac.uk/departments/mec/staff/academic-visitors/tony-croft/).

Advances in Applied Mathematics

Series Editor: Daniel Zwillinger

Introduction to Quantum Control and Dynamics
Domenico D'Alessandro

Handbook of Radar Signal Analysis
Bassem R. Mahafza, Scott C. Winton, Atef Z. Elsherbeni

Separation of Variables and Exact Solutions to Nonlinear PDEs
Andrei D. Polyanin, Alexei I. Zhurov

Boundary Value Problems on Time Scales, Volume I
Svetlin Georgiev, Khaled Zennir

Boundary Value Problems on Time Scales, Volume II
Svetlin Georgiev, Khaled Zennir

Observability and Mathematics
Fluid Mechanics, Solutions of Navier-Stokes Equations, and Modeling
Boris Khots

Handbook of Differential Equations, Fourth Edition
Daniel Zwillinger, Vladimir Dobrushkin

Experimental Statistics and Data Analysis for Mechanical and Aerospace Engineers
James Middleton

Advanced Engineering Mathematics with MATLAB®, Fifth Edition
Dean G. Duffy

Handbook of Fractional Calculus for Engineering and Science
Harendra Singh, H. M. Srivastava, Juan J. Nieto

Advanced Engineering Mathematics
A Second Course with MATLAB®
Dean G. Duffy

Quantum Computation
Helmut Bez and Tony Croft

Computational Mathematics
An Introduction to Numerical Analysis and Scientific Computing with Python
Dimitrios Mitsotakis

https://www.routledge.com/Advances-in-Applied-Mathematics/book-series/CRCADVAPPMTH?pd=publis hed,forthcoming&pg=1&pp=12&so=pub&view=list

Quantum Computation

Helmut Bez and Tony Croft

CRC Press
Taylor & Francis Group
Boca Raton London New York

CRC Press is an imprint of the
Taylor & Francis Group, an **informa** business

A CHAPMAN & HALL BOOK

First edition published 2023
by CRC Press
6000 Broken Sound Parkway NW, Suite 300, Boca Raton, FL 33487-2742

and by CRC Press
4 Park Square, Milton Park, Abingdon, Oxon, OX14 4RN

CRC Press is an imprint of Taylor & Francis Group, LLC

© 2023 Helmut Bez and Tony Croft

Library of Congress Cataloging-in-Publication Data

Names: Bez, H. E., author. | Croft, Tony, 1957- author.
Title: Quantum computation/Helmut Bez and Tony Croft.
Description: First edition. | Boca Raton : CRC Press, [2023] | Series:
  Advances in applied mathematics | Includes bibliographical references
  and index.
Identifiers: LCCN 2022039303 (print) | LCCN 2022039304 (ebook) | ISBN
  9781032206486 (hbk) | ISBN 9781032206493 (pbk) | ISBN 9781003264569
  (ebk)
Subjects: LCSH: Quantum computing.
Classification: LCC QA76.889 .B49 2023 (print) | LCC QA76.889 (ebook) |
  DDC 004.1--dc23/eng/20221104
LC record available at https://lccn.loc.gov/2022039303
LC ebook record available at https://lccn.loc.gov/2022039304

ISBN: 978-1-032-20648-6 (hbk)
ISBN: 978-1-032-20649-3 (pbk)
ISBN: 978-1-003-26456-9 (ebk)

DOI: 10.1201/9781003264569

Typeset in LM Roman
by KnowledgeWorks Global Ltd.

Publisher's note: This book has been prepared from camera-ready copy provided by the authors.

*For the many happy and productive years we had together,
Helmut Bez wishes to dedicate his contribution to this book to
his late wife Carys.*

*For countless hours of patience whilst this book has been written,
Tony Croft wishes to dedicate his contribution to his wife Kate.*

Contents

Preface

In this book the elements of quantum computation are presented by comparison, where possible, with the digital case. The path from the introduction of the axioms of quantum computation to a discussion of algorithms that demonstrate its potential is kept short – the intention being to move the reader, having no previous knowledge of quantum computation, rapidly to a point where more advanced aspects may be investigated.

Quantum computation has evolved from the foundations of digital computer science and quantum mechanics and, whilst it is possible to 'understand' quantum computation without physics, the approach does require a number of 'leaps of faith' by the learner – and it depends upon what one means by 'understand'. Here an attempt is made to remove some of the mystery from the processes of quantum computation by considering the elementary quantum physics on which it is based. In particular, we

1. introduce the Stern-Gerlach experiment to provide support for the axioms of quantum computation, and

2. present the measurement process in terms of the von Neumann-Lüders postulate of quantum mechanics, rooting measurement in quantum computation directly to a fundamental axiom of quantum physics.

The target readership of this book comprises those wishing to acquaint themselves with quantum computation and have some mathematical skills and some understanding of digital computation.

There are, currently, two main approaches to quantum computation, specifically:

1. quantum-gate computing,

2. adiabatic quantum computing – or quantum annealing.

Quantum-gate computing is easier to relate to the digital case, and many of the world's leading computer companies are pursuing this approach. For these reasons, this text relates to the gate model of quantum computation.

There is little that is new in this book; it comprises the work of others compiled to assist the intended readership, i.e., advanced undergraduate and masters students in mathematics or digital computation or physics, in acquiring a sufficient understanding of quantum computation to enable more advanced study of the topic to be undertaken.

Readers with a strong background in mathematics may wish to start their reading at Part II, using Part I as reference material. Those knowledgeable in quantum mechanics may wish to skip the first chapter of Part II and the material on quantum measurement. In addition Chapter 18 of Part II, relating to junk removal from general circuits, may be omitted on first reading.

Part I is intended to provide a detailed account of all the mathematical tools necessary to understand Part II and enable the reader to progress further with their studies. Successful students need to become very fluent in the manipulation of sets and functions, topics described in Chapters 1 and 2. A vector space representation is required to encapsulate the properties of quantum mechanical systems and the generally accepted choice for the number field over which the vector space is defined is the set of complex numbers. Thus Chapter 3 provides an extensive introduction to this topic and the essential results required later in the book. Thereafter, Part I progresses through topics in linear algebra, including details of vector and matrix algebra, vector spaces, inner products, linear operators, eigenvalues and eigenvectors and tensor product spaces. Quantum computation requires extensive use of a wide range of mathematical notation and to assist the reader navigate through this, a comprehensive list of symbols is provided at the beginning of the book. Throughout, an abundance of worked examples and exercises (many with solutions at the back of the book) provide opportunities for the reader to practice and develop the required fluency. Four Appendices provide supplementary information, for reference where necessary, on probability, trigonometric functions and identities, coordinate system and field axioms. Armed with all the mathematical tools provided in Part I, the reader will be well-equipped to tackle topics in quantum measurement, quantum information processing and the algorithms designed to demonstrate the potential power of quantum computation in Part II.

Finally, we hope you learn a great deal from this book, and come to share our enthusiasm for quantum computation.

Helmut Bez
Tony Croft
December 2022

Acknowledgements

The authors gratefully acknowledge the most helpful comments and suggestions received from the following colleagues who commented on various drafts of chapters in the book: Dr Francis Duah, Dr James Flint, Professor Michael Grove, Dr Glynis Hoenselaers, Professor Matthew Inglis, Professor Barbara Jaworski, Dr Ciarán Mac an Bhaird.

Symbols

Symbol Description

Sets and numbers

$\in$	is an element of set
$\lvert A\rvert$	cardinality of set A
$\cap$	set intersection
$\cup$	set union
$\overline{A}$	is the complement of set A
$\subset$	is a proper subset of
$\subseteq$	is a subset of
$\supset$	is a proper superset of
$\supseteq$	is a superset of
$\mathbb{E}$	universal set
$\emptyset$	empty set
$\mathbb{R}$	set of real numbers
$\mathbb{R}^+$	set of positive real numbers
$[0,p)$	interval $0 \le x < p$, for $x \in \mathbb{R}$
$\mathbb{N}$	set of natural numbers $0,1,2,\ldots$
$\mathbb{Z}$	set of integers $\ldots,-2,-1,0,1,2,\ldots$
$\mathbb{Q}$	set of rational numbers $\frac{p}{q}$ with $p,q \in \mathbb{Z}, q \neq 0$
$\mathbb{C}$	set of complex numbers
$A \times B$	Cartesian product of sets A and B
$\mathbb{R}^2$	Cartesian product $\mathbb{R} \times \mathbb{R}$
$\mathbb{R}^n$	Cartesian product $\mathbb{R} \times \mathbb{R} \times \ldots \times \mathbb{R}$
$\mathbb{C}^2$	Cartesian product $\mathbb{C} \times \mathbb{C}$
$\mathbb{C}^n$	Cartesian product $\mathbb{C} \times \mathbb{C} \times \ldots \times \mathbb{C}$
$A\backslash B$	the set A with all elements of set B removed
$\mathbb{R}\backslash 0$	the set $\mathbb{R}$ with 0 removed
$\mathbb{B}$	the set of binary digits $\{0,1\}$
$\mathbb{B}^k$	the Cartesian product $\mathbb{B}\times\mathbb{B}\times\ldots\times\mathbb{B}$
$\{0,1\}^k$	the Cartesian product $\mathbb{B}\times\mathbb{B}\times\ldots\times\mathbb{B}$
0^n	an n-tuple of zeros, $(0,0,0,\ldots,0)$, also $(000\ldots0)$
$\sum_{i=1}^n x_i$	the sum $x_1 + x_2 + \ldots + x_n$
δ_{ij}	Kronecker delta $= \begin{cases} 1 & i=j \\ 0 & i \neq j \end{cases}$
e	exponential constant, $\mathrm{e} = 2.718\ldots$
$\equiv$	is identically equal to
$\sim$	is equivalent to

Functions

$f : A \to B$	function mapping set A to set B
$(f \circ g)(x)$	composition $f(g(x))$
$\mathcal{F}(A,B)$	set of all functions from A to B

Complex numbers

$\mathrm{i}^2 = -1$	i the imaginary unit
$a + \mathrm{i}b$	Cartesian form of z
$z = r\angle\theta$	polar form $r(\cos\theta + \mathrm{i}\sin\theta)$
$z = r\mathrm{e}^{\mathrm{i}\theta}$	exponential form
z^*	complex conjugate of z
$\lvert z\rvert$	modulus of complex number z
$\mathrm{Re}(z)$	real part of z
$\mathrm{Im}(z)$	imaginary part of z
$\arg(z)$	argument of z

Vectors

$\overrightarrow{OP}$	position vector of point P
$\{i,j\}$	standard orthonormal basis in $\mathbb{R}^2$
$\{i,j,k\}$	standard orthonormal basis in $\mathbb{R}^3$
$\lVert u\rVert$	norm, modulus, magnitude of vector u
$\lvert\ \rangle$	ket
$\langle\ \rvert$	bra
$\lvert 0\rangle$	ket $0 = \begin{pmatrix} 1 \\ 0 \end{pmatrix}$
$\lvert 1\rangle$	ket $1 = \begin{pmatrix} 0 \\ 1 \end{pmatrix}$
$u \cdot v$	scalar (dot) product of vectors u and v
$\langle u,v\rangle$	inner product of vectors u and v
$\langle u\vert v\rangle$	inner product of vectors u and v
$u \times v$	vector product of vectors u and v
$u \otimes v$	tensor product of vectors u and v
$\hat{n}$	unit vector in the direction of n

Matrices

$I_{n \times n}$	$n \times n$ identity matrix
I_n	$n \times n$ identity matrix, $I_{n \times n}$
A^T	transpose of matrix A
A^{-1}	inverse of matrix A
$A^\dagger$	conjugate transpose of matrix A
$|A|$	$\det A$ = determinant of matrix A
diag{ }	diagonal matrix

Tensors

$u \otimes v$	tensor product of vectors u and v
$|\psi\rangle \otimes |\phi\rangle$	tensor product of tensors $|\psi\rangle$ and $|\phi\rangle$
$|\psi\rangle|\phi\rangle$	alternative to $|\psi\rangle \otimes |\phi\rangle$
$|\psi \ \phi\rangle$	alternative to $|\psi\rangle \otimes |\phi\rangle$
$\mathcal{A} \otimes \mathcal{B}$	tensor product of linear operators $\mathcal{A}$ and $\mathcal{B}$
$\bigotimes_{i=1}^{n} \mathcal{A}$	n-fold tensor product of the linear operator $\mathcal{A}$
$B_{\mathbb{C}^2}$	$\{|0\rangle, |1\rangle\}$ basis states of $\mathbb{C}^2$
$B_{\otimes^2 \mathbb{C}^2}$	$\{|00\rangle, |01\rangle, |10\rangle, |11\rangle\}$ basis states of $\mathbb{C}^2 \otimes \mathbb{C}^2 = \otimes^2 \mathbb{C}^2$
$B_{\mathbb{C}^2 \otimes \mathbb{C}^2}$	same as $B_{\otimes^2 \mathbb{C}^2}$

Boolean algebra

$\wedge$	Boolean operator *and*
$\vee$	Boolean operator *or*
$\oplus$	Boolean operator 'exclusive-or' (*xor*)
$\overline{x}$	*not x* when $x \in \mathbb{B}$

Counting and probability

$n!$	factorial n $= n \times (n-1) \times \cdots \times 3 \times 2 \times 1$
^nP_r	the number of permutations of r objects chosen from n distinct objects
^nC_r	the number of combinations of r objects chosen from n distinct objects
$\binom{n}{r}$	same as ^nC_r

Group theory

$(G, \circ)$	The set G with binary operation $\circ$ satisfying the group axioms
$H \leq G$	H is a subgroup of G
$H \triangleleft G$	H is a normal subgroup of G
G/H	The quotient or factor group
$(\mathbb{Z}/n\mathbb{Z}, +)$	The additive group of integers modulo n
$(\mathbb{Z}_n, +)$	(alternative: The additive group of integers modulo n)
$(\mathbb{Z}/n\mathbb{Z})^\times$	The multiplicative group of integers modulo n
$U(2)$	unitary group on $\mathbb{C}^2$

Abbreviations

iff	if and only if
wrt	with respect to

Part I

Mathematical Foundations for Quantum Computation

1

Mathematical preliminaries

1.1 Objectives

Quantum computation is multidisciplinary drawing as it does from computer science, physics and mathematics. Even when considering solely mathematical aspects of quantum computation, the student will need to use a diverse range of tools and techniques from seemingly different branches of mathematics, a characteristic which makes the subject particularly demanding for the novice. An extensive mathematical toolkit is therefore required for a thorough study of the subject. This toolkit is built-up gradually throughout Part I of this book. However, right from the start, there is a need to have some exposure to the vocabulary of mathematical structures that will follow. The primary objective of this chapter is to provide such a vocabulary in a less than formal way, with more formal treatments being detailed in subsequent chapters. A second objective is to describe and give practice using some preliminary techniques, such as modular arithmetic, working with Boolean connectives, and equivalence relations that will be important as you move through the book.

We shall first consider 'sets'. These are fundamental mathematical structures used as building blocks to construct more advanced and useful mathematical entities. Essentially, a set is a collection of objects called elements. Often these elements will be numbers. As we progress through the book, we shall introduce different types of set elements such as functions, vectors and matrices. Once we have defined sets of interest, they can be endowed with additional properties and become yet more advanced mathematical structures, for example groups, vector spaces and more. All of these objects are important in the study of quantum computation. This chapter introduces terminology associated with sets. It describes several ways in which we can perform calculations, or algebra, on sets and details laws which set algebra must obey. It also provides an informal introduction to some of the important mathematical structures that will be required later. It then goes on to discuss number systems and combinatorics. We shall see that a thorough knowledge of the binary number system, which uses just the digits 0 and 1, and the various ways in which these numbers can be manipulated is essential for an understanding of quantum computation. Combinatorics is that branch of mathematics that deals with counting and forms the basis of much of probability theory. These topics will be drawn on throughout the book.

Finally, the reader's attention is drawn to several Appendices which review some basic aspects of probability, essential trigonometric identities, Cartesian, plane polar and spherical polar coordinate systems, and the field axioms.

DOI: 10.1201/9781003264569-1

1.2   Definitions and notation

Definition 1.1 Set

   A **set** *is any collection of objects, states, letters, numbers, etc.*

One way of describing a set is to name it and then list its members, or **elements**, enclosed in braces { }. For example, we can define a set labelled A by

$$A = \{a, b, c\}.$$

The set A has three elements.

   A set S, say, which has a finite number of elements is called a **finite set**. The number of elements in any finite set is called its **cardinality**, written $|S|$.

   The set of **integers**, written $\mathbb{Z}$, which has an infinite number of elements, is given by

$$\mathbb{Z} = \{\ldots, -3, -2, -1, 0, 1, 2, 3, \ldots\}.$$

Another way of describing a set is to use descriptive words, for example '$\mathbb{N}$ is the set of counting numbers, $0, 1, 2, \ldots$'. We use the symbols $\in$ and $\notin$ to mean 'is an element of', or 'is not an element of', respectively, so for example $3 \in \mathbb{Z}$, $\frac{2}{3} \notin \mathbb{Z}$.

   A third way of describing a set is to define a rule by which all members of a set can be found. Consider the following notation:

$$A = \{n : n \in \mathbb{Z} \text{ and } n \geq 10\}.$$

This is read as 'A is the set of values n such that n is in the set $\mathbb{Z}$ (i.e., the set of integers), and which are greater than or equal to 10'. Thus $A = \{10, 11, 12, 13, \ldots\}$. Some sets of numbers that are of particular interest in the study of quantum computation are

$$\mathbb{B} = \{0, 1\} \qquad \text{the set of binary digits}$$

$$\mathbb{N} = \text{ the set of natural numbers, } 0, 1, 2, \ldots$$

$$\mathbb{Z} = \text{ the set of integers, } \ldots, -3, -2, -1, 0, 1, 2, 3, \ldots$$

$$\mathbb{Q} = \text{ the set of rational numbers, } \frac{p}{q}, \text{ where } p, q \in \mathbb{Z}, q \neq 0$$

$$\mathbb{R} = \text{ the set of all real numbers}$$

$$\mathbb{R}^+ = \text{ the set of all positive real numbers}$$

$$\mathbb{C} = \text{ the set of all complex numbers.}$$

The set of real numbers $\mathbb{R}$ contains integers, rational numbers and irrational numbers (numbers that are not rational, such as π, $\sqrt{2}$). Most of these numbers are familiar from school days. However, quantum computation requires a much larger set of numbers than is provided by $\mathbb{R}$ alone. This necessitates the introduction of the set $\mathbb{C}$ of so-called **complex numbers**. These are numbers which have both real and imaginary components. We study these in detail in Chapter 3. Complex numbers are required to describe the state of a quantum system.

   The symbol $\backslash$ is used to mean 'remove from the set' so that, for example, $\mathbb{R}\backslash 0$ means all real numbers except the value 0.

The set of all elements of interest at any particular time is known as the **universal set**, written herein as $\mathbb{E}$. A set which has no elements is called an **empty set**, written $\emptyset$ or $\{\ \}$.

The set of real numbers $\mathbb{R}$ together with the operations of addition and multiplication is an algebraic structure known as a **field**, a term which will be used repeatedly throughout the book. Loosely speaking this means that the real numbers obey rules, known as **field axioms** (see Appendix D), that enable us to add, subtract, multiply and divide by any non-zero number, in the manner in which we are already very familiar. Similarly, the sets of rational numbers, $\mathbb{Q}$, and complex numbers, $\mathbb{C}$, each form a field. Technicalities of a field need not concern us further, and whenever we refer to a field generally, it may be helpful for you to have the fields $\mathbb{R}$ or $\mathbb{C}$ in mind.

A rule by which we can take two elements of a set and combine them to produce another element in the same set is called a **binary operation**. Familiar binary operations are addition and multiplication in the set of real numbers, but there are many more, as we shall see.

We can perform calculations on whole sets using the operations intersection, union, exclusive-or and complement which are defined as follows:

Definition 1.2 Intersection

$$A \cap B = \{x : x \in A \text{ and } x \in B\}.$$

Thus $x \in A \cap B$ if x is in both A and B. If the intersection of A and B is the empty set, i.e., $A \cap B = \emptyset$, then A and B have no elements in common and are said to be **disjoint**.

Definition 1.3 Union

$$A \cup B = \{x : x \in A \text{ or } x \in B \text{ or in both}\}.$$

Thus $x \in A \cup B$ if x is in A or B or in both of A and B. Technically this operation is called an **inclusive-or** because it includes the possibility that x is in both sets.

Definition 1.4 Exclusive-or (*xor*)

$$A \oplus B = \{x : x \in A \text{ or } x \in B \text{ but not in both}\}.$$

Thus $x \in A \oplus B$ if x is in just one of A or B.

Definition 1.5 Complement

$$\overline{A} = \{x : x \in \mathbb{E} \text{ but } x \notin A\}.$$

Thus the complement of A comprises all those elements of the universal set that are not in A.

Two sets are said to be **equal** if they have precisely the same elements. If all the elements of a set A are also elements of a set B, and provided A and B are not the same, we say that A is a **proper subset** (or simply a **subset**) of B, written $A \subset B$. If we wish to allow the possibility that A and B may have precisely the same elements, we write $A \subseteq B$. Likewise, if the set A includes all elements of B, and provided $A \neq B$, then we say A is a **proper superset** of B, written $A \supset B$. Allowing for the possibility of A and B to be the same, we can write $A \supseteq B$. By convention, the empty set $\emptyset$ is a subset of any set A, that is $\emptyset \subset A$. If a set is non-empty, it is regarded as a subset of itself, i.e., $A \subseteq A$.

Exercises

1.1 Write down all the subsets of $\mathbb{B} = \{0, 1\}$.

1.2 The set of all subsets of a given set A is called the **power set** of A. How many elements are in the power set of $\mathbb{B} = \{0, 1\}$?

1.3 A set has three elements. How many elements are in its power set?

1.4 The cardinality of a set is n. How many elements are in its power set?

1.5 Explain why the set of even integers is a subset of $\mathbb{Z}$. Is the set of odd integers a subset of $\mathbb{Z}$?

1.3 Venn diagrams

Venn diagrams provide a graphical way of picturing sets. The sets are usually drawn as circles from which various properties can be deduced. Figure 1.1 shows sets A and B, their intersection $A \cap B$, their union $A \cup B$ and the complement of A.

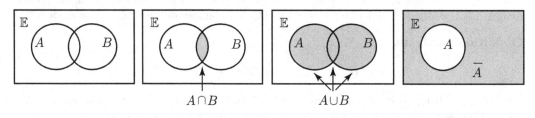

FIGURE 1.1
Venn diagrams used to depict sets and set operations: intersection, union and complement.

1.4 Laws of set algebra

The operations $\cap$, $\cup$ and complement can be used to define new sets. It is possible to show that the following laws hold when manipulating such operations (Table 1.1). In many cases these laws are obvious from an inspection of an appropriate Venn diagram. The laws of Boolean algebra, which we state shortly, are analogous to these. The further laws in Table 1.2 can be derived from those in Table 1.1.

1.5 Boolean algebra

Boolean algebra is concerned with the manipulation of **Boolean variables**. These are variables that can take only one of two values. Examples include TRUE or FALSE, and UP

TABLE 1.1

The laws of set algebra

$A \cup B = B \cup A$ $A \cap B = B \cap A$	commutative laws
$A \cup (B \cup C) = (A \cup B) \cup C$ $A \cap (B \cap C) = (A \cap B) \cap C$	associative laws
$A \cap (B \cup C) = (A \cap B) \cup (A \cap C)$ $A \cup (B \cap C) = (A \cup B) \cap (A \cup C)$	distributive laws
$A \cup \emptyset = A$ $A \cap \mathbb{E} = A$	identity laws
$A \cup \overline{A} = \mathbb{E}$ $A \cap \overline{A} = \emptyset$ $\overline{\overline{A}} = A$	complement laws

TABLE 1.2

Laws derivable from Table 1.1

$A \cup (A \cap B) = A$ $A \cap (A \cup B) = A$	absorption law
$(A \cap B) \cup (A \cap \overline{B}) = A$ $(A \cup B) \cap (A \cup \overline{B}) = A$	minimisation laws
$\overline{A \cup B} = \overline{A} \cap \overline{B}$ $\overline{A \cap B} = \overline{A} \cup \overline{B}$	De Morgan's laws

or DOWN, and ON or OFF. If A is a binary digit, that is it can take the values 0 or 1, then A is a Boolean variable. Given two or more Boolean variables, A, B, C, ... we can combine them to produce Boolean expressions using binary operations called **logical connectives**. These are defined in Chapter 2 when we explain Boolean logic gates. For now it suffices to say that the logical connectives *and* ($\wedge$), *or* ($\vee$) and *not* ($\bar{\ }$) can be manipulated with laws directly analogous to the laws of set algebra if we interpret $\vee$ as $\cup$, $\wedge$ as $\cap$, 1 as the universal set and 0 as the empty set. Thus we have the laws in Table 1.3. Further laws given in Table 1.4 can be derived from those in Table 1.3. In Section 1.5.1 we discuss another logical connective, the 'exclusive-or', denoted $\oplus$.

TABLE 1.3

The laws of Boolean algebra

$A \vee B = B \vee A$ $A \wedge B = B \wedge A$	commutative laws
$A \vee (B \vee C) = (A \vee B) \vee C$ $A \wedge (B \wedge C) = (A \wedge B) \wedge C$	associative laws
$A \wedge (B \vee C) = (A \wedge B) \vee (A \wedge C)$ $A \vee (B \wedge C) = (A \vee B) \wedge (A \vee C)$	distributive laws
$A \vee 0 = A$ $A \wedge 1 = A$	identity laws
$A \vee \overline{A} = 1$ $A \wedge \overline{A} = 0$ $\overline{\overline{A}} = A$	complement laws

TABLE 1.4

Laws derivable from Table 1.3

$A \vee (A \wedge B) = A$	absorption law
$A \wedge (A \vee B) = A$	
$(A \wedge B) \vee (A \wedge \overline{B}) = A$	minimisation laws
$(A \vee B) \wedge (A \vee \overline{B}) = A$	
$\overline{A \vee B} = \overline{A} \wedge \overline{B}$	De Morgan's laws
$\overline{A \wedge B} = \overline{A} \vee \overline{B}$	
$A \vee 1 = 1$	
$A \wedge 0 = 0$	

1.5.1 The exclusive-or operator (xor)

We now focus on the **exclusive-or** operator, written $\oplus$, which is applied to a pair of Boolean variables. This binary operation is immensely important in defining Boolean functions and in all quantum algorithms. Thus, familiarity with its manipulation is essential.

Definition 1.6 Exclusive-or operator $\oplus$

The exclusive-or operator, written $\oplus$ and also referred to as xor, is defined on the Boolean variables x and y so that $x \oplus y = 1$ whenever x or y (but not both) equals 1, and is zero otherwise (Table 1.5).

TABLE 1.5

Definition of $\oplus$

$0 \oplus 0$	$=$	0
$0 \oplus 1$	$=$	1
$1 \oplus 0$	$=$	1
$1 \oplus 1$	$=$	0

Example 1.5.1

Show that if $x \in \mathbb{B}$ then:

   (a)   if $x = 0$, $1 \oplus x = 1$.

   (b)   if $x = 1$, $1 \oplus x = 0$.

Solution

(a) if $x = 0$, we see directly from Table 1.5 that $1 \oplus x = 1 \oplus 0 = 1$.

(b) if $x = 1$, then $1 \oplus x = 1 \oplus 1 = 0$, as required. Expressions of the form $1 \oplus x$ will occur frequently in the quantum algorithms in Chapter 22. It is important to note that whatever the value of $x \in \mathbb{B}$, $1 \oplus x$ gives the other value.

Example 1.5.2

Show that if $x \in \mathbb{B}$ then $x \oplus x = 0$.

Solution

Given that $x \in \mathbb{B} = \{0, 1\}$ then either $x = 0$ or $x = 1$.

Suppose $x = 0$. Then $x \oplus x = 0 \oplus 0 = 0$.

Suppose $x = 1$. Then $x \oplus x = 1 \oplus 1 = 0$.

Thus, whatever x, $x \oplus x = 0$.

Results such as this will be used repeatedly when simplifying Boolean expressions used in quantum gates. For example, it follows immediately that, for $x_1, x_2 \in \mathbb{B}$

$$(x_1 \wedge x_2) \oplus (x_1 \wedge x_2) = 0.$$

Exercises

1.6 Let $x \in \mathbb{B} = \{0, 1\}$. Let $s \in \mathbb{B} = \{0, 1\}$.

(a) Show that $x \oplus x = 0$.

(b) Show that $x \oplus 0 = x$.

(c) Show that $x \oplus s \oplus s = x$.

(d) Show that $x \oplus x \oplus s = s$.

1.6 Groups

The mathematical structure known as a **group** pervades quantum computation. Groups are built using a set, G say, and a binary operation, $\circ$ say, used to combine elements of that set. In practice, the binary operation will often be $+$, $\times$ or $\oplus$.

Definition 1.7 Group

*A **group** is a set G with a binary operation which assigns to each pair of elements, $a, b \in G$, another element $a \circ b$ which is also in G for which the following **group axioms** hold:*

*1. given $a, b \in G$, then $a \circ b \in G$ (**closure**)*

*2. for any $a, b, c \in G$ then $(a \circ b) \circ c = a \circ (b \circ c)$ (**associativity**)*

*3. there exists an **identity** or **unit** element $e \in G$ such that $e \circ a = a \circ e = a$ for all $a \in G$.*

*4. for any $a \in G$ there exists an element a^{-1} also in G such that $a \circ a^{-1} = a^{-1} \circ a = e$. The element a^{-1} is called the **inverse** of a.*

We will usually denote a group by $(G, \circ)$. On occasions the group operation will be replaced by juxtaposition of elements, for example writing $a \circ b$ as simply ab. This is particularly so (but is not restricted to the case) when the binary operation is multiplication. Note that a^{-1} does not mean the reciprocal of a nor a power of a. It is purely a notation for the inverse element of the group.

Many familiar sets and binary operations can be readily identified as groups. Consider the following examples.

Example 1.6.1 The group $(\mathbb{Z}, +)$

Consider the set of integers $\mathbb{Z}$ with the operation of addition, $+$. Clearly, adding any two integers results in another integer, so $\mathbb{Z}$ is closed under addition. Associativity is obvious: $p + (q + r) = (p + q) + r$ for any $p, q, r \in \mathbb{Z}$. The identity element is 0 because adding zero to any integer does not change it. Finally, the inverse of any integer p is $-p$ since $p + (-p) = 0$, and $-p \in \mathbb{Z}$. Hence the structure $(\mathbb{Z}, +)$ is a group.

Definition 1.8 Commutative or Abelian Group
If $(G, \circ)$ is a group, and additionally the operation $\circ$ is commutative, that is $a \circ b = b \circ a$ for all $a, b \in G$, then $(G, \circ)$ is said to be a **commutative group** *or an* **Abelian group**.

Clearly, because addition of integers is commutative $(\mathbb{Z}, +)$ is Abelian.

Definition 1.9 Subgroup
If $(G, \circ)$ is a group, then if a subset H of G is itself a group with the same operation it is called a **subgroup**, *written $H \leq G$.*

It is straightforward to check that the set of even integers form a subgroup of $(\mathbb{Z}, +)$.

Exercises

   1.7   Explain why the set of odd integers does not form a subgroup of $(\mathbb{Z}, +)$.

The following examples illustrate some more groups. Further group theory follows in Chapter 8.

Example 1.6.2 The permutation group P_3

Consider the set of **permutations** or **arrangements** of three objects a, b, c. These are listed:

$$abc, acb, cba, bac, cab, bca.$$

To study these further, we need to agree a convention to keep track of changes in the ordering. There are several ways of doing this, but suppose we let P_{12} indicate the operation 'interchange the first and second letters'. Then P_{12} changes abc to bac, written

$$P_{12} : abc \rightarrow bac.$$

Similarly, $P_{12} : cab \rightarrow acb$ and so on.

   Permutations can be applied repeatedly. The so-called **composition** of permutations $P_{23}P_{12}$ will be interpreted as applying P_{12} first followed by applying P_{23}. Thus

$$P_{23}P_{12} : abc \rightarrow bac \rightarrow bca.$$

Observe that the permutation bca cannot be obtained from abc by a single, simple interchange of two elements, but can be achieved introducing a cyclic rotation operation so that the first element moves to the third position, the third to the second and the second to the first, which we shall denote by P_{132} (the 132 indicating $1 \rightarrow 3 \rightarrow 2 \rightarrow 1$). Thus

$$P_{132} : abc \rightarrow bca.$$

Similarly, the operation P_{123} moves the first element to the second position, the second to the third and the third to the first:

$$P_{123} : abc \to cab.$$

The operation labelled '1' defined as $1 : abc \to abc$ shall mean leave all elements alone. Thus there are six operations required to achieve all possible permutations:

$$1, P_{12}, P_{13}, P_{23}, P_{132}, P_{123}.$$

The six operations thus defined can be applied repeatedly as in the example $P_{23}P_{12}$ above. In doing so, we shall always apply the rightmost operation first. The result of combining operations in this way can be placed in a so-called **Cayley table** (Table 1.6). When reading the table, it is important to note that the operation in the header row is applied first. Observe, for example, that $P_{23}P_{12} = P_{132}$. It is straightforward to check that the set of permutations of abc with the operation of composition as defined above satisfies all the axioms required to form a group. The identity element of the group is 1. This group is referred to as the **permutation group** P_3. Permutations of more than three objects can be used to define further permutation groups, P_4, $P_5, \ldots$ in a similar fashion.

TABLE 1.6

Cayley table for the permutation group P_3

	1	P_{12}	P_{13}	P_{23}	P_{132}	P_{123}
1	1	P_{12}	P_{13}	P_{23}	P_{132}	P_{123}
P_{12}	P_{12}	1	P_{132}	P_{123}	P_{13}	P_{23}
P_{13}	P_{13}	P_{123}	1	P_{132}	P_{23}	P_{12}
P_{23}	P_{23}	P_{132}	P_{123}	1	P_{12}	P_{13}
P_{132}	P_{132}	P_{23}	P_{12}	P_{13}	P_{123}	1
P_{123}	P_{123}	P_{13}	P_{23}	P_{12}	1	P_{132}

Exercises

1.8 Referring to Example 1.6.2 use the Cayley table to write down the inverse of each of the elements of the group. Which elements are self-inverse ?

In the following example we shall meet another group which although superficially looks quite different has a Cayley table with an identical structure to the permutation group of Example 1.6.2. This leads to the concept of **isomorphism** which is a way of comparing apparently dissimilar groups. Permutations can also be regarded as functions. This aspect is discussed in Chapter 2.

Example 1.6.3 The symmetric group S_3

Consider the various symmetry operations on the equilateral triangle shown in Figure 1.2 which result in the triangle being in the same position albeit with the vertex labels altered. So, for example the operation 'reflect the triangle in the line L_1' is denoted by S_{12}. The operation S_{123} is a clockwise rotation through $2\pi/3$. The operation S_{132} is an anti-clockwise rotation through $2\pi/3$. We can apply these symmetry operations sequentially and the results

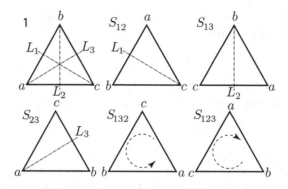

FIGURE 1.2
The symmetric group S_3: symmetries of an equilateral triangle.

of doing this are given in Table 1.7. Note when reading the table, the operation in the header row is applied first. Observe that the structure of the Cayley table for S_3 is identical to that of the permutation group P_3 in Example 1.6.2. We say that there is an **isomorphism** between P_3 and S_3, that is there is a one-to-one correspondence between the elements of the two sets, which preserves the structure of the group, i.e., P_3 and S_3 are isomorphic.

TABLE 1.7
Cayley table for the symmetric group S_3

	1	S_{12}	S_{13}	S_{23}	S_{132}	S_{123}
1	1	S_{12}	S_{13}	S_{23}	S_{132}	S_{123}
S_{12}	S_{12}	1	S_{132}	S_{123}	S_{13}	S_{23}
S_{13}	S_{13}	S_{123}	1	S_{132}	S_{23}	S_{12}
S_{23}	S_{23}	S_{132}	S_{123}	1	S_{12}	S_{13}
S_{132}	S_{132}	S_{23}	S_{12}	S_{13}	S_{123}	1
S_{123}	S_{123}	S_{13}	S_{23}	S_{12}	1	S_{132}

1.7 Cartesian product

Definition 1.10 Cartesian product
*The **Cartesian product** of two sets A and B, written $A \times B$, is the set of all ordered pairs of elements (a, b) such that $a \in A$ and $b \in B$.*

The term 'ordered pair' indicates that the order in which a and b appear is important, so that generally (a, b) is distinct from (b, a) and here a is chosen from the set A, whilst b is chosen from the set B. For example, given

$$A = \{1, 2, 3\}, \qquad B = \{d, e\}$$

then

$$A \times B = \{(1, d), (1, e), (2, d), (2, e), (3, d), (3, e)\}.$$

If A and B are finite sets with cardinalities n and m, respectively, then the cardinality of $A \times B$ is nm. Clearly the cardinality of $A \times B$ above is $3 \times 2 = 6$. The two sets involved can be the same. Consider the following example.

Example 1.7.1 The Cartesian product $\mathbb{B} \times \mathbb{B} = \mathbb{B}^2$

For the set $\mathbb{B} = \{0, 1\}$ write down the set $\mathbb{B} \times \mathbb{B}$.

Solution

The set $\mathbb{B} \times \mathbb{B}$ consists of all ordered pairs of elements (a, b) where $a \in \mathbb{B}$ and $b \in \mathbb{B}$:

$$\mathbb{B} \times \mathbb{B} = \{(0,0), (0,1), (1,0), (1,1)\}.$$

$\mathbb{B} \times \mathbb{B}$ is also written as $\mathbb{B}^2$. Observe that the cardinality of $\mathbb{B}^2$ is $2 \times 2 = 4$.

Example 1.7.2 Ordered n-tuples

We can produce the Cartesian product of more than two sets, $A_1, A_2, \ldots, A_n$, in a natural way to give an ordered n-tuple:

$$a = (a_1, a_2, \ldots, a_n)$$

where $a_i \in A_i$. The set $\mathbb{B}^n$ consists of all ordered n-tuples of the Boolean variables 0 and 1. We shall sometimes write this set as

$$\{0, 1\}^n$$

elements of which include, for example, the n-tuples

$$(0, 1, 0, 1, \ldots, 0), \quad (1, 1, 1, 1, \ldots, 0), \quad (0, 0, 0, 0, \ldots, 1).$$

Sometimes these n-tuples are written as **binary strings**, that is, strings of binary digits: $0101 \ldots 0$, $1111 \ldots 0$ and so on. An n-tuple of zeros, $(0, 0, 0, \ldots, 0)$, or $000 \ldots 0$ will also be written concisely as 0^n.

As a further example, the set $\{0, 1\}^3$ is given by

$$\{0, 1\}^3 = \{000, 001, 010, 011, 100, 101, 110, 111\}.$$

We shall see that n-tuples $\{0, 1\}^n$ arise frequently in quantum algorithms.

Example 1.7.3 The set $\mathbb{R} \times \mathbb{R} \times \mathbb{R} = \mathbb{R}^3$

The set $\mathbb{R} \times \mathbb{R} \times \mathbb{R}$, written $\mathbb{R}^3$ is the set of all ordered triples (x, y, z) of real numbers. Later we shall see that such an ordered triple can be thought of as an object called a **row vector** or **column vector** in $\mathbb{R}^3$, that is,

$$(x, y, z) \qquad \text{or} \qquad \begin{pmatrix} x \\ y \\ z \end{pmatrix}.$$

Collections of vectors like these, when they are provided with some specific rules for addition and multiplication by real numbers are called **vector spaces over the field** $\mathbb{R}$, as detailed in Chapter 6.

Example 1.7.4 The set $\mathbb{C} \times \mathbb{C} = \mathbb{C}^2$

The set $\mathbb{C} \times \mathbb{C} = \mathbb{C}^2$ is the set of all ordered pairs of complex numbers. (For those readers unfamiliar with complex numbers, details are given in Chapter 3. A complex number takes the form $a + b\mathrm{i}$, with $a, b \in \mathbb{R}$ and the imaginary unit i is defined such that $\mathrm{i}^2 = -1$.) Depending upon the context, we might choose to write elements of this set as row or column vectors. Thus the following are both elements of $\mathbb{C}^2$:

$$(3 + 4\mathrm{i}, -11 + 2\mathrm{i}), \qquad \begin{pmatrix} 2 - 5\mathrm{i} \\ 1 + 3\mathrm{i} \end{pmatrix}.$$

Collections of vectors of complex numbers, when provided with specific rules for addition and multiplication by complex numbers, are called **vector spaces over the field** $\mathbb{C}$. We shall see that such vectors are used to represent the state of a quantum system.

Example 1.7.5

A **matrix** (plural, matrices) is a mathematical object built from rows of row vectors, or columns of column vectors. So

$$\begin{pmatrix} 4 & -3 & 2 \\ 2 & 2 & -1 \end{pmatrix}, \qquad \begin{pmatrix} 1 & 0 & 0 \\ 0 & 1 & 0 \\ 0 & 0 & 1 \end{pmatrix}, \qquad \begin{pmatrix} 4 + \mathrm{i} & 7 + 2\mathrm{i} \\ 2 - 6\mathrm{i} & 1 - \mathrm{i} \end{pmatrix}$$

are examples of matrices. Sets of matrices endowed with additional rules for addition and multiplication by a real or complex number become vector spaces. These spaces are crucial for modelling evolution of quantum states with time, and for modelling the measurement of observables. Matrices are studied in detail in Chapters 5 and 7.

Exercises

   1.9   What is the cardinality of the set $\mathbb{B}^2$?

  1.10   What is the cardinality of the set $\mathbb{B}^n$?

1.8   Number bases

The **decimal number system**, known as **base 10**, uses the ten digits $0, 1, 2, \ldots, 8, 9$. However, applications in computer science and engineering make use of other number systems, particularly in bases 2, 8 and 16. Such number systems are referred to as **binary, octal** and **hexadecimal**, respectively. The binary system, with 2 as its base, uses only the two digits 0 and 1. These are called binary digits or **bits**. So a bit takes its value from the set $\mathbb{B} = \{0, 1\}$. Consider, for example, the binary number 1101_2 where the subscript 2 indicates the base. This means that as we move from the right to the left, the position of each digit represents an increasing power of 2. So,

$$\begin{aligned} 1101_2 &= 1(2^3) + 1(2^2) + 0(2^1) + 1(2^0) \\ &= 8 + 4 + 0 + 1 \\ &= 13_{10}. \end{aligned}$$

A **digital register** is a physical entity capable of representing, at any given time, a binary number. Whereas a 1 bit digital register is capable of representing just one of the numbers 0 or 1, a 2-bit register can represent $00, 01, 10$ and 11, that is any element of $\{0,1\}^2$. Likewise, an n-bit register is capable of representing any one of the 2^n binary numbers $000\cdots 0, \ldots, 111\cdots 1$ in $\{0,1\}^n$.

Example 1.8.1

When performing calculations on quantum states, we shall frequently come across objects of the form
$$|10\rangle \quad |101\rangle \quad |10011\rangle \quad \text{or} \quad |11110\rangle$$
where strings of binary digits are parenthesised by $|\ \rangle$. To avoid writing lengthy strings of zeros and ones, we sometimes express these binary strings of digits as their decimal equivalents. With this convention, you should confirm that
$$|10\rangle = |2\rangle, \quad |101\rangle = |5\rangle \quad |10011\rangle = |19\rangle \quad \text{and} \quad |11110\rangle = |30\rangle.$$

1.9 Modular arithmetic

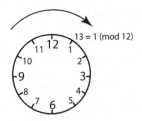

FIGURE 1.3
Counting modulo 12.

Modular arithmetic is a method of counting with integers, $\ldots, -3, -2, -1, 0, 1, 2, 3, \ldots,$ where instead of this sequence continuing forever, it 'wraps around' once a particular value, called the **modulus**, is reached. Probably the most familiar example is arithmetic modulo 12, used in an everyday clock. When counting hours beyond 12 o'clock, it is usual to start again at 1, so that 13 looks to be the same time as 1. (Figure 1.3.) We say that 13 **is congruent to** 1 modulo 12 and write this as
$$13 \equiv 1 \ (\mathrm{mod}\, 12).$$
This is equivalent to saying that
$$13 - 1 \text{ is divisible by } 12.$$
Similarly $15 \equiv 3 \ (\mathrm{mod}\, 12)$ since $15 - 3$ is divisible by 12. It is usual to regard 12, as 0, so that when working in mod 12, we use the digits $0, 1, 2, \ldots, 11$. More generally, we have the following definition:

Definition 1.11 Congruence
*We say that a is **congruent** to b modulo n, written $a \equiv b \ (mod\, n)$, if*
$$a - b \text{ is divisible by } n, \qquad \text{or} \qquad a - b = kn \text{ for some } k \in \mathbb{Z}.$$

Note that 'is congruent to' is a so-called **equivalence relation** as we shall show in Section 1.10.

Example 1.9.1 Addition modulo 2 and the exclusive-or operation $\oplus$.

We shall see that use of addition modulo 2 is commonplace in quantum computation. Working in modulo 2 we use only the digits 0 and 1, and hence there is an immediate relationship between binary arithmetic and counting modulo 2. We can visualise this in Figure 1.4. Observe that $2 \equiv 0 \,(\mathrm{mod}\,2)$ and consequently that $1 + 1 = 0 \,(\mathrm{mod}\,2)$.

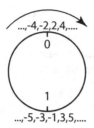

FIGURE 1.4
Counting modulo 2.

We can represent the results of addition modulo 2 in Table 1.8. The $+ \,(\mathrm{mod}\,2)$ operation is particularly important in quantum computation.

TABLE 1.8
Addition modulo 2

$+$ or $\oplus$	0	1
0	0	1
1	1	0

It is equivalent to the **exclusive-or** operation, denoted $\oplus$, on Boolean variables. Recall that the exclusive-or takes the value 1 when either, but not both, of the operands equals 1 and is 0 otherwise. Thus

$$\text{if } a, b \in \mathbb{B} \quad \text{then} \quad a \oplus b = a + b \,(\mathrm{mod}\,2).$$

Example 1.9.2

We can extend the addition process in Example 1.9.1:

$$\text{if } x_i \in \mathbb{B} \quad \text{then} \quad x_1 \oplus x_2 \oplus \ldots \oplus x_k = x_1 + x_2 + \ldots + x_k \,(\mathrm{mod}\,2).$$

For example,

$$1 \oplus 0 \oplus 1 \oplus 1 \oplus 1 \oplus 1 = 1 + 0 + 1 + 1 + 1 + 1$$
$$= 5$$
$$= 1 (\mathrm{mod}\,2).$$

TABLE 1.9

Multiplication modulo 2

$\times$	0	1
0	0	0
1	0	1

Example 1.9.3 Multiplication modulo 2

Multiplication modulo 2 is used in some quantum algorithms. Table 1.9 defines this. Multiplication modulo 2 is equivalent to the *and* operation $\wedge$ on Boolean variables which is defined such that

$$0 \wedge 0 = 0, \quad 0 \wedge 1 = 0, \quad 1 \wedge 0 = 0, \quad 1 \wedge 1 = 1.$$

Observe that the result of calculating $x \wedge y$ when $x, y \in \mathbb{B}$, is 1 if and only if $x = y = 1$.

Example 1.9.4

Consider the expression

$$(0 \wedge s_1) \oplus (1 \wedge s_2) \qquad \text{where } s_1, s_2 \in \mathbb{B}.$$

Find the values of s_1 and s_2 such that this expression equals zero.

Solution

Table 1.10 shows the possible combinations of values of s_1 and s_2 and the results of calculating $(0 \wedge s_1)$, $(1 \wedge s_2)$ and $(0 \wedge s_1) \oplus (1 \wedge s_2)$.

TABLE 1.10

Table for Example 1.9.4

s_1	s_2	$0 \wedge s_1$	$1 \wedge s_2$	$(0 \wedge s_1) \oplus (1 \wedge s_2)$
0	0	0	0	0
0	1	0	1	1
1	0	0	0	0
1	1	0	1	1

Observe from the table that the only combinations of s_1 and s_2 resulting in 0 are $s_1 = s_2 = 0$ and $s_1 = 1, s_2 = 0$. When we study quantum algorithms in Chapter 23, we shall need to solve equations such as

$$(0 \wedge s_1) \oplus (1 \wedge s_2) = 0.$$

Exercises

1.11   Consider the expression

$$(1 \wedge s_1) \oplus (1 \wedge s_2) \qquad \text{where } s_1, s_2 \in \mathbb{B}.$$

Find the values of s_1 and s_2 such that this expression equals zero.

Example 1.9.5 Addition modulo 2: the bit-wise sum

For $x \in \mathbb{B}^k$ and $y \in \mathbb{B}^k$, where $x = (x_1, x_2, \ldots, x_k)$, $y = (y_1, y_2, \ldots, y_k)$, we define the **bit-wise** $\oplus$ operator as

$$x \oplus y = (x_1 \oplus y_1, x_2 \oplus y_2, \ldots, x_k \oplus y_k).$$

This is just the exclusive-or operator applied to corresponding components. For example, suppose

$$x = (1, 0, 1, 1, 1, 1)$$
$$y = (1, 1, 0, 0, 1, 1)$$

then

$$x \oplus y = (1 \oplus 1, 0 \oplus 1, 1 \oplus 0, 1 \oplus 0, 1 \oplus 1, 1 \oplus 1)$$
$$= (0, 1, 1, 1, 0, 0).$$

This operation may be seen written vertically: consider the next example.

Example 1.9.6 Addition modulo 2: the bit-wise sum

Find $111 \oplus 101$ where $\oplus$ represents bit-wise sum.

Solution

Corresponding bits are added modulo 2:

$$\begin{array}{r} 111 \quad \oplus \\ 101 \\ \hline 010 \end{array}$$

(Finally, note that the bit-wise sum is not the same as addition modulo 2 when carry digits are used).

Exercises

1.12 Let $x \in \mathbb{B}^2 = \{0, 1\}^2$. Let $s \in \mathbb{B}^2 = \{0, 1\}^2$. Suppose $x = (1, 1) \equiv 11$. Suppose $s = (1, 0) \equiv 10$. Verify the following:

   (a) Show that $x \oplus x = 00$,
   (b) Show that $x \oplus 0 = x$ (here $0 \equiv 00 \equiv (0, 0)$),
   (c) Show that $x \oplus s \oplus s = x$,
   (d) Show that $x \oplus x \oplus s = s$.

Example 1.9.7 The additive group of integers modulo 2

Consider addition modulo 2 of the integers in the set $\{0, 1\}$. Table 1.8 shows this addition rule. It is straightforward to show that this set with the operation of addition modulo 2 satisfies the axioms for a group. This group is often written as $(\mathbb{Z}/2\mathbb{Z}, +)$ or simply $\mathbb{Z}/2\mathbb{Z}$, and is referred to as the additive group of integers modulo 2.

Exercises

1.13 State the identity element of the group $(\mathbb{Z}/2\mathbb{Z}, +)$.

1.14 Produce a group table for $(\mathbb{Z}/3\mathbb{Z}, +)$. What is the identity element in this group?

Example 1.9.8 The multiplicative group of integers modulo 8

Table 1.11 shows a multiplication rule for elements in the set $\{1, 3, 5, 7\}$ modulo 8. It is

TABLE 1.11
Cayley table for
the group
$(\mathbb{Z}/8\mathbb{Z})^{\times}$

×	1	3	5	7
1	1	3	5	7
3	3	1	7	5
5	5	7	1	3
7	7	5	3	1

straightforward to show that this set with the operation of multiplication modulo 8 satisfies the rules for a group. This group is often written $(\mathbb{Z}/8\mathbb{Z})^{\times}$ and is referred to as the multiplicative group of integers modulo 8. The elements of $(\mathbb{Z}/8\mathbb{Z})^{\times}$ are those integers in the interval $1 \leq n \leq 7$ which are coprime to 8, that is, their only common factor with 8 is 1.

1.10 Relations, equivalence relations and equivalence classes

Given sets A and B, we can define relationships between elements in the two sets. Specifically, we consider the ordered pair (a, b) with $a \in A$, $b \in B$, i.e., $(a, b) \in A \times B$, and ask whether or not a is related to b.

Definition 1.12 Relation
*For the ordered pair $(a, b) \in A \times B$, a **relation** R is such that either*

  1. a is related to b, written $a\,R\,b$, or

  2. a is not related to b, written $a\,\not{R}\,b$.

Consider the following example.

Example 1.10.1

The sets A and B are defined as $A = \{2, 3, 5\}$, $B = \{3, 6, 7\}$. For $a \in A$ and $b \in B$, we shall say a is related to b if a is a factor of b. The statement 'is a factor of' is the relation, R, say. We can then write expressions such as $3\,R\,6$ because 3 is a factor of 6. Likewise $5\,\not{R}\,3$ since 5 is not a factor of 3. We can imagine this relation as shown in Figure 1.5.

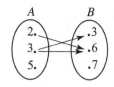

FIGURE 1.5
The arrows show the relation R: 'is a factor of', between elements of two sets.

It is also helpful to think of the relation as a rule which sends, via the arrows, (some) elements of A, the 'input', to (some) elements of B, the 'output'. Note that not all elements of A and B are necessarily involved. We can also think of a relation as defining a subset of the Cartesian product $A \times B$. In this case the subset, $\hat{R}$ say, is $\{(2,6),(3,3),(3,6)\}$.

Example 1.10.2

Consider a set A and the ordered pair (a_1, a_2), with $a_1, a_2 \in A$, which is an element of the Cartesian product $A \times A$. There are many ways that a_1 and a_2 may, or may not, be related. Suppose, for example, that A is the set of integers between 1 and 4. The Cartesian product $A \times A$ is then the set of ordered pairs

$$\{(1,1),(1,2),(1,3),(1,4),(2,1),(2,2),(2,3),(2,4),$$

$$(3,1),(3,2),(3,3),(3,4),(4,1),(4,2),(4,3),(4,4)\}.$$

We consider possible relationships within each ordered pair, (a_1, a_2). We can ask, for example, whether a_1 is a multiple of a_2. So, for the element $(4,2)$ it is clearly the case that the first element is a multiple of the second. The statement 'is a multiple of' is a relation, R, say. Having defined the relation R, we can write $4 R 2$. Clearly, for the element $(1,3)$, the first element is not a multiple of the second and so $1 \not R 3$. We can imagine the relation R as shown in Figure 1.6.

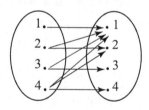

FIGURE 1.6
The arrows show the relation R: 'is a multiple of', between elements of two sets.

With any given relation, it is the case that $a_1 R a_2$ or $a_1 \not R a_2$, that is, any two elements are either related or they are not. The relation R defines a subset, $\hat{R}$ say, of $A \times A$:

$$\hat{R} = \{(a_1, a_2) : a_1 R a_2\}.$$

For the case above, it is straightforward to check that

$$\hat{R} = \{(1,1),(2,1),(2,2),(3,1),(3,3),(4,1),(4,2),(4,4)\}$$

which is a subset of the Cartesian product $A \times A$.

In the study of functions (Chapter 2), relations which will become particularly relevant are 'one-to-one' and 'two-to-one' relations defined on the Cartesian product $A \times B$, such as those depicted in Figure 1.7. We shall see in Section 1.11 that one-to-one relations on $A \times A$ represent arrangements or permutations of the elements of the set A.

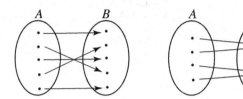

FIGURE 1.7
A one-to-one and a two-to-one relation.

1.10.1 Equivalence relations

A particular type of relation known as an equivalence relation is significant in quantum computation.

Definition 1.13 Equivalence relation
*A relation defined on the Cartesian product $A \times A$ is called an **equivalence relation**, denoted $\sim$ say, if it satisfies the following:*

    1. $a \sim a$ for every $a \in A$ (referred to as reflexivity)

    2. if $a \sim b$ then $b \sim a$ (referred to as symmetry)

    3. if $a \sim b$ and $b \sim c$ then $a \sim c$,   (referred to as transitivity)

Definition 1.14 Equivalence class
*Given an equivalence relation, $\sim$, the **equivalence class** of any element $a \in A$, written $[a]$, is the set of all elements in A to which a is related:*

$$[a] = \{b \in A : a \sim b\}$$

The set of equivalence classes is denoted by $A/\sim$.

Example 1.10.3 The integers modulo 4, ($\mathbb{Z}/4\mathbb{Z}$)

In this example we consider the integers modulo 4, as depicted in Figure 1.8.
    Working in modulo 4 recall that we use the integers 0, 1, 2, 3 and then below 0 and above 3 we 'wrap around' as indicated. Consider the congruence relation

$$a \equiv b \;(\text{mod } 4).$$

Recall that $a \equiv b \;(\text{mod } 4)$ if $a - b$ is divisible by 4. Inspection of Figure 1.8 readily reveals that

$$6 \equiv 2 \;(\text{mod } 4), \qquad\qquad 17 \equiv 9 \;(\text{mod } 4)$$

and so on. Equivalently, $6 - 2$ is divisible by 4, and $17 - 9$ likewise. To confirm that this is indeed an equivalence relation, note that

$$a \equiv a \bmod 4 \qquad \text{since } a - a = 0 \text{ is divisible by 4 (reflexivity)}$$

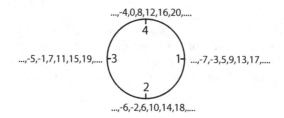

FIGURE 1.8
Integers modulo 4.

Also, if $a - b$ is divisible by 4, then $b - a = -(a - b)$ is divisible by 4 (symmetry). Finally, if $a - b$ and $b - c$ are both divisible by 4, then we can write

$$a - b = 4k, \quad b - c = 4\ell, \qquad \text{for some } k, \ell \in \mathbb{Z}.$$

Adding these two expressions:

$$a - c = 4(k + \ell)$$

which is divisible by 4 since $k + \ell \in \mathbb{Z}$. This confirms transitivity, and hence, the relation 'is congruent to, mod 4' is an equivalence relation. Observe that the relation divides the integers into equivalence classes as is clearly illustrated in Figure 1.8.

$$[0] = \ldots, -4, 0, 4, 8, 12, 16, \ldots \qquad [1] = \ldots, -3, 1, 5, 9, 13, 17, \ldots$$

$$[2] = \ldots, -2, 2, 6, 10, 14, 18, \ldots \qquad [3] = \ldots, -1, 3, 7, 11, 15, \ldots$$

The set of equivalence classes is often written $\mathbb{Z}/4\mathbb{Z}$. Every integer is in one and only one equivalence class. We say that the equivalence relation **partitions** the integers into equivalence classes. So equivalence classes are distinct.

Example 1.10.4

If $\sim$ is the equivalence relation $a \equiv b \pmod{12}$, then the equivalence classes are:

$$[0] = \{\ldots, 0, 12, 24, \ldots\}, \qquad [1] = \{\ldots, 1, 13, 25, \ldots\},$$

$$[2] = \{\ldots, 2, 14, 26, \ldots\}, \qquad [3] = \{\ldots, 3, 15, 27, \ldots\}, \qquad \text{and so on.}$$

Definition 1.15 Equivalence classes and the partition of a set
Given an equivalence relation $\sim$ on $A \times A$, denote the equivalence class of an element $a \in A$ by $[a]$. Then for any $b \in A$, either $[a] \cap [b] = \emptyset$ or $[a] = [b]$. So equivalence classes are distinct entities and the equivalence relation can be thought of as 'dividing up' the set into these distinct classes, hence the notation $A/\sim$.

Example 1.10.5 Equivalence classes and a projective space

Consider the xy plane, $\mathbb{R}^2$, in Figure 1.9. We can define a relation, $\sim$, between two points (x, y) and (x^*, y^*) in the plane by

$$(x, y) \sim (x^*, y^*) \quad \text{if the points lie on the same straight line through the origin.}$$

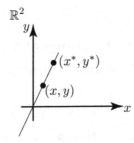

FIGURE 1.9

(x, y) and (x^*, y^*) lie on the same line through the origin.

The relation $\sim$ is not an equivalence relation (see the Exercises below). We now remove the origin and consider the set $\mathbb{R}^2 \setminus \{(0,0)\}$. It is straightforward to show that, with this restriction, $\sim$ is an equivalence relation. The set of equivalence classes, denoted $\mathbb{R}^2 \setminus \{(0,0)\}/\sim$, in this context often referred to as a **projective space**, is the set of straight lines through the origin. Any two points are regarded as equivalent if they lie on the same straight line through the origin. We shall refer to the lines as **rays** in Chapter 13. Points on the same line, that is on the same ray, are in the same equivalence class (Figure 1.10).

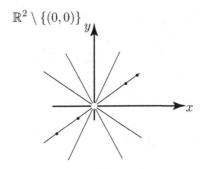

FIGURE 1.10

Points • on the same line or same ray are in the same equivalence class.

Exercises

1.15 Consider the set $\mathbb{R}^2$ (i.e., without removal of the origin). Show that the relation $(x, y) \sim (x^*, y^*)$ if the points lie on the same straight line through the origin, is not an equivalence relation because the transitivity requirement fails.

1.16 Consider the set $\mathbb{R}^2 \setminus \{(0,0)\}$ and the equivalence relation: $(x, y) \sim (x^*, y^*)$ if the points lie on the same straight line through the origin. Show that this is equivalent to writing $(x, y) \sim (x^*, y^*)$ if and only if $(x^*, y^*) = \lambda(x, y)$ for some $\lambda \in \mathbb{R} \setminus \{0\}$.

1.11 Combinatorics – permutations and combinations

Combinatorics is that branch of mathematics concerned with counting.

Definition 1.16 Permutation
A **permutation** *of n distinct objects is an arrangement of those objects.*

For example, with two objects, a and b, there are just two permutations

$$ab, ba.$$

With three distinct objects, a, b and c, there are six permutations

$$abc, acb, bac, bca, cab, cba.$$

More generally, there are $n!$ permutations of n distinct objects.

Example 1.11.1

Consider the set of strings of two digit binary numbers $\mathbb{B}^2 = \{00, 01, 10, 11\}$ and the one-to-one relation shown in Figure 1.11.

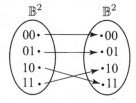

FIGURE 1.11
A relation as a permutation of binary strings.

We can think of this relation as a permutation of the set of four objects $\{00, 01, 10, 11\}$, specifically the permutation $00, 01, 11, 10$. We have seen that there are $n!$ permutations of n distinct objects and consequently there $4! = 24$ different permutations of the elements of set $\mathbb{B}^2$. Calculations such as this will be required when we meet functions defined on $\mathbb{B}^2$ in Chapter 2.

If only r of the n distinct objects are selected for arrangement, there are $\dfrac{n!}{(n-r)!}$ permutations. This is often written ^nP_r.

For example, given the objects a, b, c above, if we select just two at a time, there are $^3P_2 = \dfrac{3!}{(3-2)!} = 6$ permutations:

$$ab, ba, ac, ca, bc, cb.$$

On the other hand, given four distinct objects, which have $4! = 24$ permutations, taking just two at a time, gives $\dfrac{4!}{(4-2)!} = 12$ permutations.

Suppose now that not all the objects are distinct. If out of the n objects, n_1 are identical and of type 1, n_2 are identical and of type 2, and so on, the number of permutations using all n objects are $\dfrac{n!}{n_1!n_2!\dots n_k!}$.

Combinations are closely related to permutations.

Definition 1.17 Combination
A **combination** *is a selection of r objects from n distinct objects. When making the selection, the order in which we write the objects down is of no consequence.*

For example, the combination ab is the same as the combination ba. The number of combinations is often written ^nC_r, or sometimes $\binom{n}{r}$ and is given by $\dfrac{n!}{(n-r)!r!}$.

1.12 End-of-chapter exercises

1. What is the cardinality of the power set of the set $\mathbb{B}^n$?

2. By drawing Venn diagrams, verify De Morgan's laws:
$$\overline{A \cap B} = \overline{A} \cup \overline{B}, \qquad \overline{A \cup B} = \overline{A} \cap \overline{B}.$$

3. Use the laws of set algebra to simplify the following expressions:
   (a) $(A \cap B) \cup (A \cap \overline{B})$
   (b) $A \cup (\overline{A} \cap \overline{B})$.

4. Consider the set $\mathbb{Z}$ and the equivalence relation $R : a \equiv b \pmod 6$. Write down an expression for each of the equivalence classes. Confirm that for $x, y \in \mathbb{Z}$ with $x \neq y$, $[x] \cap [y] = \emptyset$ or $[x] = [y]$.

5. Consider the set of all triangles in the plane. Two triangles are *similar* if their corresponding angles are the same. Show that 'is similar to' is an equivalence relation.

6. Consider the set of points in $\mathbb{R}^2$ and the relation $(x, y) \sim (x^*, y^*)$ if and only if $y - x = y^* - x^*$. Show that $\sim$ is an equivalence relation and give a geometric interpretation of the equivalence classes.

7. Consider the set $\mathbb{B} = \{0, 1\}$.
   (a) Performing addition modulo 2, construct an addition table.
   (b) Performing multiplication modulo 2, construct a multiplication table.
   (c) State the additive identity element.
   (d) State the multiplicative identity element.
   (e) Show that the set $\mathbb{B} = \{0, 1\}$ with the above operations $+$ and $\times$ satisfies the axioms of a field (see Appendix D). This is an example of a **finite field** often denoted $\mathbb{F}_2$ or $\mathrm{GF}(2)$.

8. Let $x \in \mathbb{B}^n = \{0, 1\}^n$. Let $s \in \mathbb{B}^n = \{0, 1\}^n$.
   (a) Show that $x \oplus x = 0$.
   (b) Show that $x \oplus 0 = x$.
   (c) Show that $x \oplus s \oplus s = x$.
   (d) Show that $x \oplus x \oplus s = s$.

9. Let $x, y \in \mathbb{B}$.

    (a) Show that
    $$x \oplus y = (x \vee y) \wedge (\overline{x} \vee \overline{y}).$$

    (b) Show that
    $$x \oplus y = (x \wedge \overline{y}) \vee (\overline{x} \wedge y).$$

10. Given an ordered n-tuple $a \in \mathbb{B}^n$ where $a = (a_1, a_2, \ldots, a_n)$ and $a_i \in \mathbb{B}$, then the **Hamming weight** of a is defined to be the number of 1's in a. Show that this is equivalent to adding the terms of a. That is
    $$\text{Hamming weight of } a = \sum_{i=1}^{n} a_i$$

    Calculate the Hamming weight of $a \in \mathbb{B}^6$ when $a = (1, 0, 1, 1, 1, 1)$.

2

Functions and their application to digital gates

2.1 Objectives

A **function** is a special type of relation between elements of two sets. The first set can be thought of as providing the input to the function, and the second set as providing the output. In this sense, a function is used to process data.

When we refer to a digital computer's 'state' at any time t, we are referring to the digital data in its registers. A program moves the computer through a sequence of states, these changes being realised through the application of classical Boolean gates. Mathematically, Boolean gates are functions which act on input data, i.e., strings of bits, to produce output data, also strings of bits, and in doing so the state of the system evolves.

The objective of this chapter is to introduce terminology and notation associated with functions. We show how functions can be represented with pictures, tables and mathematical formulae. Functions defined on the set $\{0, 1\}$ are of particular relevance in quantum computation so these are our primary focus. We make particular reference to so-called bijective functions and inverse functions. These play a central role in the description of reversible digital gates and also quantum gates which are used to process quantum information.

2.2 Introductory definitions and terminology

Definition 2.1 Function, or mapping
*Consider two sets A and B. A **function**, or **mapping**, f say, is a relation or a rule which assigns to each element of A just one element of B.*

Here we have labelled the function f, but other letters, including Greek characters, e.g., π, σ, will also be used. Schematically we can represent a function as shown in the example of Figure 2.1 in which the function f maps the element a_1 to b_3, a_2 to b_4 and so on.

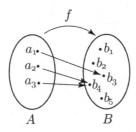

FIGURE 2.1
Two sets and the function $f : A \rightarrow B$. Each element of A maps to just one element in B.

DOI: 10.1201/9781003264569-2

We write $f : A \to B$, and $f(a_1) = b_3$, $f(a_2) = b_4$ and so on. We refer to a_1, a_2, a_3 as **inputs** to, or **arguments** of, the function, and b_3, b_4 as **outputs**. The set A is called the **domain** of the function. The set B is called the **co-domain**. The set of all the elements in B which are mapped to by the function is called the **range** of the function. In this case the range, i.e., the set of outputs, is $\{b_3, b_4\}$. Note that the range is not necessarily the same as the co-domain as not all elements of B need to be mapped onto.

Definition 2.2 Onto or surjective function
If all the elements of the co-domain of a function are mapped onto by some element in the domain, then the function is said to be **onto** *or* **surjective**.

Note that some elements of A might map to the same element of B as occurs in Figure 2.1.

Definition 2.3 One-to-one or injective function
If each element in the range is mapped to by just one element in the domain, then the function is said to be **one-to-one** *or* **injective**. *A function which is not one-to-one is referred to as* **many-to-one**.

Definition 2.4 Bijective function
If a function is both surjective and injective it is said to be **bijective**. *When this happens, there is a* **one-to-one correspondence** *between elements in A and elements in B.*

Bijective functions play a central role in the description of reversible digital and quantum gates which are used to process quantum information.

Boolean functions are particularly relevant to the study of quantum computation. These are functions where values in both the domain and the co-domain are chosen from the set of binary digits $\mathbb{B} = \{0, 1\}$. Consider the following example.

Example 2.2.1 A function $f : \mathbb{B} \to \mathbb{B}$

Consider a function $f : \mathbb{B} \to \mathbb{B}$, an example of which is shown in Figure 2.2. The domain of f is the set $\mathbb{B} = \{0, 1\}$, as is the co-domain. The element 1 in the domain is mapped to 0 in the co-domain. We write $f(1) = 0$. The element 0 is mapped to 1. We write $f(0) = 1$. Functions of this type are ubiquitous in digital computation. In Section 2.4, we will consider several such functions from $\mathbb{B}$ to $\mathbb{B}$.

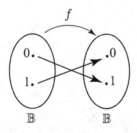

FIGURE 2.2
An example of a function $f : \mathbb{B} \to \mathbb{B}$.

Example 2.2.2 Set permutations as functions

Suppose $A = \{1, 2, 3, 4, 5\}$. Consider the function $\pi : A \to A$ depicted in Figure 2.3. Observe carefully that this function is one-to-one and onto, i.e., a bijection, and its effect is to permute the elements of the set A. We can think of the function as effecting the permutation:

$$12345 \rightarrow 23154$$

Indeed we say that a permutation is a bijection from A to itself. Set permutations arise in the study of group theory (Chapter 8). In this example $\pi(1) = 2$, $\pi(2) = 3$, $\pi(3) = 1$, $\pi(4) = 5$, $\pi(5) = 4$.

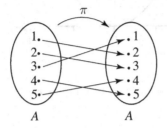

FIGURE 2.3
An example of a function $\pi : A \rightarrow A$ which permutes a set.

There are several common ways of writing down set permutations. One such way is **cycle notation** which would represent this permutation as

$$(123)(45)$$

which means $1 \rightarrow 2$, $2 \rightarrow 3$, $3 \rightarrow 1$; $4 \rightarrow 5$, $5 \rightarrow 4$. In cycle notation an element which gets mapped to itself is usually omitted from the list.

Exercises

2.1   Use cycle notation to represent the bijection $\sigma(1) = 2$, $\sigma(2) = 3$, $\sigma(3) = 4$, $\sigma(4) = 5$, $\sigma(5) = 1$.

2.2   Draw a set diagram which depicts the permutation $(12)(345)$.

Example 2.2.3 Real-valued functions and graphs

Consider the function $f : \mathbb{R} \rightarrow \mathbb{R}$ defined by $f(x) = x^2$, or equivalently $f : x \rightarrow x^2$. The domain of this function is the set of real numbers $\mathbb{R}$. This set provides the input. For example if $x = 3$ the output is $f(3) = 3^2 = 9$. If $x = -2$, the output is $f(-2) = (-2)^2 = 4$. The input-output pairs $(-2, 4)$, $(3, 9)$ can be plotted as points using Cartesian coordinates (x, y) and by joining the points we obtain the **graph** of the function illustrated in Figure 2.4.

Exercises

2.3   Show that the function $f : \mathbb{R} \rightarrow \mathbb{R}$, $f(x) = x^2$, is neither one-to-one nor surjective.

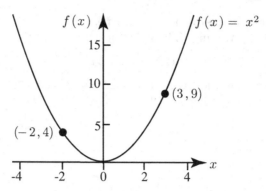

FIGURE 2.4
Graph of the function $f : \mathbb{R} \to \mathbb{R}$ defined by $f(x) = x^2$.

2.4   Show that the function $f : \mathbb{R} \to \mathbb{R}$, $f(x) = x^3$, is both one-to-one and surjective, and is therefore bijective.

Definition 2.5 Composite function
Consider two functions $f : A \to B$ and $g : B \to C$. Starting with $x \in A$, then $f(x) \in B$. We can then use g to map $f(x)$ to $g(f(x))$ as shown in Figure 2.5. The function $g(f(x))$, also written $(g \circ f)(x)$, or simply $g \circ f$, is called the **composite function** *or* **composition** *of f and g. In other words*

$$(g \circ f) : A \to C, \qquad (g \circ f)(x) = g(f(x))$$

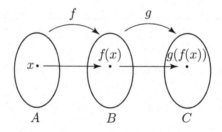

FIGURE 2.5
The composite function $g(f(x))$ applies f first and then g.

Composition of functions obeys the **associative law**, that is, if $f : A \to B$, $g : B \to C$, $h : C \to D$ then

$$h \circ (g \circ f) = (h \circ g) \circ f.$$

So, provided the order of h, g and f is maintained, it does not matter which composite function $g \circ f$ or $h \circ g$ is evaluated first.

Definition 2.6 Inverse function
When a function is bijective there is a one-to-one correspondence between its inputs and outputs. Consequently we can define a function, labelled f^{-1} and called the **inverse function**, *which reverses the rule given by f. This is depicted in Figure 2.6. Note that $f^{-1}(f(x)) = x$, that is $(f^{-1} \circ f)(x) = x$. We then say that f is* **invertible** *or* **reversible**.

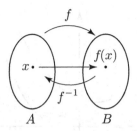

FIGURE 2.6
The inverse function f^{-1}, when such exists, reverses the process in f.

Exercises

2.5 Is the function f in Example 2.2.1 invertible ?

2.3 Some more functions $f : \mathbb{R} \to \mathbb{R}$

There are several common functions for which both the domain and co-domain are the set of real numbers $\mathbb{R}$. We shall refer to these as **real-valued functions**. This section provides a brief review of the functions we will need in this book.

Polynomial functions $f : \mathbb{R} \to \mathbb{R}$ take the form

$$f(x) = a_n x^n + a_{n-1} x^{n-1} + a_{n-2} x^{n-2} + \ldots + a_2 x^2 + a_1 x + a_0$$

where $a_i \in \mathbb{R}$, $n \in \mathbb{N}$, $i = 0, \ldots, n$. The value of n is referred to as the **degree** of the polynomial. A polynomial of degree 0 is a **constant function**. The graph of a real-valued function is obtained by plotting and joining points with coordinates $(x, f(x))$. It is straightforward to produce graphs of these functions using a graphing calculator, software or one of the many available on-line graph plotting tools. The graphs of some low degree polynomials are shown in Figure 2.7. Note that the graph of a polynomial of degree 1, i.e., $f(x) = a_0 + a_1 x$, is a straight line. The graph of a constant function (degree 0) is a horizontal line.

An **exponential function** $f : \mathbb{R} \to \mathbb{R}$ has the form

$$f(x) = a^x$$

where a is a positive real constant. In quantum computing the value of a is commonly 2. The graph of $f(x) = 2^x$, for $x \geq 0$, is shown in Figure 2.8. Another commonly used value for a is the **exponential constant** $e = 2.718\ldots$ in which case $f(x) = e^x$ is referred to as **the exponential function**. Values of exponential functions can be obtained using a calculator. In order to demonstrate exponential growth, the value of a must be greater than 1.

The **logarithm to base 2** function, $f : \mathbb{R}^+ \to \mathbb{R}$ denoted by $f(x) = \log_2 x$, has a value y such that $x = 2^y$. Values of $\log_2 x$ can be found using the formula $\log_2 x = \dfrac{\log_{10} x}{\log_{10} 2}$. The values of $\log_{10}$ can be obtained using a calculator. The graph of $\log_2 x$ is shown in Figure 2.8. Note that the domain of this function is restricted to $\mathbb{R}^+$; the logarithm of 0 and of negative numbers is not defined when using real numbers.

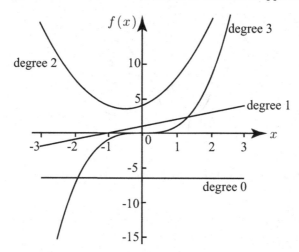

FIGURE 2.7
Some typical polynomial functions of low degree.

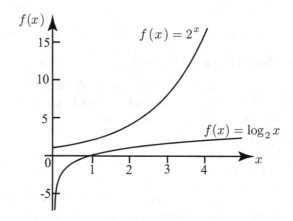

FIGURE 2.8
The functions defined by $f(x) = 2^x$ and $f(x) = \log_2 x$.

The **trigonometric functions**, $f : \mathbb{R} \to \mathbb{R}$, $f(x) = \sin x$ and $f(x) = \cos x$, follow from their respective trigonometric ratios (see Appendix B). (Likewise for the function $f(x) = \tan x$ though some domain restriction is needed because the tangent function is undefined when $x = (2k + 1)\frac{\pi}{2}$, $k \in \mathbb{Z}$.) In general, values of these functions can be obtained using a calculator, but some common special cases are given in the Appendix.

As usual, the graphs of the trigonometric functions are obtained by calculating, plotting and joining points with coordinates $(x, f(x))$. Graphs of $f(x) = \sin x$ and $f(x) = \cos x$ are shown in Figure 2.9. There will be a need to write expressions involving trigonometric functions in different yet equivalent forms. To do this use is made of trigonometric identities also given in Appendix B.

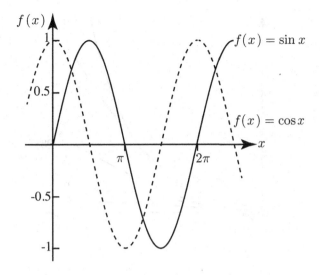

FIGURE 2.9
The functions $f(x) = \sin x$ and $f(x) = \cos x$.

Exercises

2.6 Show that the functions $f(x) = \sin x$ and $f(x) = \cos x$ shown in Figure 2.9 are neither one-to-one nor surjective.

2.7 Using software, or otherwise, plot a graph of $f(x) = \cos 2x$. Determine whether this function is bijective.

2.8 Plot a graph of $f(x) = \cos \frac{\theta}{2}$ for $0 \leq \theta \leq \pi$. Deduce that if $r = \cos \frac{\theta}{2}$ for $0 \leq \theta \leq \pi$, then $0 \leq r \leq 1$.

2.9 Plot a graph of $f(x) = \sin \frac{\theta}{2}$ for $0 \leq \theta \leq \pi$. Deduce that if $r = \sin \frac{\theta}{2}$ for $0 \leq \theta \leq \pi$, then $0 \leq r \leq 1$.

2.10 Use a trigonometric identity (Appendix B) to show that if

$$x = \sin\frac{\theta}{2}\cos\varphi, \quad y = \sin\frac{\theta}{2}\sin\varphi, \quad z = \cos\frac{\theta}{2}$$

then $x^2 + y^2 + z^2 = 1$.

2.3.1 The relative growth of functions

Complexity theory is concerned with the efficiency of algorithms. One way in which efficiency is measured is by estimating the time required to perform a calculation as the number of inputs, x, increases. Observe from Figure 2.10 that as x increases some functions grow much faster than others. An understanding of the relative growth of these functions is essential to an understanding of the assessment of algorithm efficiency – both in the digital and quantum domains. It is important to recognise that exponential functions grow much more rapidly than polynomial and logarithmic ones as x increases.

To emphasise further the importance of relative growth consider the function values in Table 2.1 and observe how incredibly rapidly the exponential function, 2^x, grows even compared to the rapidly growing quadratic x^2.

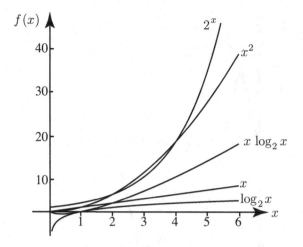

FIGURE 2.10
Comparing the relative growth of the functions $\log_2 x$, x, $x \log_2 x$, x^2 and 2^x.

TABLE 2.1
Table showing the relative growth of
common functions

	$x = 10$	$x = 100$	$x = 1000$
$\log_2 x$	3.3	6.6	10.0
$x \log_2 x$	33.2	664.4	9965.8
x^2	100	10 000	1 000 000
2^x	1024	10^{30}	10^{301}

2.4   The Boolean functions $f : \mathbb{B} \to \mathbb{B}$

There are just four Boolean functions $f : \mathbb{B} \to \mathbb{B}$ where $\mathbb{B}$ is the set of binary digits $\{0, 1\}$. In this section we define the four functions and illustrate several results which will be useful when we meet Deutsch's quantum algorithm in Chapter 22. We denote the set containing all four functions $f : \mathbb{B} \to \mathbb{B}$ by $\mathcal{F}(\mathbb{B}, \mathbb{B})$.

Example 2.4.1 The digital *not* function

The function $f : \mathbb{B} \to \mathbb{B}$, illustrated in Figure 2.11, maps the binary digit 1 to 0, and the binary digit 0 to 1, that is $f(1) = 0$, $f(0) = 1$.

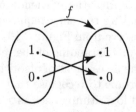

FIGURE 2.11
The digital *not* function $f : \mathbb{B} \to \mathbb{B}$, $f(x) = 1 \oplus x$.

There are several ways of expressing this function. We could write a mathematical formula:

$$f(x) = \begin{cases} 0 & \text{when} & x = 1 \\ 1 & \text{when} & x = 0 \end{cases}.$$

An alternative way is to tabulate the inputs and outputs as shown in Table 2.2. It is easily

TABLE 2.2
Representing the
digital *not* function
f by its inputs and
outputs

input x	output
1	0
0	1

seen from Figure 2.11 that this function is bijective: there is a one-to-one correspondence between values in the domain and the co-domain. This function is referred to as a **digital *not* gate** or a **digital *not* function**. On occasions we will write not_D to distinguish it from the quantum equivalent not_Q. We will indicate the *not* function using an overbar so that this function can be written $f(x) = \overline{x}$, that is $\overline{1} = 0$, and $\overline{0} = 1$. Further, you should verify that the *not* function f can be expressed using the exclusive-or operator as $f(x) = 1 \oplus x$. It can be represented diagrammatically as in Figure 2.12.

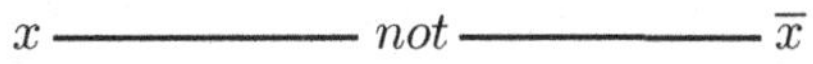

$$x \text{———} not \text{———} \overline{x}$$

FIGURE 2.12
The digital *not* function.

Exercises

2.11   Given $f : \mathbb{B} \to \mathbb{B}$, $f(x) = 1 \oplus x$, verify that $f(0) = 1 \oplus 0 = 1$ and $f(1) = 1 \oplus 1 = 0$.

There are just three further functions $f : \mathbb{B} \to \mathbb{B}$. Two of these are represented by constant functions, giving a constant (i.e., the same) output whatever the value of the input. One is the **identity function** $i(x)$ (or sometimes we shall write $i_D(x)$), which produces the same output as input. Let us label these four functions as f_0, f_1, f_2, f_3, (with f_2 being the digital *not* function of Example 2.4.1).

In summary

$$f_0 : \mathbb{B} \to \mathbb{B}, \quad f_0(x) = 0 \quad \text{for all } x \in \mathbb{B}$$

$$f_1 : \mathbb{B} \to \mathbb{B}, \quad f_1(x) = i(x) = x \quad \text{for all } x \in \mathbb{B}$$

$$f_2 : \mathbb{B} \to \mathbb{B}, \quad f_2(x) = \overline{x} \quad \text{for all } x \in \mathbb{B}$$

$$f_3 : \mathbb{B} \to \mathbb{B}, \quad f_3(x) = 1 \quad \text{for all } x \in \mathbb{B}.$$

We can record the possible inputs and outputs of these functions in Tables 2.3 ... 2.5. (The digital *not* function, f_2, has been defined already in Table 2.2.)

TABLE 2.3
The constant
function $f_0(x) = 0$.

input x	output 0
1	0
0	0

TABLE 2.4
The identity function
$f_1(x) = i(x) = x$.

input x	output x
1	1
0	0

TABLE 2.5
The constant
function $f_3(x) = 1$.

input x	output 1
1	1
0	1

The functions f_2 and f_1 in Tables 2.2 and 2.4 are said to be **balanced** in that the function outputs the same number of 0's as 1's. Functions f_0 and f_3 are said to be **unbalanced** and are constant. Determining whether a given but unknown function $f : \mathbb{B} \to \mathbb{B}$ is balanced or constant is an important challenge for quantum algorithms as we shall see in Chapter 22.

Exercises

2.12   Which of the four functions $f : \mathbb{B} \to \mathbb{B}$ cannot be inverted ?

The following three examples illustrate useful manipulations required in the development of quantum algorithms that we study in Chapter 22.

Example 2.4.2

Suppose f is one of the four functions $f : \mathbb{B} \to \mathbb{B}$ but we know not which. The exclusive-or operator is defined on the Boolean variables x and y so that $x \oplus y = 1$ whenever x or y (but not both) equals 1, and is zero otherwise.

(a) Suppose that $f(0) \oplus f(1) = 0$. Show that f must be one of the two constant functions.

(b) Suppose that $f(0) \oplus f(1) = 1$. Show that f must be one of the two balanced functions.

Solution

(a) The inputs and outputs of the four functions, $f_0, \ldots, f_3$, together with the result of calculating $f(0) \oplus f(1)$, are shown in Table 2.6. Recall that $f(0) \oplus f(1)$ is 1 whenever $f(1)$ or $f(0)$ (but not both) equals 1.

TABLE 2.6

input to f	f_0	f_1	f_2	f_3
1	0	1	0	1
0	0	0	1	1
$f(0) \oplus f(1)$	0	1	1	0

From the table, it is clear that if $f(0) \oplus f(1) = 0$, then f must be one of f_0 or f_3, that is, it is a constant function.

(b) Likewise, if $f(0) \oplus f(1) = 1$, then f must be one of f_1 or f_2, that is, it is a balanced function.

Example 2.4.3

Consider the four functions $f : \mathbb{B} \to \mathbb{B}$. Show that in all cases

$$(-1)^{f(0)}(-1)^{f(0)\oplus f(1)} = (-1)^{f(1)}$$

Solution

Table 2.6 is extended in Table 2.7 to include the rows $(-1)^{f(0)}$, $(-1)^{f(1)}$, $(-1)^{f(0)\oplus f(1)}$, $(-1)^{f(0)}(-1)^{f(0)\oplus f(1)}$ from which the result $(-1)^{f(0)}(-1)^{f(0)\oplus f(1)} = (-1)^{f(1)}$ is clear.

TABLE 2.7

input to f	f_0	f_1	f_2	f_3
1	0	1	0	1
0	0	0	1	1
$f(0) \oplus f(1)$	0	1	1	0
$(-1)^{f(0)}$	1	1	-1	-1
$(-1)^{f(1)}$	1	-1	1	-1
$(-1)^{f(0)\oplus f(1)}$	1	-1	-1	1
$(-1)^{f(0)}(-1)^{f(0)\oplus f(1)}$	1	-1	1	-1

Example 2.4.4

Suppose f is one of the four functions $f : \mathbb{B} \to \mathbb{B}$. For $x \in \mathbb{B}$ show that whatever the output $f(x)$,

$$f(x) \oplus 0 = f(x)$$

and

$$f(x) \oplus 1 = \overline{f(x)}$$

where the overline $^{\overline{}}$ denotes the *not* function.

Solution

The value of $f(x)$ is either 0 or 1. Consider Table 2.8 from which it is clear that $f(x) \oplus 0$ has the same value as $f(x)$. From Table 2.9, it is clear that $f(x) \oplus 1$ is $\overline{f(x)}$.

TABLE 2.8

$f(x)$	0	$f(x) \oplus 0$
1	0	1
0	0	0

TABLE 2.9

$f(x)$	1	$f(x) \oplus 1$
1	1	0
0	1	1

Exercises

2.13   For $f : \mathbb{B} \to \mathbb{B}$ show that if $x \in \mathbb{B}$ then $f(x) \oplus f(x)$ is always zero.

2.14   For $f : \mathbb{B} \to \mathbb{B}$ show that if $x \in \mathbb{B}$, $f(x) \oplus 0 = f(x)$.

2.15   Show that not all four functions $f : \mathbb{B} \to \mathbb{B}$ are bijective.

2.16   Write down explicit expressions for all four functions in the set $\mathcal{F}(\mathbb{B}, \mathbb{B})$.

Example 2.4.5 The functions $f : \mathbb{B} \to \mathbb{B}$ with binary operation $\oplus$ form a group

Show that the set of functions $f_i : \mathbb{B} \to \mathbb{B}$, $i = 0, \ldots, 3$, defined as $f_0(x) = 0$, $f_1(x) = x$, $f_2(x) = \bar{x}$, $f_3(x) = 1$, together with the exclusive-or operator, $\oplus$, form a group, where $(f \oplus g)(x)$ is defined as

$$(f \oplus g)(x) = f(x) \oplus g(x).$$

Solution

The effect of applying the operation $\oplus$ to the four functions is shown in Table 2.10. For example, consider $(f_2 \oplus f_1)(x)$ which is defined to be $f_2(x) \oplus f_1(x)$. When $x = 0$, $(f_2 \oplus f_1)(0) = f_2(0) \oplus f_1(0) = 1 \oplus 0 = 1$. Likewise, when $x = 1$, $(f_2 \oplus f_1)(1) = f_2(1) \oplus f_1(1) = 0 \oplus 1 = 1$. Thus $(f_2 \oplus f_1)(x) = 1 = f_3(x)$. You should verify the other results for yourself.

TABLE 2.10

The operation $\oplus$
applied to the set of
functions $f_i : \mathbb{B} \to \mathbb{B}$

$\oplus$	f_0	f_1	f_2	f_3
f_0	f_0	f_1	f_2	f_3
f_1	f_1	f_0	f_3	f_2
f_2	f_2	f_3	f_0	f_1
f_3	f_3	f_2	f_1	f_0

Clearly the given set of functions is closed under the defined operation $\oplus$. The identity element is f_0, and each element is its own inverse (e.g., observe $f_2 \oplus f_2 = f_0$). This group is denoted $(\mathcal{F}(\mathbb{B}, \mathbb{B}), \oplus)$ and is used in the quantisation of Boolean functions.

2.5   Functions defined on Cartesian products

Suppose the domain of a function is a set of ordered pairs, (a, b) with $a \in A$ and $b \in B$, i.e., the domain is the Cartesian product $A \times B$. We can define a function, f, which maps each ordered pair in $A \times B$ to just one element in a third set C, $f : A \times B \to C$. The terminology introduced in Section 2.2 is precisely the same. We can think of this function in several ways. It could be tabulated. Alternatively, it could be depicted in either of the ways shown in Figure 2.13. The co-domain C could itself be a Cartesian product, as we shall see shortly.

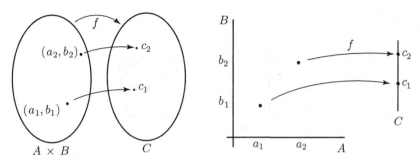

FIGURE 2.13
The function $f : A \times B \to C$, defined on a Cartesian product.

2.5.1 The Boolean functions $f : \mathbb{B} \times \mathbb{B} \to \mathbb{B}$

We now consider specific cases where the domain of the function is the Cartesian product $\mathbb{B} \times \mathbb{B} = \mathbb{B}^2$ and the co-domain is $\mathbb{B}$, that is $f : \mathbb{B} \times \mathbb{B} \to \mathbb{B}$. We denote the set of all such functions by $\mathcal{F}(\mathbb{B}^2, \mathbb{B})$. The first example makes use of the logical connective exclusive-or:

Definition 2.7 Exclusive-or, $\oplus$
*Suppose $x, y \in \mathbb{B}$. The binary operator xor, (**exclusive-or**), written $\oplus$, is equal to 1 when either but not both of x and y are equal to 1, and is zero otherwise:*

$$0 \oplus 0 = 0$$
$$0 \oplus 1 = 1$$
$$1 \oplus 0 = 1$$
$$1 \oplus 1 = 0.$$

Example 2.5.1 The *xor* gate or classical *cnot* function $f(x, y) = x \oplus y$

Consider the function f defined, using the exclusive-or $\oplus$, as

$$f : \mathbb{B} \times \mathbb{B} \to \mathbb{B}$$

$$f(x, y) = x \oplus y = \begin{cases} 0 & \text{when} & x = 0, \ y = 0 \\ 1 & \text{when} & x = 0, \ y = 1 \\ 1 & \text{when} & x = 1, \ y = 0 \\ 0 & \text{when} & x = 1, \ y = 1 \end{cases}$$

This function is depicted in Figure 2.14 and tabulated in Table 2.11.

TABLE 2.11
The function $f : \mathbb{B} \times \mathbb{B} \to \mathbb{B}$,
$f(x, y) = x \oplus y$

input (x, y)	output $x \oplus y$
(0,0)	0
(0,1)	1
(1,0)	1
(1,1)	0

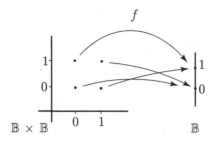

FIGURE 2.14
A pictorial representation of the function $f : \mathbb{B} \times \mathbb{B} \to \mathbb{B}$, $f(x, y) = x \oplus y$.

Observe that the output of this function is 1 when either but not both of x and y is 1, and that it is balanced. This function is an **exclusive-or gate** (*xor*) (sometimes called a **classical** *cnot* **function**) represented diagrammatically as in Figure 2.15.

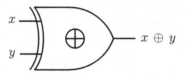

FIGURE 2.15
Diagrammatic representation of an *xor* gate or classical *cnot* function.

It is important for what follows to note that this function is not one-to-one. For example, both $(0, 1)$ and $(1, 0)$ get mapped to 1. Thus it is not invertible because we cannot recover the input values by knowing the output value. This has implications for reversibility which we shall discuss shortly.

Exercises

2.17   Use Table 2.11 to show that the binary operation $\oplus$ is commutative, that is $x \oplus y = y \oplus x$.

There are $2^4 = 16$ functions in the set $\mathcal{F}(\mathbb{B}^2, \mathbb{B})$, that is, of the type $f : \mathbb{B} \times \mathbb{B} \to \mathbb{B}$. Each of these functions accepts two Boolean variables x and y, combining these in various ways to produce a single Boolean variable as output. Mathematically, combination is achieved using further logical connectives: *and* $\wedge$, *or* $\vee$ and *not* $^-$ which are defined below:

Definition 2.8 Conjunction: and, $\wedge$
*Suppose $x, y \in \mathbb{B}$. The binary operator and, (**conjunction**), written $\wedge$, is equal to 1 when*

both x and y are equal to 1, and is zero otherwise:

$$0 \wedge 0 = 0$$
$$0 \wedge 1 = 0$$
$$1 \wedge 0 = 0$$
$$1 \wedge 1 = 1.$$

Definition 2.9 Disjunction: or, $\vee$
*The binary operator or, (**disjunction**), written $\vee$, is equal to 1 when either or both x and y are equal to 1, and is zero otherwise:*

$$0 \vee 0 = 0$$
$$0 \vee 1 = 1$$
$$1 \vee 0 = 1$$
$$1 \vee 1 = 1.$$

Definition 2.10 negation: not, $\overline{x}$
*The unary operator not, (**negation**), written $\overline{x}$, is equal to 1 when $x = 0$ and equal to 0 when $x = 1$:*

$$\overline{0} = 1$$
$$\overline{1} = 0.$$

It is important to note that just two connectives, conjunction $\wedge$ and negation $^-$, are sufficient to write any binary expression in an equivalent form. Thus we say that the set $\{\wedge, ^-\}$ is **functionally complete** or **universal**.

TABLE 2.12
The complete set of functions $f : \mathbb{B} \times \mathbb{B} \to \mathbb{B}$

x	y	f_0	f_1	f_2	f_3	f_4	f_5	f_6	f_7	f_8	f_9	f_{10}	f_{11}	f_{12}	f_{13}	f_{14}	f_{15}
0	0	0	0	0	0	0	0	0	0	1	1	1	1	1	1	1	1
0	1	0	0	0	0	1	1	1	1	0	0	0	0	1	1	1	1
1	0	0	0	1	1	0	0	1	1	0	0	1	1	0	0	1	1
1	1	0	1	0	1	0	1	0	1	0	1	0	1	0	1	0	1

The sixteen functions in $\mathcal{F}(\mathbb{B}^2, \mathbb{B})$ are defined by their inputs and outputs given in Table 2.12. Each can be expressed mathematically using combinations of $\wedge$, $\vee$ and $^-$ as shown in Table 2.13. Observe that f_6 in the table is the *xor* gate already discussed in Example 2.5.1. It is important to note that none of these functions are one-to-one. Knowing the output, it is impossible to determine the input. Consequently it is not possible to reverse the effects of these functions – they are not invertible. We see how to overcome this limitation shortly.

2.5.2 The Boolean functions $f : \mathbb{B}^2 \to \mathbb{B}^2$.

In the following example both the domain and the co-domain of the given function are the Cartesian product $\mathbb{B} \times \mathbb{B} = \mathbb{B}^2$. So the input to the function is an ordered pair of Boolean variables, likewise the output. We saw in Example 2.5.1 that the classical Boolean *cnot* gate was not invertible. In the following example, we see how this situation is remedied.

TABLE 2.13

Functions $f : \mathbb{B} \times \mathbb{B} \to \mathbb{B}$ expressed
mathematically

		logic description
$f_0(x,y) =$	0	
$f_1(x,y) =$	$x \wedge y$	*and*
$f_2(x,y) =$	$x \wedge \overline{y}$	
$f_3(x,y) =$	x	
$f_4(x,y) =$	$\overline{x} \wedge y$	
$f_5(x,y) =$	y	
$f_6(x,y) =$	$x \oplus y$	*xor*
$f_7(x,y) =$	$x \vee y$	*or*
$f_8(x,y) =$	$\overline{x \vee y}$	*not or (nor)*
$f_9(x,y) =$	$\overline{x \oplus y}$	
$f_{10}(x,y) =$	$\overline{y}$	
$f_{11}(x,y) =$	$x \vee \overline{y}$	
$f_{12}(x,y) =$	$\overline{x}$	
$f_{13}(x,y) =$	$\overline{x} \vee y$	
$f_{14}(x,y) =$	$\overline{x \wedge y}$	*not and (nand)*
$f_{15}(x,y) =$	1	

Example 2.5.2 The digital Feynman *cnot* gate.

Consider the function $F_D : \mathbb{B}^2 \to \mathbb{B}^2$ defined by

$$F_D(x,y) = (x, x \oplus y).$$

known as a digital Feynman *cnot* gate. (Here, we are using the letter F for a Feynman
gate and the subscript D to indicate we are considering the digital version. Later we will
introduce F_Q for its quantum emulation.) This function is sometimes written as

$$cnot : \mathbb{B}^2 \to \mathbb{B}^2, \quad cnot(x,y) = (x, x \oplus y).$$

It is tabulated in Table 2.14 and represented diagrammatically in Figure 2.16. An alternative
diagrammatic form – Feynman's form – is shown in Figure 2.17. Observe from the table
that this function is bijective.

TABLE 2.14

The function $F_D : \mathbb{B}^2 \to \mathbb{B}^2$,
$F_D(x,y) = (x, x \oplus y)$

input (x,y)	output $(x, x \oplus y)$
(0,0)	(0,0)
(0,1)	(0,1)
(1,0)	(1,1)
(1,1)	(1,0)

Unlike the earlier classical *cnot* gate, the Feynman *cnot* gate is reversible: we can recover the input values if we know the output values, that is, the gate is invertible. You
should verify this yourself. The Feynman *cnot* gate can also be represented as $F_D(x,y) =$
$(f_3(x,y), f_6(x,y))$ of Tables 2.12 and 2.13.

It is worth noting that *cnot* is sometimes referred to as a 'controlled not' function. If
the first argument of the function is 0, the second does not change; if the first argument is
1, we apply a *not* to the second argument.

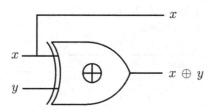

FIGURE 2.16
Diagrammatic representation of the Feynman *cnot* gate.

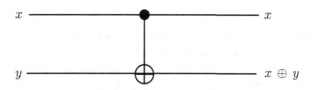

FIGURE 2.17
Feynman's form of the digital invertible *cnot* gate: input (x, y), output $(x, x \oplus y)$.

Example 2.5.3 The identity function on $\mathbb{B}^2$.

Consider the function $i : \mathbb{B}^2 \to \mathbb{B}^2$ defined by

$$i(x, y) = (x, y).$$

Because the output of the function $i(x, y) = (x, y)$ is the same as the input, this is referred

TABLE 2.15
The identity function
$i : \mathbb{B}^2 \to \mathbb{B}^2$, $i(x, y) = (x, y)$

input (x, y)	output (x, y)
(0,0)	(0,0)
(0,1)	(0,1)
(1,0)	(1,0)
(1,1)	(1,1)

to as the **identity function** on $\mathbb{B}^2$. This function is tabulated in Table 2.15. Note that using Tables 2.12 and 2.13, we can write $i(x, y) = (f_3(x, y), f_5(x, y))$.

Example 2.5.4

Consider the function $f : \mathbb{B}^2 \to \mathbb{B}^2$ defined by

$$f(x, y) = (x \vee y, x).$$

Note that this is the function $f(x, y) = (f_7, f_3)$ from Tables 2.12 and 2.13. Show that this function is not bijective.

TABLE 2.16

The function $f : \mathbb{B}^2 \to \mathbb{B}^2$,
$f(x, y) = (x \vee y, x)$

input (x, y)	output $(x \vee y, x)$
(0,0)	(0,0)
(0,1)	(1,0)
(1,0)	(1,1)
(1,1)	(1,1)

Solution

We tabulate the function as shown in Table 2.16. Observe from the table that the element $(1, 1)$ in the range of the function is mapped to by both $(1, 0)$ and $(1, 1)$ in the domain. Hence, this function cannot be injective, and therefore, it cannot be bijective. A consequence is that this function is not reversible: knowing the output $(1, 1)$, we cannot recover the input.

We note that all Boolean functions $f : \mathbb{B}^2 \to \mathbb{B}^2$ may be represented in the form $f(x, y) = (f_i(x, y), f_j(x, y))$ for some f_i and f_j, the functions $\mathbb{B}^2 \to \mathbb{B}$ in Tables 2.12 and 2.13. This gives us a total of $4^2 \times 4^2 = 4^4 = 256$ functions in the set $\mathcal{F}(\mathbb{B}^2, \mathbb{B}^2)$. However, it can be shown that only $4! = 24$ of these are invertible. The invertible ones are identified in Proposition 2.1.

Proposition 2.1 *The invertible functions* $f : \mathbb{B}^2 \to \mathbb{B}^2$ *are the pairs:*

$$(f_3, f_5), \ (f_3, f_6), \ (f_3, f_9), \ (f_3, f_{10}),$$

$$(f_5, f_6), \ (f_5, f_9), \ (f_5, f_{12}),$$

$$(f_6, f_{10}), \ (f_6 f_{12}),$$

$$(f_9, f_{10}), \ (f_9, f_{12}),$$

$$(f_{10}, f_{12})$$

and the same pairs in contrary order. Explicitly, the pairs are

$$(x, y), \ (x, x \oplus y), \ (x, \overline{x \oplus y}), \ (x, \overline{y}),$$

$$(y, x \oplus y), \ (y, \overline{x \oplus y}), \ (y, \overline{x}),$$

$$(x \oplus y, \overline{y}), \ (x \oplus y, \overline{x}),$$

$$(\overline{x \oplus y}, \overline{y}), \ (\overline{x \oplus y}, \overline{x}),$$

$$(\overline{y}, \overline{x}).$$

Proof

The 24 invertible functions $g_k : \mathbb{B}^2 \to \mathbb{B}^2$, for $0 \leq k \leq 23$ are identified by selecting from the 256 only those which are bijective, yielding:

x	y	g_0	g_1	g_2	g_3	g_4	g_5	g_6	g_7	g_8	g_9	g_{10}	g_{11}
0	0	00	00	00	00	00	00	01	01	01	01	01	01
0	1	01	01	10	10	11	11	00	00	10	10	11	11
1	0	10	11	01	11	01	10	10	11	00	11	10	00
1	1	11	10	11	01	10	01	11	10	11	00	00	10

x	y	g_{12}	g_{13}	g_{14}	g_{15}	g_{16}	g_{17}	g_{18}	g_{19}	g_{20}	g_{21}	g_{22}	g_{23}
0	0	10	10	10	10	10	10	11	11	11	11	11	11
0	1	00	00	01	01	11	11	01	01	10	10	00	00
1	0	01	11	00	11	00	01	10	00	01	00	01	10
1	1	11	01	11	00	01	00	00	10	00	01	10	01

that is, we have

$$
\begin{aligned}
g_0 &= (f_3, f_5), & g_1 &= (f_3, f_6), & g_2 &= (f_5, f_3), & g_3 &= (f_6, f_3), & g_4 &= (f_5, f_6), \\
g_5 &= (f_6, f_5), & g_6 &= (f_3, f_9), & g_7 &= (f_3, f_{10}), & g_8 &= (f_5, f_9), & g_9 &= (f_6, f_{10}), \\
g_{10} &= (f_6, f_{12}), & g_{11} &= (f_5, f_{12}), & g_{12} &= (f_9, f_3), & g_{13} &= (f_{10}, f_3), & g_{14} &= (f_9, f_5), \\
g_{15} &= (f_{10}, f_6), & g_{16} &= (f_{12}, f_5), & g_{17} &= (f_{12}, f_6), & g_{18} &= (f_{10}, f_{12}), & g_{19} &= (f_9, f_{12}), \\
g_{20} &= (f_{12}, f_{10}), & g_{21} &= (f_{12}, f_9), & g_{22} &= (f_9, f_{10}), & g_{23} &= (f_{10}, f_9).
\end{aligned}
$$

Of all the invertible functions of the type $f : \mathbb{B}^2 \to \mathbb{B}^2$, the Feynman *cnot* gate of Example 2.5.2, $f(x, y) = (x, x \oplus y)$ is, essentially, the only one of computational significance.

Exercises

2.18   Consider the function $f : \mathbb{B}^2 \to \mathbb{B}^2$, $f(x, y) = (x, y \oplus 0)$. Tabulate input and output values and show that f is bijective.

2.19   Consider the function $f : \mathbb{B}^2 \to \mathbb{B}^2$, $f(x, y) = (x, y \oplus 1)$. Tabulate input and output values and show that f is bijective.

2.5.3   Further Boolean functions

Example 2.5.5 The digital Toffoli gate or *ccnot* gate.

Consider the three-input function $T_D : \mathbb{B}^3 \to \mathbb{B}^3$ defined by

$$T_D(x, y, z) = (x, y, z \oplus (x \wedge y))$$

known as a **digital Toffoli gate** or *ccnot* gate. This function is tabulated in Table 2.17 and is often represented as shown in Figure 2.18. If necessary, when producing such a table, intermediate steps can be introduced. So, for example, to evaluate $z \oplus (x \wedge y)$ when the input is $(1, 1, 1)$ note that $x \wedge y = 1 \wedge 1 = 1$, and then $z \oplus (x \wedge y) = 1 \oplus 1 = 0$. Careful inspection of the table reveals that this is a bijective function and hence is reversible.

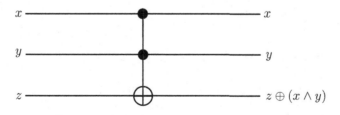

FIGURE 2.18
The digital Toffoli or *ccnot* gate.

Consider again the Toffoli gate in Example 2.5.5. By setting $z = 0$, we obtain

$$T_D(x, y, 0) = (x, y, 0 \oplus (x \wedge y)).$$

TABLE 2.17

The function $T_D : \mathbb{B}^3 \to \mathbb{B}^3$,
$T_D(x, y, z) = (x, y, z \oplus (x \wedge y))$

input (x, y, z)	output $(x, y, z \oplus (x \wedge y))$
(0,0,0)	(0,0,0)
(0,0,1)	(0,0,1)
(0,1,0)	(0,1,0)
(0,1,1)	(0,1,1)
(1,0,0)	(1,0,0)
(1,0,1)	(1,0,1)
(1,1,0)	(1,1,1)
(1,1,1)	(1,1,0)

We have seen earlier (Example 2.4.4) that $0 \oplus f(x) = f(x)$, and it follows that

$$T_D(x, y, 0) = (x, y, x \wedge y).$$

Hence, with $z = 0$, T_D enables evaluation of the *and* function $x \wedge y$. Also, with $x = y = 1$ we have

$$\begin{aligned} T_D(1, 1, z) &= (1, 1, z \oplus (1 \wedge 1)) \\ &= (1, 1, z \oplus 1) \\ &= (1, 1, \overline{z}) \end{aligned}$$

since $z \oplus 1 = \overline{z}$ (see Example 2.4.4). Hence, the Toffoli gate can be used to calculate the *not* function. We noted earlier that the set $\{\wedge, \overline{\ }\}$ is functionally complete or universal. Because the $\wedge$ and not functions can be calculated using carefully chosen values of the inputs, the Toffoli function can perform any computation on a digital computer in a reversible manner. We also have

$$T_D(x, y, 1) = (x, y, 1 \oplus (x \wedge y)) = (x, y, \overline{x \wedge y})$$

to give *nand* (x, y), that is f_{14} in Table 2.13. Further, it is straightforward to verify that $T_D(x, 1, 0) = (x, 1, x)$, $T_D(1, x, 0) = (1, x, x)$ and $T_D(x, x, 0) = (x, x, x)$, so this gate can be used to copy input values.

In summary it follows that the Toffoli function defines a complete (or universal) set for the generation of invertible (or reversible) versions of the Boolean functions $f : \mathbb{B}^2 \to \mathbb{B}$. The Toffoli function is also a significant quantum operator as we shall see in Chapter 15. It is, however, not universal for quantum computation, but from the above, it follows that a quantum computer can implement all possible digital computations.

Example 2.5.6 The digital Peres or half-adder gate.

Consider the three-input function $P_D : \mathbb{B}^3 \to \mathbb{B}^3$ defined by

$$P_D(x, y, z) = (x, y \oplus x, z \oplus (x \wedge y)).$$

This function is tabulated in Table 2.18 and often represented as in Figure 2.19. This function is known as the **digital Peres** or **half-adder gate**. It can be checked directly from the table that this is a bijective function. The Peres gate may be viewed as a product of the Toffoli gate with the invertible *cnot* gate, as we show in Section 2.7.

TABLE 2.18

The function $P_D : \mathbb{B}^3 \rightarrow \mathbb{B}^3$,

$P_D(x, y, z) = (x, y \oplus x, z \oplus (x \wedge y))$

input (x, y, z)	output $(x, y \oplus x, z \oplus (x \wedge y))$
(0,0,0)	(0,0,0)
(0,0,1)	(0,0,1)
(0,1,0)	(0,1,0)
(0,1,1)	(0,1,1)
(1,0,0)	(1,1,0)
(1,0,1)	(1,1,1)
(1,1,0)	(1,0,1)
(1,1,1)	(1,0,0)

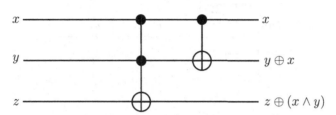

FIGURE 2.19

The digital Peres half-adder.

2.6 Further composition of functions

Example 2.6.1

Consider the function $F : \mathbb{B}^2 \rightarrow \mathbb{B}^2$ defined by

$$F(x, y) = (\overline{y}, \overline{x}).$$

Obtain an expression for the composite function $F \circ F$.

Solution

$$
\begin{aligned}
F \circ F = (F \circ F)(x, y) &= F(F(x, y)) \\
&= F(\overline{y}, \overline{x}) \\
&= (\overline{\overline{x}}, \overline{\overline{y}}) \\
&= (x, y).
\end{aligned}
$$

Note that the result of finding $F \circ F$ is the identity function on $\mathbb{B}^2$ so that $F(x, y) = (\overline{y}, \overline{x})$ is its own inverse, i.e., it is self-inverse.

Example 2.6.2

Consider the functions $F : \mathbb{B}^2 \rightarrow \mathbb{B}^2$ and $G : \mathbb{B}^2 \rightarrow \mathbb{B}^2$ defined by

$$F(x, y) = (x \oplus y, \overline{y}), \qquad G(x, y) = (\overline{x \oplus y}, \overline{y}).$$

Obtain an expression for the composite function $F \circ G$.

Solution

$$F \circ G = (F \circ G)(x, y) = F(G(x, y))$$
$$= F(\overline{x \oplus y}, \overline{y})$$
$$= ((\overline{x \oplus y}) \oplus \overline{y}, \overline{\overline{y}}).$$

Now consider just $(\overline{x \oplus y}) \oplus \overline{y}$. This function is tabulated in Table 2.19 and is equal to x. Consequently, $F \circ G$ is given by Table 2.20 and thus, in this example, as in Example 2.6.1,

TABLE 2.19
Tabulation of the function $(\overline{x \oplus y}) \oplus \overline{y}$

x	y	$x \oplus y$	$\overline{x \oplus y}$	$\overline{y}$	$(\overline{x \oplus y}) \oplus \overline{y}$
0	0	0	1	1	0
0	1	1	0	0	0
1	0	1	0	1	1
1	1	0	1	0	1

TABLE 2.20
Calculating the
composite function
$F \circ G$

(x, y)	$(F \circ G)(x, y)$
$(0, 0)$	$(0, 0)$
$(0, 1)$	$(0, 1)$
$(1, 0)$	$(1, 0)$
$(1, 1)$	$(1, 1)$

$F \circ G$ is the identity function on $\mathbb{B}^2$. You should verify that $G \circ F$ is also equal to the identity function. Observe that $F(x, y) = (x \oplus y, \overline{y})$ is the inverse of $G(x, y) = (\overline{x \oplus y}, \overline{y})$ (and vice-versa).

2.7   The Cartesian product of functions

Definition 2.11 The Cartesian product of functions
*Consider the two functions $f_1 : A_1 \to B_1$ and $f_2 : A_2 \to B_2$. Note that the two functions may be defined on different domains, A_1 and A_2. The **Cartesian product** of the two functions is another function, $f_1 \times f_2$, defined as follows:*

$$f_1 \times f_2 : A_1 \times A_2 \to B_1 \times B_2$$

$$(f_1 \times f_2)(a_1, a_2) = (f_1(a_1), f_2(a_2)).$$

Note $(f_1(a_1), f_2(a_2)) \in B_1 \times B_2$ for all $a_1 \in A_1$ and $a_2 \in A_2$.

Example 2.7.1

We have already seen the *cnot* function

$$cnot : \mathbb{B}^2 \to \mathbb{B}^2, \qquad cnot(x, y) = (x, x \oplus y)$$

and the identity function on $\mathbb{B}$,

$$i : \mathbb{B} \to \mathbb{B}, \qquad i(x) = x.$$

Note that the *cnot* and identity function are defined on different domains. Thus we can now determine the meaning of functions such as

$$cnot \times i : \mathbb{B}^2 \times \mathbb{B} \to \mathbb{B}^2 \times \mathbb{B}$$

$$(cnot \times i)((x, y), z) = (cnot(x, y), i(z))$$
$$= ((x, x \oplus y), z).$$

Strictly, the right-hand side is a vector in $\mathbb{B}^2 \times \mathbb{B}$. But since there is a natural isomorphism from $\mathbb{B}^2 \times \mathbb{B}$ to $\mathbb{B}^3$, we can readily write the result as

$$(cnot \times i)(x, y, z) = (x, x \oplus y, z).$$

Example 2.7.2 The digital Peres gate revisited

We have seen the digital Toffoli and Peres gates in Examples 2.5.5, and 2.5.6.

$$T_D : \mathbb{B}^3 \to \mathbb{B}^3, \qquad T_D(x, y, z) = ccnot(x, y, z) = (x, y, z \oplus (x \wedge y))$$

$$P_D : \mathbb{B}^3 \to \mathbb{B}^3, \qquad P_D(x, y, z) = (x, y \oplus x, z \oplus (x \wedge y)).$$

Observe that

$$\begin{aligned}
P_D(x, y, z) &= (x, y \oplus x, z \oplus (x \wedge y)) \\
&= (cnot(x, y), z \oplus (x \wedge y)) \\
&= (cnot \times i)(x, y, z \oplus (x \wedge y)) \\
&= (cnot \times i)(T_D(x, y, z)) \\
&= (cnot \times i) \circ T_D(x, y, z).
\end{aligned}$$

Thus $P_D = (cnot \times i) \circ ccnot$.

2.8 Permuting (swapping) binary digits and binary strings

The Boolean functions which we have introduced can be used to construct functions which permute, or swap, binary digits. In this section we explore several ways in which this can be achieved.

2.8.1 A classical digital circuit for swapping binary digits

The function $f : \mathbb{B}^2 \to \mathbb{B}^2$ defined by $f(x,y) = (y,x)$, which swaps the binary digits x and y, is one of the 24 invertible functions referred to in Section 2.5.2. Suppose we wish to create a classical circuit to swap digits. The classical *xor* gate introduced in Example 2.5.1 is not invertible, and to produce a classical circuit for swapping digits, it is necessary to 'signal-split', or 'fan-out' at three points of the circuit – see Figure 2.20.

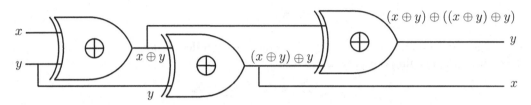

FIGURE 2.20
A classical circuit for swapping binary digits.

To verify that this circuit performs as intended work through Table 2.21 from which it follows that $(x \oplus y) \oplus y = x$ and $(x \oplus y) \oplus ((x \oplus y) \oplus y) = y$: the digits have been swapped.

TABLE 2.21
A classical circuit for swapping binary digits (see Figure 2.20)

x	y	$x \oplus y$	$(x \oplus y) \oplus y = x$	$(x \oplus y) \oplus ((x \oplus y) \oplus y) = y$
0	0	0	0	0
0	1	1	0	1
1	0	1	1	0
1	1	0	1	1

2.8.2 Swapping binary digits using the Feynman *cnot* gate

A functionally equivalent circuit to that of Figure 2.20 may be defined using the invertible Feynman *cnot* function introduced in Example 2.5.2, $cnot : \mathbb{B}^2 \to \mathbb{B}^2$, $cnot(x,y) = (x, x \oplus y)$. We define the $cnot_1$ and $cnot_2$ functions by:

$$cnot_1(x,y) = (x, y \oplus x), \quad cnot_2(x,y) = (x \oplus y, y)$$

which are shown graphically in Figure 2.21.

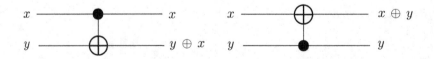

FIGURE 2.21
The digital, invertible $cnot_1$ and $cnot_2$ gates.

We can show that
$$cnot_2 \circ cnot_1 \circ cnot_2(x,y) = (y,x),$$
i.e., a digital *swap* function defined by explicitly invertible gates as shown in Figure 2.22.

Specifically,

$$cnot_1 \circ cnot_2(x,y) = cnot_1(cnot_2(x,y))$$
$$= cnot_1((x \oplus y, y))$$
$$= (x \oplus y, y \oplus (x \oplus y)).$$

Then,

$$cnot_2 \circ cnot_1 \circ cnot_2(x,y) = cnot_2(x \oplus y, y \oplus (x \oplus y))$$
$$= ((x \oplus y) \oplus (y \oplus (x \oplus y)), y \oplus (x \oplus y))$$
$$= (y, x)$$

(using Table 2.21). For brevity we often depict bit-swapping as shown in Figure 2.23.

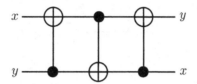

FIGURE 2.22
The *cnot* implementation of the digital 2-bit swap circuit.

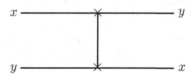

FIGURE 2.23
Alternative representation of the digital 2-bit swap gate.

2.8.3 Swapping strings of binary digits – vectorising the *swap* operator

Suppose we define the bit-swapping function $swap_{1,2}(x,y) = (y,x)$, then we can 'vectorise' this in an obvious way to create a swap function, $swap_v$ say, for binary strings.

To swap the strings $(x_1, x_2; y_1, y_2)$ to $(y_1, y_2; x_1, x_2)$ we have:

$$swap_v(x_1, x_2; y_1, y_2) = swap_{2,4} \circ swap_{1,3}(x_1, x_2; y_1, y_2)$$
$$= swap_{2,4}(y_1, x_2; x_1, y_2)$$
$$= (y_1, y_2; x_1, x_2)$$

as shown in Figure 2.24.

This extends in the obvious way to swap strings of length n; see Figure 2.25.

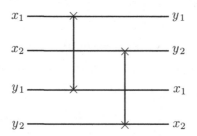

FIGURE 2.24
Vectorising the *swap* function to swap binary strings.

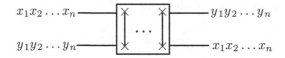

FIGURE 2.25
A *swap* function for binary strings.

2.9 Copying binary digits and binary strings

In this section, we look at several ways in which single binary digits and then strings of binary digits can be copied. Specifically, we consider signal-splitting or 'fan-out', a *dupe* gate, the *cnot* gate, the Toffoli gate and the Feynman double gate.

2.9.1 Fan-out

Figure 2.26 shows a simple circuit for copying a bit. There is clearly something physical happening at the point where the 'signal' is 'split'; this is usually referred to as 'fan-out' and taken as implicit (or assumed without comment) in digital computing. Clearly the circuit transforms $x \in \mathbb{B}$ to $(x, x) \in \mathbb{B}^2$.

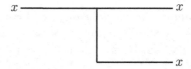

FIGURE 2.26
A fan-out circuit for copying, $x \in \mathbb{B}$.

2.9.2 A *dupe* gate

An alternative view of fan-out, more suited to comparison with the quantum case, is to make the signal-splitting process explicit with a '*dupe*' gate

$$(i, i) : \mathbb{B} \to \mathbb{B}^2$$

defined by:

$$(i, i)(x) = (i(x), i(x))$$
$$= (x, x).$$

Here $i : \mathbb{B} \to \mathbb{B}$, $i(x) = x$ for $x \in \mathbb{B}$ is the identity function. As a function $(i, i) : \mathbb{B} \to \mathbb{B}^2$ is not invertible. It is invertible if the range is defined to be the diagonal subset $\{(x, x) : x \in \mathbb{B}\}$ of $\mathbb{B}^2$. The *dupe* gate (i, i), which is functionally equivalent to the fan-out circuit above, may be represented as shown in Figure 2.27. Clearly a network of *dupe* gates enables multiple copies of the input bit to be produced – see Figure 2.28.

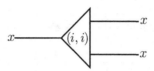

FIGURE 2.27
A *dupe* gate to copy a bit.

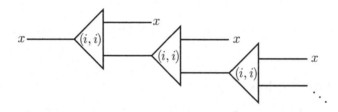

FIGURE 2.28
A network of *dupe* gates for $x \to x^k$ where $x^k = (x, x, x, \ldots) = (xxx \ldots)$.

2.9.3   The *cnot* gate.

The fan-out and *dupe* gate approaches above are non-invertible processes. At the expense of additional, or auxiliary, inputs, digital copying can be achieved using the Feynman *cnot* invertible gate of Example 2.5.2. Observe that the Feynman *cnot* gate may be expressed as a function of the *dupe* gate and the classical, non-invertible, *xor* gate – see Figure 2.29. The

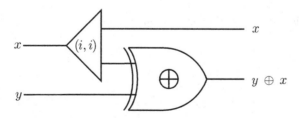

FIGURE 2.29
The *cnot* gate as a function of the *dupe* and classical non-invertible *xor* gates.

function of Figure 2.29 yields (x, x) on the input of $(x, 0)$, (since $0 \oplus x = x$), the 0 being the auxiliary input needed to copy using an invertible-digital circuit; mathematically the circuit may be expressed as:

$$(i \times xor) \circ ((i, i) \times i)(x, y) = (i \times xor)(x, x, y)$$
$$= (x, y \oplus x)$$
$$= cnot(x, y).$$

Hence, $cnot(x, 0) = (x, x)$. The circuit of *cnot* gates shown (in Feynman notation) in Figure 2.30 yields x^k (i.e., $(xxx \ldots x)$) on the input of $x, 0^{k-1}$ (i.e., $(x000 \ldots 0)$). Again we note

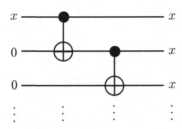

FIGURE 2.30
Copying a binary digit using the *cnot* gate.

the necessity for auxiliary inputs 0^{k-1} to implement the desired circuit using only invertible gates.

2.9.4 The Feynman double gate

Similarly, with the Feynman double gate defined by $F_D^*(x, y, z) = (x, y \oplus x, z \oplus x)$, we have

$$F_D^*(x, 0, 0) = (x, x, x)$$

as depicted in Figure 2.31.

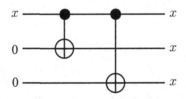

FIGURE 2.31
Copying a binary digit using the Feynman double gate.

2.9.5 The *ccnot* (Toffoli) gate

With the *ccnot* (Toffoli) gate, $T_D(x, y, z) = (x, y, z \oplus (x \wedge y))$ (Example 2.5.5) we have:

$$T_D(x, 1, 0) = (x, 1, 0 \oplus (x \wedge 1)) = (x, 1, x \wedge 1) = (x, 1, x).$$

2.9.6 Fan-out copying of binary strings

The fan-out circuit of Figure 2.32 maps the string $x_1 x_2 \in \mathbb{B}^2$ to the string $x_1 x_2 x_1 x_2 \in \mathbb{B}^4$ or, equivalently, $(x_1, x_2) \in \mathbb{B}^2$ to $((x_1, x_2), (x_1, x_2)) \in \mathbb{B}^2 \times \mathbb{B}^2$.

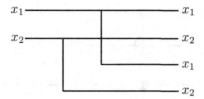

FIGURE 2.32
A fan-out Boolean copy circuit: $x_1, x_2 \in \{0, 1\}$.

2.9.7 Digital string copying with the non-reversible *copy* and reversible *swap* gates

We can represent the process represented by Figure 2.32 with the circuit shown in Figure 2.33. The *swap* gate is clearly invertible and may be implemented using the *cnot* gate shown earlier in Figure 2.22. Copying a 2-bit string x_1, x_2, in the manner of Figure 2.33, may therefore be expressed mathematically as:

$$
\begin{aligned}
swap_{23} \circ ((i, i) \times (i, i))(x_1, x_2) &= swap_{23}((i, i)(x_1), (i, i)(x_2)) \\
&= swap_{23}(x_1, x_1, x_2, x_2) \\
&= (x_1, x_2, x_1, x_2)
\end{aligned}
$$

and similarly for arbitrary strings $x_1 \cdots x_n$ of digital bits.

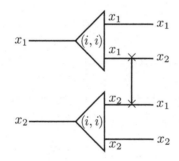

FIGURE 2.33
An alternative representation of the copy circuit of Figure 2.32.

2.9.8 Digital string copying using only reversible (the *cnot* and *swap*) gates

An alternative network, using only invertible Boolean gates, is shown in Figure 2.34. The input $(x_1, 0, x_2, 0)$ yields (x_1, x_2, x_1, x_2). We note that the use of only reversible functions again requires the input of auxiliary bits – y_1 and y_2 in this case.

2.9.9 Digital string copying using only the *cnot* gate

'Vectorising' the invertible *cnot* gate to copy strings we have:

$$
cnot(x_1, x_2 \ y_1, y_2) = (x_1, x_2, y_1 \oplus x_1, y_2 \oplus x_2)
$$

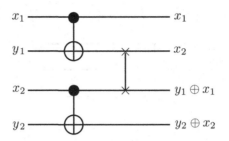

FIGURE 2.34
An invertible copy gate using the *cnot* and *swap* gates – functionally equivalent to that of Figure 2.33.

as depicted in Figure 2.35, and hence, by setting $y_1 = y_2 = 0$,

$$cnot(x_1, x_2, 0, 0) = (x_1, x_2, x_1, x_2)$$

sometimes written concisely as

$$cnot(x_1 x_2, 00) = (x_1 x_2, x_1 x_2).$$

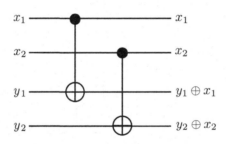

FIGURE 2.35
Copying a binary string using the *cnot* gate.

More generally
$$cnot(x_1 x_2 \cdots x_n, 0^n) = (x_1 x_2 \cdots x_n, \; x_1 x_2 \cdots x_n).$$

2.10   Periodic functions

2.10.1   Real-valued periodic functions

Definition 2.12 Periodic function
Consider a real-valued function $f : \mathbb{R} \to \mathbb{R}$. If for some value $p \in \mathbb{R} \backslash 0$, we have

$$f(x) = f(x + p)$$

for all x, then f is said to be **periodic**. *The smallest p for which this is true is called the* **period** *of f.*

The graph of such a function has a definite pattern which is repeated at regular intervals.

Example 2.10.1

The continuous function $f(x) = \sin x$ shown in Figure 2.36 has period $p = 2\pi$. Observe that 2π is the smallest interval on the x axis over which the pattern repeats.

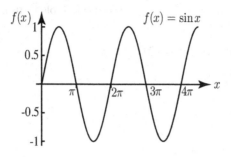

FIGURE 2.36
The function $f(x) = \sin x$ with period $p = 2\pi$.

Example 2.10.2

Figure 2.37 shows a graph of a periodic function which is discontinuous (i.e., there are jumps in the graph). The function is defined as

$$f(x) = x, \quad 0 \le x < 1, \qquad \text{period } p = 1.$$

Observe (by introducing the dashed extension to the line in Figure 2.37) that in the interval $1 \le x < 2$ the function is defined by $f(x) = x - 1$. Likewise, in the interval $2 \le x < 3$, the function is defined by $f(x) = x - 2$, and so on.

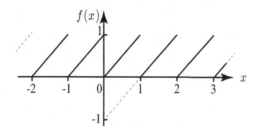

FIGURE 2.37
A discontinuous, periodic function with period $p = 1$.

Exercises

2.20 By sketching a graph, or otherwise, show that a one-to-one function cannot be periodic.

2.21   If $f : \mathbb{R} \to \mathbb{R}$ is a constant function, i.e., $f(x) = c$, for all $x \in \mathbb{R}$, show that f is periodic but has no period.

2.22   If $f : \mathbb{R} \to \mathbb{R}$ is periodic with period p show that $f(x \pm np) = f(x)$ for all $n \in \mathbb{N}$.

2.10.2   Periodic Boolean functions

Following the case of real functions, we now consider Boolean functions which exhibit periodicity.

Definition 2.13 Periodic Boolean function
The Boolean function $f : \mathbb{B}^n \to \mathbb{B}^n$ is periodic if there exists $s \in \mathbb{B}^n$, $s \neq 0^n$, such that

$$f(x) = f(x \oplus s)$$

for all $x \in \mathbb{B}^n$.

An understanding of periodic Boolean functions is essential for the development of some quantum algorithms, e.g., Simon's algorithm (see Chapter 23).

Example 2.10.3

Consider the Boolean function $f : \mathbb{B}^3 \to \mathbb{B}^3$ defined in Table 2.22. Show that this function is periodic with period $s = 001$.

TABLE 2.22
A Boolean function
$f : \mathbb{B}^3 \to \mathbb{B}^3$

x	$f(x)$
000	000
001	000
010	000
011	000
100	101
101	101
110	111
111	111

Solution

Table 2.23 shows the result of calculating $x \oplus s$ and $f(x \oplus s)$. For example, when $x = 111$, $x \oplus s = 111 \oplus 001 = 110$, and $f(110) = 111 = f(111)$. Observe, by carefully inspecting the table, that for each x, $f(x \oplus s) = f(x)$, and hence, the function is periodic with period $s = 001$.

Exercises

2.23   Consider the Boolean function $f : \mathbb{B}^3 \to \mathbb{B}^3$ defined in Table 2.24.

(a) Show that this function is many-to-one (in fact four-to-one).

TABLE 2.23

$x \oplus s$ and $f(x \oplus s)$ when $s = 001$

x	$f(x)$	$x \oplus s$	$f(x \oplus s)$
000	000	001	000
001	000	000	000
010	000	011	000
011	000	010	000
100	101	101	101
101	101	100	101
110	111	111	111
111	111	110	111

TABLE 2.24

000	111
001	010
010	010
011	111
100	010
101	111
110	111
111	010

(b) Show that this function is periodic with period $s = 011$.

(c) Show that this function is also periodic with period $s = 101$. Deduce that the period of a periodic Boolean function may not be unique.

2.11   End-of-chapter exercises

1. Explain why the relation depicted in Figure 2.38 is not a function.

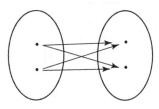

FIGURE 2.38

2. Explain why the functions $f : \mathbb{B} \to \mathbb{B}$ depicted in Figure 2.39 are not invertible.

3. If $x, y \in \mathbb{B}$ show that $x \oplus \overline{y} = \overline{x \oplus y}$.

4. Tabulate the function $f : \mathbb{B}^2 \to \mathbb{B}^2$, $f(x, y) = (y, \overline{x \oplus y})$ and determine whether it is bijective.

5. Tabulate the function $f : \mathbb{B}^2 \to \mathbb{B}^2$, $f(x, y) = (x \wedge y, \overline{x \wedge y})$ and determine whether it is one-to-one, two-to-one or neither.

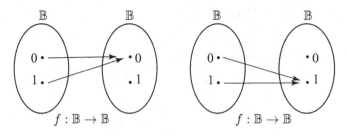

FIGURE 2.39

6. Show that the function $f : \mathbb{B}^2 \to \mathbb{B}$, $f(x,y) = (x \wedge \overline{y}) \vee (\overline{x} \wedge y)$ is equivalent to the function $f(x,y) = x \oplus y$.

7. Determine whether the function $f : \mathbb{B}^2 \to \mathbb{B}^2$, $f(x,y) = (x \oplus y, x \wedge y)$ is invertible.

8. Show that the function $f : \mathbb{B}^3 \to \mathbb{B}^3$, $f(x,y,z) = (x, x \oplus y, z \oplus (x \wedge y))$ is invertible.

9. Show the logical equivalence expressed in De Morgan's law: $\overline{x \wedge y} = \overline{x} \vee \overline{y}$.

10. Consider the three functions

$$i : \mathbb{B} \to \mathbb{B}, \ i(x) = x; \qquad F_D(x,y) : \mathbb{B}^2 \to \mathbb{B}^2, \ F_D(x,y) = (x, x \oplus y);$$

$$T_D : \mathbb{B}^3 \to \mathbb{B}^3, \ T_D(x,y,z) = (x, y, z \oplus (x \wedge y)).$$

Use the definition of the Cartesian product of two functions to show that

$$(F_D \times i) \circ T_D(x,y,z) = (F_D \times i)(x, y, z \oplus (x \wedge y))$$
$$= (F_D(x,y), i(z \oplus (x \wedge y))),$$

which is equivalent to the digital Peres gate (Example 2.5.6).

11. Consider the function $f : \mathbb{B}^3 \to \mathbb{B}^3$ tabulated in Table 2.25. State the range of this function. Explain why the co-domain is not the same as the range. Show that this function is two-to-one.

TABLE 2.25
See Q11

000	011
001	010
010	010
011	011
100	000
101	001
110	001
111	000

12. Consider the function $f : \mathbb{R} \to \mathbb{R}$. Let $x_1, x_2 \in \mathbb{R}$. Suppose $f(x_1) = f(x_2)$. Show that

(a) if $x_1 \neq x_2$ then f is many-to-one,

(b) if f is one-to-one then $x_1 = x_2$,

(c) if $f(x_1) = f(x_2)$ implies $x_1 = x_2$ then f is one-to-one.

13. Consider $f : \mathbb{R}^2 \to \mathbb{R}^2$, $f(x,y) = (x - 7y, 2x + y)$. Prove that f is a one-to-one function.

14. To demonstrate that a function $f : A \to B$ is surjective, it must be possible to choose any element $b \in B$, and then find an element $a \in A$ such that $f(a) = b$.

    (a) Show that the function $f : \mathbb{R} \to \mathbb{R}$, $f(x) = 3x + 5$ is surjective.

    (b) Show that the function $f : \mathbb{R}^+ \to \mathbb{R}$, $f(x) = \log_2 x$ is surjective.

15. A Boolean gate $G : \mathbb{B}^k \to \mathbb{B}^k$ is said to be linear if $G(a \oplus b) = G(a) \oplus G(b)$ for all $a, b \in \mathbb{B}^k$.

    (a) Show that the Feynman gate $F_D(x, y) = (x, y \oplus x)$ is linear on $\mathbb{B}^2$. Hint: you need to show $F_D(x_1, y_1) \oplus F_D(x_2, y_2) = F_D((x_1, y_1) \oplus (x_2, y_2))$.

    (b) Show that the double Feynman gate $F_D^*(x, y, z) = (x, y \oplus x, z \oplus x)$ is linear on $\mathbb{B}^3$.

16. For $x_1, x_2, x_3 \in \mathbb{B}$, by tabulating the functions show that

$$(x_1 \wedge x_2) \wedge x_3 = x_1 \wedge (x_2 \wedge x_3).$$

Deduce the associativity of $\wedge$ and note that we can simply write $x_1 \wedge x_2 \wedge x_3$ without any ambiguity.

3

Complex numbers

3.1 Objectives

The objective of this chapter is to introduce complex numbers and to develop fluency in their manipulation. It is helpful to think of a complex number as an object of the form $a + ib$ where a and b are real numbers called the real and imaginary parts of the complex number. The symbol i is referred to as the **imaginary unit** and has the unusual property that $i^2 = -1$. This is unusual in the sense that there is no real number which when squared yields a negative result, and so i is not real. Alternatively, complex numbers can be viewed as ordered pairs of real numbers, e.g. (a, b), upon which we impose rules for addition and multiplication. Both of these approaches are discussed in this chapter.

A knowledge of complex numbers is essential for understanding quantum computation. This is because the state of a quantum system is described by mathematical objects called 'vectors' that reside in a 'complex vector space'. In this chapter we shall introduce complex numbers as solutions of quadratic equations, before explaining how calculations are performed with them. Three common algebraic formats in which complex numbers can be represented are described – Cartesian, polar and exponential forms – and we explain a graphical representation known as an Argand diagram, which helps visualisation. This chapter contains essential foundations required for an understanding of complex vector spaces, tensor product spaces and quantum algorithms which follow.

3.2 Introduction to complex numbers and the imaginary number i

Suppose we wish to solve the quadratic equation $2x^2 + 2x + 5 = 0$. This equation will not factorise using real numbers. The usual procedure would be to attempt to solve it using the quadratic formula for solving $ax^2 + bx + c = 0$:

$$
\begin{aligned}
x &= \frac{-b \pm \sqrt{b^2 - 4ac}}{2a} \\
&= \frac{-2 \pm \sqrt{4 - 4(2)(5)}}{4} \\
&= \frac{-2 \pm \sqrt{-36}}{4}.
\end{aligned}
$$

Note the need to take the square root of the negative number -36. But there is no real number whose square is -36, so it is impossible to proceed without introducing a number which is not real – called an **imaginary number** – symbolised by i, with the property that $i^2 = -1$, $i = \sqrt{-1}$. With this development, we can write $\sqrt{-36}$ as $\sqrt{36} \times \sqrt{-1} = 6i$ and write

---

DOI: 10.1201/9781003264569-3

the solutions of the quadratic equation as

$$x = \frac{-2 \pm 6\mathrm{i}}{4} = \frac{-1 \pm 3\mathrm{i}}{2} = -\frac{1}{2} \pm \frac{3}{2}\mathrm{i}.$$

We can now formally write down two solutions of the quadratic equation, namely, $x = -\frac{1}{2} + \frac{3}{2}\mathrm{i}$ and $x = -\frac{1}{2} - \frac{3}{2}\mathrm{i}$. These numbers are called **complex numbers**. Observe that each is made up of two parts, a **real part**, $-\frac{1}{2}$, and an **imaginary part**, $\pm\frac{3}{2}$. In the general case we will often use the letter z to denote a complex number.

Definition 3.1 Cartesian form of a complex number
Consider the number z given by

$$z = a + \mathrm{i}b \qquad \textit{where } a \in \mathbb{R},\ b \in \mathbb{R}.$$

*Here, a is the real part of z, written $a = \mathrm{Re}(z)$, and b is the imaginary part of z, written $b = \mathrm{Im}(z)$. The form $z = a + \mathrm{i}b$ is referred to as the **Cartesian form**. We use the symbol $\mathbb{C}$ to denote the set of all complex numbers.*

Note that all real numbers are complex numbers with imaginary part b equal to zero, and so $\mathbb{R} \subset \mathbb{C}$.

Exercises

3.1 Obtain the solutions of the quadratic equation $x^2 + x + 1 = 0$ and identify their real and imaginary parts.

3.2 Given that $\mathrm{i}^2 = -1$ obtain simplified expressions for (a) i^3, (b) i^4, (c) i^5.

3.3 Given that $x = 2$ is a solution of the cubic equation $3x^3 - 11x^2 + 16x - 12 = 0$ find the remaining two solutions. (Hint: write the cubic as $(x - 2)(ax^2 + bx + c) = 0$.)

3.4 Find a quadratic equation which has solutions $z = 5 + 2\mathrm{i}$ and $z = 5 - 2\mathrm{i}$.

3.5 Evaluate

a) $\left(\dfrac{\mathrm{i}}{\sqrt{2}}\right)^2$,   b) $\left|\left(\dfrac{\mathrm{i}}{\sqrt{2}}\right)^2\right|$.

3.3 The arithmetic of complex numbers

Given two (or more) complex numbers, addition (and subtraction) is performed in a natural way by adding (or subtracting) the real and imaginary parts separately. Thus if $z_1 = 3 - 5\mathrm{i}$ and $z_2 = 8 + 4\mathrm{i}$ then

$$z_1 + z_2 = (3 - 5\mathrm{i}) + (8 + 4\mathrm{i}) = (3 + 8) + (-5\mathrm{i} + 4\mathrm{i}) = 11 - \mathrm{i}$$

$$z_1 - z_2 = (3 - 5\mathrm{i}) - (8 + 4\mathrm{i}) = (3 - 8) + (-5\mathrm{i} - 4\mathrm{i}) = -5 - 9\mathrm{i}.$$

Likewise, multiplication by a real number is performed in a natural way:

$$7z_1 = 7(3 - 5\mathrm{i}) = 21 - 35\mathrm{i}.$$

More generally, to multiply two complex numbers, we make use of the fact that $i^2 = -1$. Thus

$$z_1 z_2 = (3 - 5i)(8 + 4i) = 24 - 40i + 12i - 20i^2 = 44 - 28i.$$

In general, addition, subtraction and multiplication are achieved as follows: given complex numbers $z_1 = a_1 + b_1 i$, $z_2 = a_2 + b_2 i$, we define

$$z_1 + z_2 = (a_1 + b_1 i) + (a_2 + b_2 i) = (a_1 + a_2) + (b_1 + b_2)i$$

$$z_1 - z_2 = (a_1 + b_1 i) - (a_2 + b_2 i) = (a_1 - a_2) + (b_1 - b_2)i$$

$$z_1 z_2 = (a_1 + b_1 i)(a_2 + b_2 i) = (a_1 a_2 - b_1 b_2) + (a_1 b_2 + a_2 b_1)i.$$

Some authors prefer to introduce complex numbers as ordered pairs of real numbers, writing $a + ib$ as (a, b). Then the rules of addition, subtraction and multiplication become:

$$(a_1, b_1) + (a_2, b_2) = (a_1 + a_2, b_1 + b_2)$$

$$(a_1, b_1) - (a_2, b_2) = (a_1 - a_2, b_1 - b_2)$$

$$(a_1, b_1)(a_2, b_2) = (a_1 a_2 - b_1 b_2, a_1 b_2 + a_2 b_1).$$

Note that this approach is entirely consistent with the Cartesian form given previously.

Exercises

3.6   If $z_1 = 3 + 8i$, $z_2 = -3 - 2i$ find

    a) $z_1 + z_2$ b) $z_1 - z_2$ c) $4z_1 + 3z_2$

    d) z_1^2 e) $z_1 z_2$.

3.7   State $\mathrm{Re}(z)$ and $\mathrm{Im}(z)$ when $z = (2 + 3i)(5 - 2i)$.

Definition 3.2 Complex conjugate
*The **complex conjugate** of any complex number z, which we shall write as z^*, is found by changing the sign of the imaginary part. Thus*

$$if \quad z = a + ib \quad then \quad z^* = a - ib.$$

(Some authors write the conjugate of z as $\overline{z}$ but throughout this book we will write z^*.)

Exercises

3.8   Given $z_1, z_2 \in \mathbb{C}$, show that $(z_1 + z_2)^* = z_1^* + z_2^*$.

3.9   Given $z_1, z_2 \in \mathbb{C}$, show that $(z_1 - z_2)^* = z_1^* - z_2^*$.

Observe that if $z = a + ib$

$$z\,z^* = (a + ib)(a - ib)$$
$$= a^2 + bai - abi - b^2i^2$$
$$= a^2 + b^2 \qquad (\text{since } i^2 = -1)$$

which will always be real.

For the division of complex numbers use is made of the conjugate. Consider the following example.

Example 3.3.1

Calculate $\dfrac{z_1}{z_2}$ when $z_1 = 3 - 5i$ and $z_2 = 8 + 4i$.

Solution

$$\frac{z_1}{z_2} = \frac{3 - 5i}{8 + 4i}.$$

We now multiply both the numerator and denominator of this fraction by the complex conjugate of the denominator and then simplify the result:

$$\frac{z_1}{z_2} = \frac{3 - 5i}{8 + 4i} \times \frac{8 - 4i}{8 - 4i}$$
$$= \frac{24 - 40i - 12i - 20}{64 + 16}$$
$$= \frac{4 - 52i}{80}$$
$$= \frac{1}{20} - \frac{13}{20}i.$$

In general, division of complex numbers is achieved in the following way:
If $z_1 = a_1 + b_1i$ and $z_2 = a_2 + b_2i$ then

$$\frac{z_1}{z_2} = \frac{a_1 + b_1i}{a_2 + b_2i}$$
$$= \frac{a_1 + b_1i}{a_2 + b_2i} \times \frac{a_2 - b_2i}{a_2 - b_2i}$$
$$= \frac{a_1a_2 + b_1b_2 + i(a_2b_1 - a_1b_2)}{a_2^2 + b_2^2}.$$

Exercises

3.10   If $z_1 = 4 + 4i$, $z_2 = 3 + 5i$ find (a) $\dfrac{z_1}{z_2}$, (b) $\dfrac{z_2}{z_1}$.

3.11   If $z \in \mathbb{C}$ show that (a) $z + z^* = 2\,\mathrm{Re}(z)$,   (b) $z - z^* = 2i\,\mathrm{Im}(z)$.

3.12   Find the real and imaginary parts of

a) $\dfrac{2}{4 + i} - \dfrac{3}{2 - i}$,   b) $\dfrac{1}{i} - i$,   c) $\dfrac{1}{i^2 + i}$.

3.4   The set of all complex numbers as a field

The set of all complex numbers, $\mathbb{C}$, together with the operations of addition and multiplication as defined above form a mathematical structure called a **field**. Strictly, in order to verify this fact, there are several requirements known as **field axioms** (see Appendix D) which must be checked. Essentially, the statement that $(\mathbb{C}, +, \times)$ is a field means that we are able to add, subtract and multiply any complex number and we can divide by any non-zero complex number and still remain in the field. This field has an additive identity element, zero, which is the complex number with both real and imaginary parts equal to zero: $z = 0 + 0\mathrm{i} = 0$. The multiplicative identity element is the real number 1.

3.5   The Argand diagram and polar form of a complex number

The **Argand diagram** is a graphical representation of a complex number. Using Cartesian xy axes, the real part of the complex number $z = a + \mathrm{i}b$ is plotted on the horizontal axis (x axis) and the imaginary part on the vertical axis (y axis). The complex number z is thus represented by the point with coordinates (a, b) as shown in Figure 3.1.

Observe that by introducing polar coordinates (r, θ) (see Appendix C), we can write $a = r \cos \theta$, $b = r \sin \theta$ so that

$$z = a + \mathrm{i}b = r \cos \theta + \mathrm{i}r \sin \theta = r(\cos \theta + \mathrm{i} \sin \theta).$$

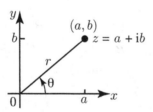

FIGURE 3.1
Argand diagram showing both Cartesian and polar forms of $z = a + \mathrm{i}b$.

Definition 3.3 Polar form of a complex number
*The form $z = r(\cos \theta + \mathrm{i} \sin \theta)$ is referred to as the **polar form** of the complex number, sometimes abbreviated to $r \angle \theta$.*

Here $r \geq 0$, a real number, is the **modulus** or **magnitude** of z, written $|z|$. The angle θ is the **argument** of z, $\arg(z)$, also commonly referred to as the **phase** or **amplitude** of the complex number. Usually, we choose θ to lie in the interval $-\pi < \theta \leq \pi$. Observe that $|z| = r = \sqrt{a^2 + b^2}$ and $\tan \theta = \dfrac{b}{a}$.

The following theorem shows how to carry out multiplication and division of complex numbers given in polar form:

Theorem 3.1 Multiplication and division in polar form

If $z_1 = r_1\angle\theta_1 = r_1(\cos\theta_1 + \mathrm{i}\sin\theta_1)$ and $z_2 = r_2\angle\theta_2 = r_2(\cos\theta_2 + \mathrm{i}\sin\theta_2)$ then

$$z_1 z_2 = r_1 r_2 \angle(\theta_1 + \theta_2)$$
$$= r_1 r_2 (\cos(\theta_1 + \theta_2) + \mathrm{i}\sin(\theta_1 + \theta_2))$$

that is, the moduli are multiplied and the arguments are added. Further,

$$\frac{z_1}{z_2} = \frac{r_1}{r_2}\angle(\theta_1 - \theta_2)$$
$$= \frac{r_1}{r_2}(\cos(\theta_1 - \theta_2) + \mathrm{i}\sin(\theta_1 - \theta_2))$$

that is, the moduli are divided and the arguments are subtracted.

Proof of these results is left as an exercise (below).

Exercises

3.13 Plot the following complex numbers on an Argand diagram and calculate the modulus and argument of each:

  a) $z = 3 + 4\mathrm{i}$, b) $z = -1 - \mathrm{i}$, c) $z = 3$,
  d) $z = 2\mathrm{i}$, e) $z = \pi + \pi\mathrm{i}$.

3.14 Find the modulus and argument of $z_1 = -\sqrt{3} + \mathrm{i}$ and $z_2 = 4 + 4\mathrm{i}$. Hence express $z_1 z_2$ and $\frac{z_1}{z_2}$ in polar form.

3.15 Use the trigonometric identities:

$$\sin(A \pm B) = \sin A \cos B \pm \cos A \sin B$$
$$\cos(A \pm B) = \cos A \cos B \mp \sin A \sin B$$

to prove that if $z_1 = r_1(\cos\theta_1 + \mathrm{i}\sin\theta_1)$ and $z_2 = r_2(\cos\theta_2 + \mathrm{i}\sin\theta_2)$ then

$$z_1 z_2 = r_1 r_2(\cos(\theta_1 + \theta_2) + \mathrm{i}\sin(\theta_1 + \theta_2))$$

and

$$\frac{z_1}{z_2} = \frac{r_1}{r_2}(\cos(\theta_1 - \theta_2) + \mathrm{i}\sin(\theta_1 - \theta_2)).$$

3.16 For any $z \in \mathbb{C}$ show that $z\,z^* = |z|^2$.

3.17 Suppose $z_1 = 3 + 2\mathrm{i}$ and $z_2 = 4 + 5\mathrm{i}$.

  (a) Find $|z_1||z_2|$ and $|z_1 z_2|$ and show that these are equal.

  (b) Find $\left|\dfrac{z_1}{z_2}\right|$ and $\dfrac{|z_1|}{|z_2|}$ and show that these are equal.

  (c) Show that the result in part (a) is true for any pair of complex numbers.

  (d) Show that the result in part (b) is true for any pair of complex numbers provided the denominator is non-zero.

3.18 Show that each of the complex numbers

$$z_1 = \frac{1}{\sqrt{3}}, \quad z_2 = -\frac{1}{\sqrt{3}}, \quad z_3 = \frac{\mathrm{i}}{\sqrt{3}}, \quad z_4 = -\frac{\mathrm{i}}{\sqrt{3}}$$

has the same magnitude.

3.6   The exponential form of a complex number

We now derive a third form in which a complex number can be expressed – the **exponential form**. The derivation requires the use of Maclaurin series which we discuss first. Maclaurin series expansions are a way of representing functions as an infinite sum of terms involving increasing powers of a variable x, say. Three commonly met expansions are the so-called **power series** for e^x, $\sin x$ and $\cos x$:

$$e^x = 1 + x + \frac{x^2}{2!} + \frac{x^3}{3!} + \frac{x^4}{4!} + \frac{x^5}{5!} + \cdots$$

$$\sin x = x - \frac{x^3}{3!} + \frac{x^5}{5!} - \cdots \qquad (x \text{ in radians})$$

$$\cos x = 1 - \frac{x^2}{2!} + \frac{x^4}{4!} - \cdots \qquad (x \text{ in radians})$$

which are valid for any real value of x. The derivation of these results can be found in most Calculus textbooks. Using the expansions for $\cos x$ and $\sin x$ and replacing x with θ, we can write

$$\cos\theta + i\sin\theta = \left(1 - \frac{\theta^2}{2!} + \frac{\theta^4}{4!} - \cdots\right) + i\left(\theta - \frac{\theta^3}{3!} + \frac{\theta^5}{5!} - \cdots\right)$$

$$= 1 + i\theta - \frac{\theta^2}{2!} - i\frac{\theta^3}{3!} + \frac{\theta^4}{4!} + \cdots$$

and from the Maclaurin series expansion of the exponential function, replacing x with $i\theta$, we note

$$e^{i\theta} = 1 + i\theta - \frac{\theta^2}{2!} - i\frac{\theta^3}{3!} + \frac{\theta^4}{4!} + \cdots$$

so that

$$e^{i\theta} = \cos\theta + i\sin\theta.$$

This result is known as **Euler's relation**. The similar relation $e^{-i\theta} = \cos\theta - i\sin\theta$ is straightforward to deduce. We thus have an alternative expression for the complex number $z = r(\cos\theta + i\sin\theta)$:

Definition 3.4 The exponential form of a complex number

$$z = re^{i\theta}$$

*is the **exponential form** of a complex number. Here r is a non-negative real number, the modulus of z, i.e., $|z|$, and usually the angle θ is chosen so that $-\pi < \theta \leq \pi$.*

When constructing quantum circuits it will be necessary to manipulate expressions involving complex numbers in polar and exponential forms. Consider the following example.

Example 3.6.1

Evaluate (a) $e^{2\pi i}$,   (b) $e^{-\pi i}$,   (c) $e^{\frac{\pi}{2}i}$.

Solution

(a) $e^{2\pi i} = \cos 2\pi + i\sin 2\pi = 1 + 0i = 1$. Alternatively observe that $e^{2\pi i}$ has modulus 1 and argument 2π and hence corresponds to the point on the Argand diagram with Cartesian coordinates $(1, 0)$.

(b) $e^{-\pi i} = \cos(-\pi) + i\sin(-\pi) = -1 + 0i = -1$.
(c) $e^{\frac{\pi}{2}i} = \cos\frac{\pi}{2} + i\sin\frac{\pi}{2} = 0 + 1i = i$.

Example 3.6.2

Use the first of Euler's relations together with the trigonometric identity $\cos^2\theta + \sin^2\theta = 1$ to show that
$$1 + e^{2i\theta} = 2(\cos^2\theta + i\sin\theta\cos\theta).$$

Solution

First note that
$$e^{2i\theta} = e^{i\theta}e^{i\theta}$$
$$= (\cos\theta + i\sin\theta)(\cos\theta + i\sin\theta)$$
$$= \cos^2\theta - \sin^2\theta + 2i\sin\theta\cos\theta.$$

Then
$$1 + e^{2i\theta} = 1 + \cos^2\theta - \sin^2\theta + 2i\sin\theta\cos\theta$$
$$= 2\cos^2\theta + 2i\sin\theta\cos\theta$$
$$= 2(\cos^2\theta + i\sin\theta\cos\theta)$$

as required.

Exercises

3.19   Evaluate (a) $e^{-2\pi i}$,   (b) $e^{\pi i}$,   (c) $e^{-\pi i/2}$,   (d) $e^{-5\pi i/2}$.

3.20   Evaluate (a) $e^{2\pi i/3}$,   (b) $e^{4\pi i/3}$,   (c) $e^{-2\pi i/3}$,   (d) $e^{-4\pi i/3}$.

3.21   Show that $1 - e^{2i\theta} = 2(\sin^2\theta - i\cos\theta\sin\theta)$.

3.22   Express $z = e^{1+i\pi/2}$ in the form $a + ib$.

Theorem 3.2 Multiplication and division in exponential form

Given two complex numbers $z_1 = r_1 e^{i\theta_1}$ and $z_2 = r_2 e^{i\theta_2}$, their product and quotient are found from
$$z_1 z_2 = r_1 r_2 e^{i(\theta_1 + \theta_2)}, \qquad \frac{z_1}{z_2} = \frac{r_1}{r_2} e^{i(\theta_1 - \theta_2)}.$$

Proof
$$z_1 z_2 = (r_1 e^{i\theta_1})(r_2 e^{i\theta_2}) = r_1 r_2 e^{i(\theta_1 + \theta_2)}$$
$$\frac{z_1}{z_2} = \frac{r_1 e^{i\theta_1}}{r_2 e^{i\theta_2}} = \frac{r_1}{r_2} e^{i(\theta_1 - \theta_2)}.$$

Example 3.6.3 Multiplication by a complex number of modulus 1: global phase

Show that multiplying an arbitrary complex number $z_1 = r_1 e^{i\theta_1}$ by any complex number with modulus equal to 1, does not change its modulus.

Solution

Let $z_2 = 1e^{i\theta_2}$ be any complex number with modulus 1. Then

$$z_2 z_1 = (1e^{i\theta_2})(r_1 e^{i\theta_1}) = r_1 e^{i(\theta_2 + \theta_1)}.$$

Observe that whilst the argument has changed, the modulus is unaltered. Multiplication of a complex number by another with modulus 1, i.e., $e^{i\phi}$ say, will be important when we introduce **global phase factors** in Chapter 13.

Example 3.6.4 Relative phase of two complex numbers

Consider the two complex numbers $z_1 = r_1 e^{i\theta_1}$ and $z_2 = r_2 e^{i\theta_2}$ shown on the Argand diagram in Figure 3.2. The angle between the two complex numbers represented in this way is $\theta_1 - \theta_2$. This quantity, or sometimes $e^{i(\theta_1 - \theta_2)}$, is referred to as the **relative phase** of the two complex numbers. Observe that

$$\frac{z_1}{z_2} = \frac{|z_1|}{|z_2|} e^{i(\theta_1 - \theta_2)}$$

and hence the relative phase can be expressed as

$$e^{i(\theta_1 - \theta_2)} = \frac{z_1/z_2}{|z_1|/|z_2|}.$$

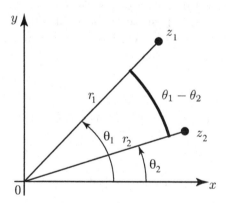

FIGURE 3.2
The angle $\theta_1 - \theta_2$ between the two complex numbers is the relative phase.

Exercises

3.23 Use Euler's relations to express $\cos\theta$ and $\sin\theta$ in terms of exponential functions.

3.24   Show that $\dfrac{1}{\cos\theta + \mathrm{i}\sin\theta} = \cos\theta - \mathrm{i}\sin\theta$.

3.25   Express $1 + \mathrm{e}^{2\mathrm{i}\omega t}$ in the form $a + \mathrm{i}b$.

3.26   Calculate the relative phase of $z_1 = 2\mathrm{e}^{\mathrm{i}\pi/3}$ and $z_2 = 5\mathrm{e}^{\mathrm{i}\pi/6}$.

3.7   The Fourier transform: an application of the exponential form of a complex number

The exponential form of a complex number is an essential ingredient of Fourier series, the Fourier transform and the discrete Fourier transform. These tools are used in the study of quantum algorithms.

Definition 3.5 Discrete Fourier transform

Suppose we have a continuous function $f(t)$ which we **sample** *at intervals T. We obtain a sequence of n sample values $f(0)$, $f(T)$, $f(2T),\ldots,f((n-1)T)$. Without loss of generality assume $T = 1$, so that we obtain the samples*

$$f(0), f(1), f(2), \ldots, f(n-1), \qquad \text{that is} \quad f(j), \ j = 0, 1, 2, \ldots, n-1.$$

The **discrete Fourier transform** *(d.f.t.) is another sequence $\hat{f}(k)$, $k = 0, 1, 2, \ldots, n-1$, defined by*

$$\hat{f}(k) = \frac{1}{\sqrt{n}} \sum_{j=0}^{n-1} f(j)\mathrm{e}^{-2\pi \mathrm{i}kj/n} \qquad \text{for } k = 0, 1, 2, \ldots, n-1.$$

There are several variants of this definition. Some authors include a factor of $\frac{1}{n}$ rather than $\frac{1}{\sqrt{n}}$ used here. Others use a positive exponential instead of the negative one used above. What is critical is that you know which formula is being used and that you apply it consistently.

Example 3.7.1

Find the discrete Fourier transform of the sequence $f(j) = 1, 1, 3, 1$.

Solution

Here the number of terms in the sequence, n, is 4. We evaluate the d.f.t. for each value of k in turn using $\hat{f}(k) = \dfrac{1}{\sqrt{n}} \displaystyle\sum_{j=0}^{n-1} f(j)\mathrm{e}^{-2\pi \mathrm{i}kj/n}$.

$\underline{k = 0}$

$$\hat{f}(0) = \frac{1}{\sqrt{4}} \sum_{j=0}^{3} f(j) = \frac{1}{2}(1 + 1 + 3 + 1) = 3.$$

$\underline{k = 1}$

$$\hat{f}(1) = \frac{1}{\sqrt{4}} \sum_{j=0}^{3} f(j)\mathrm{e}^{-\pi \mathrm{i}j/2} = \frac{1}{2}(1 + \mathrm{e}^{-\pi \mathrm{i}/2} + 3\mathrm{e}^{-\pi \mathrm{i}} + \mathrm{e}^{-3\pi \mathrm{i}/2}) = \frac{1}{2}(1 - \mathrm{i} - 3 + \mathrm{i}) = -1.$$

$\underline{k=2}$

$$\hat{f}(2) = \frac{1}{\sqrt{4}} \sum_{j=0}^{3} f(j)e^{-\pi i j} = \frac{1}{2}(1 + e^{-\pi i} + 3e^{-2\pi i} + e^{-3\pi i}) = \frac{1}{2}(1 - 1 + 3 - 1) = 1.$$

$\underline{k=3}$

$$\hat{f}(3) = \frac{1}{\sqrt{4}} \sum_{j=0}^{3} f(j)e^{-3\pi i j/2} = \frac{1}{2}(1 + e^{-3\pi i/2} + 3e^{-3\pi i} + e^{-9\pi i/2}) = \frac{1}{2}(1 + i - 3 - i) = -1.$$

Finally, $\hat{f}(k) = 3, -1, 1, -1$.

Example 3.7.2 The Hadamard transform

Consider the sequence of just two terms $f(0), f(1)$. Find its discrete Fourier transform.

Solution

From the formula for the d.f.t. with $n = 2$, we have

$$\hat{f}(k) = \frac{1}{\sqrt{2}} \sum_{j=0}^{1} f(j)e^{-\pi i k j} = \frac{1}{\sqrt{2}}(f(0) + f(1)e^{-\pi i k}).$$

$\underline{k=0}$

$$\hat{f}(0) = \frac{1}{\sqrt{2}}(f(0) + f(1))$$

$\underline{k=1}$

$$\hat{f}(1) = \frac{1}{\sqrt{2}}(f(0) + f(1)e^{-\pi i}) = \frac{1}{\sqrt{2}}(f(0) - f(1)).$$

Hence, the sequence $f(0), f(1)$ is transformed by the discrete Fourier transform to the sequence $\frac{1}{\sqrt{2}}(f(0) + f(1))$, $\frac{1}{\sqrt{2}}(f(0) - f(1))$. This particular version of the Fourier transform is referred to as the **Hadamard transform** and is fundamental in quantum computation.

Exercises

3.27   By writing $\lambda = e^{2\pi i/n}$ show that the d.f.t. can be expressed as:

$$\hat{f}(k) = \frac{1}{\sqrt{n}} \sum_{j=0}^{n-1} f(j)\lambda^{-kj} \qquad \text{for } k = 0, 1, 2, \ldots, n-1.$$

3.28   The inverse discrete Fourier transform of the sequence $\hat{f}(k)$, $k = 0, 1, 2, \ldots, n-1$ is given by

$$f(j) = \frac{1}{\sqrt{n}} \sum_{k=0}^{n-1} \hat{f}(k)e^{2\pi i k j/n} \qquad \text{for } j = 0, 1, 2, \ldots, n-1.$$

Calculate the inverse d.f.t of $\hat{f}(k) = 1, 6 + i, -9, 6 - i$.

3.29   Given a sequence of three terms $f(0)$, $f(1)$, $f(2)$ find an explicit expression for each term, $\hat{f}(k)$, in the discrete Fourier transform.

3.8 End-of-chapter exercises

1. Express the following complex numbers in Cartesian form:

   (a) $\dfrac{1}{2+3i}$

   (b) $\dfrac{1}{x-iy}$.

2. Show that $\dfrac{1}{\cos\theta - i\sin\theta} = \cos\theta + i\sin\theta$.

3. Show that for any $z_1, z_2 \in \mathbb{C}$, $(z_1 z_2)^* = z_1^* z_2^*$.

4. Suppose $z_1 = x_1 + iy_1 = r_1 e^{i\theta_1}$, $z_2 = x_2 + iy_2 = r_2 e^{i\theta_2}$. Show that

   $$|z_1 + z_2|^2 = |z_1|^2 + |z_2|^2 + |z_1|\,|z_2|(e^{i(\theta_2 - \theta_1)} + e^{i(\theta_1 - \theta_2)})$$

   and that this can also be written, using Euler's relations, as

   $$|z_1 + z_2|^2 = |z_1|^2 + |z_2|^2 + 2|z_1|\,|z_2|\cos(\theta_2 - \theta_1).$$

5. Consider the complex number $z = re^{i\theta}$. Show that the effect of multiplication by i is an anti-clockwise rotation by $\frac{\pi}{2}$ on the Argand diagram.

6. Show that the set of complex numbers $\mathbb{C}$ with the operation of addition is a group $(\mathbb{C}, +)$.

7. Show that the subset of complex numbers $\{1, -1, i, -i\}$ with the operation of multiplication is a group.

8. **De Moivre's theorem:** Use Euler's relation $e^{i\theta} = \cos\theta + i\sin\theta$ to show that when $n \in \mathbb{N}$

   $$(\cos\theta + i\sin\theta)^n = \cos n\theta + i\sin n\theta.$$

9. If $z = \dfrac{e^{i\theta}}{\sqrt{2}}$ show that $|z^2| = \dfrac{1}{2}$.

10. By writing z in polar form as $z = r(\cos\theta + i\sin\theta)$ and using De Moivre's theorem (see Question 8 above) find the three solutions of the equation $z^3 = 1$. (Hint: write $1 = 1(\cos 2k\pi + i\sin 2k\pi)$, $k = 0, 1, 2$.)

11. Show that if $\theta = \dfrac{\pi}{2}$ then $z = \dfrac{e^{i\theta}}{\sqrt{2}} = \dfrac{i}{\sqrt{2}}$.

4

Vectors

4.1 Objectives

Mathematical representations of the state of a quantum system are provided by vectors. These also hold information about the probability that the system will be in a particular state once a measurement has been made. In this chapter we shall introduce the qubit, a quantum bit, which is represented as a vector and show how qubits are manipulated. We show how two or more vectors can be multiplied by scalars (real or complex numbers) and added together to yield new vectors. Such combinations lead to the important concept of linear superposition and pave the way for a more general treatment in Chapter 6 Vector spaces. Finally, we introduce the 'inner product' of two vectors, which in turn leads to the important concepts of norm and orthogonality.

4.2 Vectors: preliminary definitions

Definition 4.1 Row vector.
An ordered set of n real numbers, u_i, $i = 1, \ldots, n$, written in the pattern or **array**

$$u = (u_1, u_2, \ldots, u_n)$$

is referred to as an n-dimensional **row** *vector of real numbers.*

Some authors indicate vectors using a bold font as in **u**. We shall avoid this practice except in some situations where there is ambiguity. The vector u is an element of the set $\mathbb{R}^n$. So, for example, the vector given by the ordered pair of real numbers

$$u = (8, -19)$$

is an element of the Cartesian product $\mathbb{R} \times \mathbb{R} = \mathbb{R}^2$. The ordered triple

$$v = (3.1, -2.8, 0)$$

is a row vector in $\mathbb{R}^3$.

Definition 4.2 Column vector.
An ordered set of n real numbers, v_i, $i = 1, \ldots, n$, written in the pattern or **array**

$$v = \begin{pmatrix} v_1 \\ v_2 \\ \vdots \\ v_n \end{pmatrix}$$

is referred to as an n-dimensional **column** *vector of real numbers.*

DOI: 10.1201/9781003264569-4

For example,

$$v = \begin{pmatrix} 3.2 \\ -9.0 \\ 0 \\ 7.5 \end{pmatrix}$$

is a column vector in $\mathbb{R}^4$. In quantum computation the entries, or **elements**, in row and column vectors will generally be complex numbers, so for example

$$c = (2 + i, 5 + 3i) \qquad \text{and} \qquad d = \begin{pmatrix} 5 - i \\ -9 - 2i \end{pmatrix}$$

are vectors in $\mathbb{C}^2$. Whether we work with row vectors or column vectors is usually determined by the context.

When working with vectors in $\mathbb{R}^n$, elements of the underlying field, in this case $\mathbb{R}$, are referred to as **scalars**. Likewise, when working with vectors in $\mathbb{C}^n$, the term scalar means a complex number, i.e., an element of $\mathbb{C}$.

If every element of a vector in $\mathbb{R}^n$ or $\mathbb{C}^n$ is the scalar 0, we refer to it as the **zero vector** $0 = (0, 0, 0, \ldots, 0)$. Note there is some potential confusion between the scalar 0 and the zero vector 0 so the reader needs to be aware of the context and choose accordingly.

In some quantum algorithms the underlying set will be $\mathbb{B} = \{0, 1\}$. The following two vectors are elements of $\mathbb{B}^4$ and $\mathbb{B}^3$, respectively:

$$x = (1, 0, 1, 1), \qquad y = \begin{pmatrix} 0 \\ 1 \\ 1 \end{pmatrix}.$$

Quantum computation uses a special notation – known as **Dirac bra-ket** notation – to denote the state of a quantum system. A **ket**, symbolised by $| \ \rangle$, can be regarded as a column vector. For example, we can write d above as

$$|d\rangle = \begin{pmatrix} 5 - i \\ -9 - 2i \end{pmatrix}.$$

Likewise, a **bra**, written $\langle \ |$, can be regarded as a row vector. For example, we can write c above as

$$\langle c| = (2 + i, 5 + 3i).$$

We will have occasion to write a specific ket as a bra and care must be taken to observe the following convention which, though strange at first, lays the groundwork for inner products which will follow. The bra $\langle d|$ corresponding to the ket $|d\rangle$ is found by forming a row vector using the same elements but taking the complex conjugate of each. Thus we can write

$$|d\rangle = \begin{pmatrix} 5 - i \\ -9 - 2i \end{pmatrix}, \qquad \langle d| = (5 + i, -9 + 2i)$$

and say that the bra $\langle d|$ is the **Hermitian conjugate** or **conjugate transpose** of the ket $|d\rangle$.

In quantum computation the two column vectors, in $\mathbb{C}^2$, $\begin{pmatrix} 1 \\ 0 \end{pmatrix}$ and $\begin{pmatrix} 0 \\ 1 \end{pmatrix}$ are used frequently and so we introduce the following notation and terminology.

Definition 4.3 $|0\rangle$ *and* $|1\rangle$

$$\text{ket } 0 = |0\rangle = \begin{pmatrix} 1 \\ 0 \end{pmatrix} \qquad \text{ket } 1 = |1\rangle = \begin{pmatrix} 0 \\ 1 \end{pmatrix}.$$

Exercises

4.1 Write down the bra corresponding to the ket $|\psi\rangle = \frac{1}{\sqrt{3}} \begin{pmatrix} i \\ 1+i \end{pmatrix}$.

4.2 Write down the ket corresponding to the bra $\langle \phi | = (\frac{1}{\sqrt{2}}, \frac{1}{\sqrt{2}})$.

4.3 Graphical representation of two- and three-dimensional vectors

When working with two- and three-dimensional vectors of real numbers it is often helpful to use a graphical representation in which **directed line segments** are used to represent the vectors. A directed line segment is just a segment of a straight line with an arrow to indicate its direction. With such a representation it is straightforward to introduce the length or **magnitude** of a vector, the angle between two vectors, and **orthogonality** which occurs when two vectors are perpendicular. These notions readily generalise to higher dimensions and are important in quantum computation as we shall see.

Consider the xy Cartesian coordinate system and the point, P, with coordinates (x_1, y_1), shown in Figure 4.1 where it is denoted $P(x_1, y_1)$.

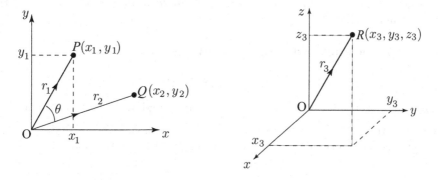

FIGURE 4.1
Graphical representation of position vectors in $\mathbb{R}^2$ and $\mathbb{R}^3$.

It is natural to draw a directed line segment from the origin to the point P. This directed line segment is referred to as the **position vector**, r_1, of P and we can write $r_1 = \overrightarrow{OP} = \begin{pmatrix} x_1 \\ y_1 \end{pmatrix}$. We say that x_1 and y_1 are the components of vector r_1 *relative* to the given xy coordinate system. Likewise, the position vector of the point Q, with coordinates (x_2, y_2), is $r_2 = \overrightarrow{OQ} = \begin{pmatrix} x_2 \\ y_2 \end{pmatrix}$. The position vector of the point $R(x_3, y_3, z_3)$ in three-dimensional space is then $r_3 = \begin{pmatrix} x_3 \\ y_3 \\ z_3 \end{pmatrix}$. Two remarks are worth making here: if we use a different coordinate system these components may change. Also, whilst we have introduced the graphical representation using column vectors, we could equally have used row vectors such as $r_1 = (x_1, y_1)$. There is then potential for some confusion because (x_1, y_1) refers to both the coordinates of a point and the position vector of that point. This distinction is of no great concern in what follows.

With these graphical representations in mind we can readily introduce the 'length' of a vector r of real numbers, written $|r|$, by making use of Pythagoras' theorem:

$$\text{length of } r_1 = |r_1| = \sqrt{x_1^2 + y_1^2}, \quad \text{length of } r_2 = |r_2| = \sqrt{x_2^2 + y_2^2}$$

$$\text{length of } r_3 = |r_3| = \sqrt{x_3^2 + y_3^2 + z_3^2}.$$

In the context of a more general column or row vector of real or complex numbers, u say, we have the following via a generalisation of Pythagoras' theorem:

Definition 4.4 Magnitude or norm of a vector
   The **magnitude** *or* **norm**, *written* $\|u\|$, *of a vector of real or complex numbers:*

$$\text{if } u = (u_1, u_2, \ldots, u_n), \quad \|u\| = \sqrt{|u_1|^2 + |u_2|^2 + \ldots + |u_n|^2}.$$

Note that it is necessary to take the modulus of each complex component prior to squaring it. This ensures that the norm is a real number. We shall discuss this aspect further when we study inner products in Section 4.6.

If $\|u\| = 1$ then u is referred to as a **unit vector**. A unit vector is frequently, but not always, indicated using the hat symbol, as in $\hat{r}$. Thus far, the two- and three-dimensional vectors of real numbers shown have been tied to the origin. In general this is not necessary and so-called **free vectors** can be placed at any location provided their lengths and directions are maintained.

Example 4.3.1 The standard orthonormal basis in $\mathbb{R}^2$ and in $\mathbb{R}^3$

Figure 4.2a shows vectors drawn from the origin to the points $(1,0)$ and $(0,1)$. We shall denote these as $i = \begin{pmatrix} 1 \\ 0 \end{pmatrix}$ and $j = \begin{pmatrix} 0 \\ 1 \end{pmatrix}$. Clearly i and j are unit vectors along the x and y axes and they are mutually perpendicular or orthogonal. As we shall see, these two vectors can be used to construct any other vector in $\mathbb{R}^2$. We refer to i and j as the **standard orthonormal basis** of $\mathbb{R}^2$.

Moving to three dimensions the standard orthonormal basis in $\mathbb{R}^3$ is $i = \begin{pmatrix} 1 \\ 0 \\ 0 \end{pmatrix}, j = \begin{pmatrix} 0 \\ 1 \\ 0 \end{pmatrix}$ and $k = \begin{pmatrix} 0 \\ 0 \\ 1 \end{pmatrix}$ (Figure 4.2b).

Example 4.3.2

Consider the ket $|\psi\rangle = \begin{pmatrix} \frac{1}{\sqrt{2}} \\ \frac{1}{\sqrt{2}} \end{pmatrix}$. Show that this is a unit vector.

Solution

$$\| \, |\psi\rangle \, \| = \sqrt{\left(\frac{1}{\sqrt{2}}\right)^2 + \left(\frac{1}{\sqrt{2}}\right)^2} = \sqrt{\frac{1}{2} + \frac{1}{2}} = 1$$

thus $|\psi\rangle$ is a unit vector.

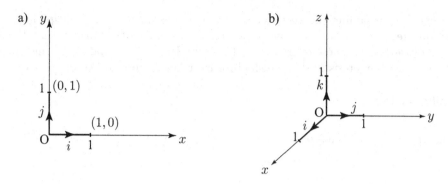

FIGURE 4.2
a) The standard orthonormal basis in $\mathbb{R}^2$, and b) in $\mathbb{R}^3$.

Exercises

4.3 Show that the bra $\langle\phi| = (\frac{\sqrt{3}}{2}, \frac{1}{2})$ is a unit vector.

4.4 Show that both ket 0 and ket 1 are unit vectors.

4.5 Draw ket 0 and ket 1 as vectors in the xy Cartesian coordinate system. Show that these vectors are orthogonal (i.e., they are perpendicular).

4.6 Find the norm of the vector $(8, -1, 2)$.

4.7 Find the norm of the vector $(i, 1 + i)$.

4.8 Show that the vector $|\psi\rangle = \begin{pmatrix} \frac{1}{\sqrt{5}} \\ \frac{2}{\sqrt{5}} \end{pmatrix}$ has unit norm.

4.4 Vector arithmetic: addition, subtraction and scalar multiplication

Vectors having the same number of components, n say, can be added or subtracted in an element-by-element fashion to yield another vector of the same size. Any vector can be multiplied by a scalar (a real or complex number) by multiplying each element of the vector by that number, again yielding another vector of the same size.

Definition 4.5 Addition and multiplication by a scalar
Given vectors u and v with components u_i and v_i, respectively, $(i = 1, \ldots, n)$, the sum $u + v$ is a vector with components $u_i + v_i$. Given a scalar k, the multiple ku is a vector with components ku_i.

For example, if $u, v \in \mathbb{C}^2$ and given by

$$u = (1 + i, 3 - 2i) \quad \text{and} \quad v = (7 - 2i, 8 - 3i)$$

then

$$u + v = (8 - i, 11 - 5i), \quad 5u = (5 + 5i, 15 - 10i), \quad 2iv = (4 + 14i, 6 + 16i).$$

This property, that adding two vectors of real or complex numbers, or multiplying a vector by a scalar, results in other vectors of the same size and nature means that the set of vectors is **closed** under these operations. Closure under addition and scalar multiplication is an important characteristic that we shall return to in Chapter 6 Vector spaces.

Example 4.4.1

Consider the real two-dimensional vectors $a = \begin{pmatrix} 3 \\ 4 \end{pmatrix}$ and $b = \begin{pmatrix} 2 \\ 1 \end{pmatrix}$ shown in Figure 4.3.

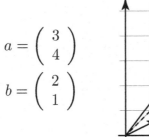
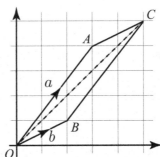

FIGURE 4.3
Vector addition.

Their sum, $a + b$, calculated by adding corresponding components, is $\begin{pmatrix} 5 \\ 5 \end{pmatrix}$. Geometrically, this result can be obtained by translating b so that its tail coincides with the head of a (that is $\overrightarrow{AC}$ in Figure 4.3). The third side of the triangle thus produced, that is OC, represents $a + b$. This is known as the **triangle law** (or sometimes the **parallelogram law**) of vector addition. Now consider the difference $a - b = \begin{pmatrix} 3 \\ 4 \end{pmatrix} - \begin{pmatrix} 2 \\ 1 \end{pmatrix} = \begin{pmatrix} 1 \\ 3 \end{pmatrix}$. Study Figure 4.3 and note that $a - b$ is represented by the vector from the head of b to the head of a, that is $\overrightarrow{BA}$.

Example 4.4.2

Using vector addition, scalar multiplication and the triangle law of addition any vector in $\mathbb{R}^2$ can be constructed from the standard orthonormal basis i and j. For example

$$\begin{pmatrix} 3 \\ 4 \end{pmatrix} = 3 \begin{pmatrix} 1 \\ 0 \end{pmatrix} + 4 \begin{pmatrix} 0 \\ 1 \end{pmatrix}.$$

Vectors in $\mathbb{R}^3$ can, likewise, be constructed from the basis vectors i, j and k.

Given any two vectors, u, v say, (of the same size), and any two scalars, k, ℓ, we can use the operations of scalar multiplication and addition to give a new vector. This leads to the concept of superposition.

Definition 4.6 Superposition
A **superposition** *of the vectors u and v is a* **linear combination**

$$w = ku + \ell v$$

where k and ℓ are real or complex numbers. Generally, given n vectors u_i, $i = 1, \ldots, n$ having the same size, the expression

$$\sum_{i=1}^{n} k_i u_i$$

where $k_i \in \mathbb{C}$, is a linear combination. (Note the u_i here are each vectors, and not the components of a vector u.)

Superpositions of vectors u_i play an essential role in quantum computation.

Exercises

4.9 Consider $u, v \in \mathbb{C}^n$ and $k \in \mathbb{C}$. Show from the definitions of vector addition and scalar multiplication that:

    a) $u + v = v + u$, i.e., vector addition is commutative,

    b) $(u + v) + w = u + (v + w)$ i.e., vector addition is associative,

    c) $0u = 0$ where on the left 0 is the scalar zero and on the right 0 is the zero vector,

    d) $1u = u$.

4.10 Express the vectors $\frac{1}{\sqrt{2}} \begin{pmatrix} 1 \\ 1 \end{pmatrix}$ and $\frac{1}{\sqrt{2}} \begin{pmatrix} -1 \\ 1 \end{pmatrix}$ as a superposition of the kets $|0\rangle$ and $|1\rangle$.

4.11 Express the vectors $\begin{pmatrix} 1 \\ 0 \end{pmatrix}$ and $\begin{pmatrix} 0 \\ 1 \end{pmatrix}$ as a superposition of the kets $|\psi\rangle$ and $|\phi\rangle$ where $|\psi\rangle = \frac{1}{\sqrt{2}} \begin{pmatrix} 1 \\ 1 \end{pmatrix}$ and $|\phi\rangle = \frac{1}{\sqrt{2}} \begin{pmatrix} 1 \\ -1 \end{pmatrix}$.

4.12 Consider $x, y \in \mathbb{B}^4$ given by

$$x = (1, 1, 0, 1), \quad y = (0, 1, 1, 1).$$

Using arithmetic modulo 2, show that $x + y = (1, 0, 1, 0)$. Show that this is equivalent to bit-wise application of the exclusive-or operator $\oplus$.

4.5 Qubits represented as vectors

A **qubit** or **quantum bit** is a mathematical representation of the state of the simplest quantum system. Chapter 12 discusses qubits in detail. One way in which a qubit can be described is by a linear combination of the two column vectors $\begin{pmatrix} 1 \\ 0 \end{pmatrix}$ and $\begin{pmatrix} 0 \\ 1 \end{pmatrix}$. We make specific use of the ket notation introduced in Section 4.2 and write $|0\rangle = \begin{pmatrix} 1 \\ 0 \end{pmatrix}$ and $|1\rangle = \begin{pmatrix} 0 \\ 1 \end{pmatrix}$ and their more general linear combination, $|\psi\rangle$ say, as

$$|\psi\rangle = \alpha_0 |0\rangle + \alpha_1 |1\rangle$$

where α_0 and α_1 are any complex numbers such that $|\alpha_0|^2 + |\alpha_1|^2 = 1$. (The reason for this requirement will become apparent in due course). Thus a qubit has an infinity of possible states which are linear combinations (also known as blends, or superpositions) of the vectors $|0\rangle$ and $|1\rangle$.

Example 4.5.1

The vector

$$|\psi\rangle = \frac{1}{\sqrt{2}}|0\rangle + \frac{1}{\sqrt{2}}|1\rangle$$

is a superposition of the kets $|0\rangle$ and $|1\rangle$. Note that, using the rules of scalar multiplication and vector addition, we could write this vector as

$$|\psi\rangle = \frac{1}{\sqrt{2}}\begin{pmatrix} 1 \\ 0 \end{pmatrix} + \frac{1}{\sqrt{2}}\begin{pmatrix} 0 \\ 1 \end{pmatrix} = \begin{pmatrix} \frac{1}{\sqrt{2}} \\ 0 \end{pmatrix} + \begin{pmatrix} 0 \\ \frac{1}{\sqrt{2}} \end{pmatrix} = \begin{pmatrix} \frac{1}{\sqrt{2}} \\ \frac{1}{\sqrt{2}} \end{pmatrix}.$$

You should verify that this ket has norm 1 (which is a consequence of requiring $|\alpha_0|^2 + |\alpha_1|^2 = 1$, as above).

4.5.1   Combining kets and functions $f : \mathbb{B} \to \mathbb{B}$

When we study quantum algorithms, we will need to be familiar with the manipulation of kets containing functions, specifically the four functions $f : \mathbb{B} \to \mathbb{B}$ defined in Section 2.4. Here we prove an important result. Working through the Proof will provide excellent practice in the required manipulations.

Lemma 4.1 *If* $f : \mathbb{B} \to \mathbb{B}$ *then*

$$|f(x)\rangle - |1 \oplus f(x)\rangle = (-1)^{f(x)}(|0\rangle - |1\rangle), \qquad x \in \mathbb{B}.$$

Proof

(a) Consider the balanced, identity function $f(x) = x$, $x \in \mathbb{B}$. Clearly $f(0) = 0, f(1) = 1$. So, if $x = 0$, then $f(x) = 0$ and

$$\begin{aligned}
|f(x)\rangle - |1 \oplus f(x)\rangle &= |f(0)\rangle - |1 \oplus f(0)\rangle \\
&= |0\rangle - |1 \oplus 0\rangle \\
&= |0\rangle - |1\rangle \\
&= (-1)^0(|0\rangle - |1\rangle) \\
&= (-1)^{f(x)}(|0\rangle - |1\rangle) \qquad \text{as required.}
\end{aligned}$$

Likewise, if $x = 1$, then $f(x) = 1$ and

$$\begin{aligned}
|f(x)\rangle - |1 \oplus f(x)\rangle &= |f(1)\rangle - |1 \oplus f(1)\rangle \\
&= |1\rangle - |1 \oplus 1\rangle \\
&= |1\rangle - |0\rangle \\
&= (-1)^1(|0\rangle - |1\rangle) \\
&= (-1)^{f(x)}(|0\rangle - |1\rangle) \qquad \text{as required.}
\end{aligned}$$

and therefore the Lemma is true when $f(x) = x$.

(b) Consider the balanced, *not* function $f(x) = \overline{x}$, $x \in \mathbb{B}$. Then $f(0) = 1, f(1) = 0$.

$$|f(x)\rangle - |1 \oplus f(x)\rangle = |\overline{x}\rangle - |1 \oplus \overline{x}\rangle$$
$$= \begin{cases} |1\rangle - |0\rangle & \text{if } x = 0 \\ |0\rangle - |1\rangle & \text{if } x = 1 \end{cases}$$
$$= (-1)^{f(x)}(|0\rangle - |1\rangle)$$

and again, the Lemma is true.

(c) Consider the constant function $f(x) = 1$, $x \in \mathbb{B}$.

$$|f(x)\rangle - |1 \oplus f(x)\rangle = |1\rangle - |1 \oplus 1\rangle$$
$$= |1\rangle - |0\rangle$$
$$= (-1)^{f(x)}(|0\rangle - |1\rangle).$$

(d) Consider the constant function $f(x) = 0$, $x \in \mathbb{B}$.

$$|f(x)\rangle - |1 \oplus f(x)\rangle = |0\rangle - |1 \oplus 0\rangle$$
$$= |0\rangle - |1\rangle$$
$$= (-1)^{f(x)}(|0\rangle - |1\rangle)$$

as required. ∎

4.6 The inner product (scalar product, dot product)

'Multiplication' of vectors is important in the study of quantum computation. There are several ways that the product of two vectors can be defined depending upon the application that we are interested in. We focus here on the **inner product**. The inner product is often referred to as the **scalar product** or **dot product**, particularly when the elements of the vectors are real and the vectors are two- or three-dimensional. In quantum computation the vectors under consideration will often have complex elements.

4.6.1 The inner product in $\mathbb{R}^2$ and $\mathbb{R}^3$

We begin our study of inner products with vectors in $\mathbb{R}^2$ and $\mathbb{R}^3$. In fact there are many different inner products on $\mathbb{R}^2$ and $\mathbb{R}^3$ and we focus here on the so-called **Euclidean inner product** because it possesses a helpful geometrical interpretation.

Definition 4.7 Euclidean inner product in $\mathbb{R}^2$ and $\mathbb{R}^3$
Given two vectors u and v we denote their Euclidean inner product, or scalar product, by $\langle u, v \rangle$ (some authors write $u \cdot v$) and define it to be

$$\langle u, v \rangle = \|u\| \|v\| \cos \theta$$

where θ is the angle between u and v (Figure 4.4).

By rearrangement, this definition provides a way of writing the cosine of the angle, and hence the angle between the two vectors:

$$\cos \theta = \frac{\langle u, v \rangle}{\|u\| \|v\|}.$$

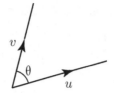

FIGURE 4.4
The scalar product can be used to determine θ.

Further, observe that

$$\langle u, u \rangle = \|u\|\|u\| = \|u\|^2$$

(since $\cos 0 = 1$) and hence the Euclidean inner product can be used to define the modulus or norm of a vector.

Definition 4.8 The Euclidean norm

$$\|u\| = \sqrt{\langle u, u \rangle}.$$

With $\langle u, v \rangle$ defined in this way it is possible to show that the inner product of vectors u, v, w in $\mathbb{R}^2$ and $\mathbb{R}^3$ has the following properties :

1. $\langle u, v \rangle = \langle v, u \rangle$,   (commutativity)
2. $k\langle u, v \rangle = \langle ku, v \rangle = \langle u, kv \rangle$,   where $k \in \mathbb{R}$
3. $\langle (u + v), w \rangle = \langle u, w \rangle + \langle v, w \rangle$,   (distributivity)
4. $\langle u, u \rangle \geq 0$ and $\langle u, u \rangle = 0$ if and only if $u = 0$ (positivity).

We now demonstrate an alternative but equivalent formula for the Euclidean inner product in $\mathbb{R}^2$, one which we shall generalise shortly to $\mathbb{R}^n$.

Theorem 4.1 *Given two vectors $u = (u_1, u_2)$ and $v = (v_1, v_2)$, their scalar product $\langle u, v \rangle = \|u\|\|v\| \cos \theta$, where θ is the angle between u and v, is given by*

$$\langle u, v \rangle = u_1 v_1 + u_2 v_2.$$

Proof
Using the cosine rule, referring to Figure 4.4,

$$\|u - v\|^2 = \|u\|^2 + \|v\|^2 - 2\|u\|\,\|v\| \cos \theta$$
$$= \|u\|^2 + \|v\|^2 - 2\langle u, v \rangle.$$

Writing $u = (u_1, u_2)$, $v = (v_1, v_2)$ then

$$\|(u_1 - v_1, u_2 - v_2)\|^2 = \|(u_1, u_2)\|^2 + \|(v_1, v_2)\|^2 - 2\langle u, v \rangle$$
$$(u_1 - v_1)^2 + (u_2 - v_2)^2 = u_1^2 + u_2^2 + v_1^2 + v_2^2 - 2\langle u, v \rangle$$
$$u_1^2 + v_1^2 - 2u_1 v_1 + u_2^2 + v_2^2 - 2u_2 v_2 = u_1^2 + u_2^2 + v_1^2 + v_2^2 - 2\langle u, v \rangle$$
$$-2u_1 v_1 - 2u_2 v_2 = -2\langle u, v \rangle$$

so that

$$\langle u, v \rangle = u_1 v_1 + u_2 v_2$$

as required. Further, observe that $\langle u, u \rangle = u_1^2 + u_2^2 = \|u\|^2$. Note that this is consistent with the definition of norm introduced in Section 4.3. ∎

Exercises

4.13   Calculate the Euclidean inner product of the vectors $u = (4, -2)$ and $v = (3, -3)$ and hence determine the angle between u and v.

4.14   Show that the Euclidean inner product on $\mathbb{R}^3$ defined by $\langle u, v \rangle = u \cdot v = \|u\| \|v\| \cos \theta$ can be expressed as $\langle u, v \rangle = u_1 v_1 + u_2 v_2 + u_3 v_3 = \sum_{i=1}^{3} u_i v_i$.

4.15   Show that the Euclidean inner product of $u = (5, -3, 2)$ and $v = (-2, 4, 1)$ is -20.

We are now in a position to generalise the previous results to vectors in $\mathbb{R}^n$.

Definition 4.9 Inner product in $\mathbb{R}^n$
Consider row vectors in $\mathbb{R}^n$. Suppose $u = (u_1, u_2, \ldots, u_n)$ and $v = (v_1, v_2, \ldots, v_n)$. Their inner product, written $\langle u, v \rangle$ is defined to be

$$\langle u, v \rangle = u_1 v_1 + u_2 v_2 + \ldots + u_n v_n = \sum_{i=1}^{n} u_i v_i$$

the result being a scalar. We generalise the idea of the cosine of an angle between two vectors, $u, v \in \mathbb{R}^n$, and hence the angle between them, by defining this to be

$$\cos \theta = \frac{\langle u, v \rangle}{\|u\| \|v\|}.$$

It can be shown that this is a well-defined quantity.

It is important to point out again that there are different ways in which an inner product on $\mathbb{R}^n$ can be defined. Consequently the norms of vectors and the angles between them depend upon the choice of inner product. Strictly speaking, when referring to the norm and angle between vectors we should state which inner product is being used. The inner product in $\mathbb{R}^n$ has a number of properties: if $u = (u_1, u_2, \ldots, u_n)$, $v = (v_1, v_2, \ldots, v_n)$, $w = (w_1, w_2, \ldots, w_n)$ where $u_i, v_i, w_i \in \mathbb{R}$, the inner product, $\langle u, v \rangle = \sum_{i=1}^{n} u_i v_i$ satisfies:

1. $\langle u, v \rangle = \langle v, u \rangle$,   (commutativity)

2. $k\langle u, v \rangle = \langle ku, v \rangle = \langle u, kv \rangle$,   where $k \in \mathbb{R}$

3. $\langle (u + v), w \rangle = \langle u, w \rangle + \langle v, w \rangle$,   (distributivity)

4. $\langle u, u \rangle \geq 0$, and $\langle u, u \rangle = 0$ if and only if $u = 0$ (positivity).

Properties (2) and (3) together mean that the scalar product is linear. Commutativity can be used to show that $\langle u, (v + w) \rangle = \langle u, v \rangle + \langle u, w \rangle$. It is very helpful to gain familiarity with properties such as (2) and (3) because we draw upon similar properties to establish the tensor product in Chapter 10.

We now turn our attention to finding the inner product of two vectors of complex numbers. A word of warning is necessary. Conventions differ among computer scientists, physicists, engineers and mathematicians as to how the inner product is defined. We shall give two definitions below, starting with the one used frequently by mathematicians.

4.6.2   The inner product on $\mathbb{C}^n$: Definition 1

Definition 4.10 The inner product on $\mathbb{C}^n$

Consider two row vectors in $\mathbb{C}^n$. Suppose $u = (u_1, u_2, \ldots, u_n)$ and $v = (v_1, v_2, \ldots, v_n)$ where $u_i, v_i \in \mathbb{C}$. We define the inner product (often referred to as the **standard Hermitian inner product***) written as $\langle u, v \rangle$ to be*

$$\langle u, v \rangle = u_1 v_1^* + u_2 v_2^* + \ldots + u_n v_n^* = \sum_{i=1}^{n} u_i v_i^*$$

where $*$ denotes complex conjugate.

Observe this definition is very similar to the previous one used for vectors of real numbers but now the operation requires taking the complex conjugate of the elements in the second vector.

Example 4.6.1

If $u = (1 + 2i, 3 - 2i)$ and $v = (5 - i, 7 + 4i)$ find $\langle u, v \rangle$, $\langle v, u \rangle$, $\langle u, u \rangle$ and $\langle v, v \rangle$.

Solution

$$\begin{aligned}
\langle u, v \rangle &= (1 + 2i)(5 - i)^* + (3 - 2i)(7 + 4i)^* \\
&= (1 + 2i)(5 + i) + (3 - 2i)(7 - 4i) \\
&= (5 - 2 + i + 10i) + (21 - 8 - 12i - 14i) \\
&= 16 - 15i.
\end{aligned}$$

$$\begin{aligned}
\langle v, u \rangle &= (5 - i)(1 + 2i)^* + (7 + 4i)(3 - 2i)^* \\
&= (5 - i)(1 - 2i) + (7 + 4i)(3 + 2i) \\
&= (5 - 2 - i - 10i) + (21 - 8 + 12i + 14i) \\
&= 16 + 15i.
\end{aligned}$$

$$\begin{aligned}
\langle u, u \rangle &= (1 + 2i)(1 + 2i)^* + (3 - 2i)(3 - 2i)^* \\
&= (1 + 2i)(1 - 2i) + (3 - 2i)(3 + 2i) \\
&= (1 + 4) + (9 + 4) \\
&= 18.
\end{aligned}$$

$$\begin{aligned}
\langle v, v \rangle &= (5 - i)(5 - i)^* + (7 + 4i)(7 + 4i)^* \\
&= (5 - i)(5 + i) + (7 + 4i)(7 - 4i) \\
&= (25 + 1) + (49 + 16) \\
&= 91.
\end{aligned}$$

We make two important observations about these answers. Firstly, the inner product is not commutative, that is $\langle u, v \rangle \neq \langle v, u \rangle$. In fact observe that $\langle u, v \rangle = \langle v, u \rangle^*$. Secondly, the inner product of any non-zero vector with itself is always a real, non-negative number.

We summarise several properties of this inner product below:
if $u = (u_1, u_2, \ldots, u_n)$, $v = (v_1, v_2, \ldots, v_n)$, $w = (w_1, w_2, \ldots, w_n)$ where $u_i, v_i, w_i \in \mathbb{C}$, the inner product,

$$\langle u, v \rangle = \sum_{i=1}^{n} u_i v_i^*$$

satisfies:

1. $\langle u, v \rangle = \langle v, u \rangle^*$,
2. $k\langle u, v \rangle = \langle ku, v \rangle = \langle u, k^*v \rangle$,   where $k \in \mathbb{C}$
3. $\langle (u + v), w \rangle = \langle u, w \rangle + \langle v, w \rangle$,   (distributivity)
4. $\langle u, u \rangle \geq 0$ and $\langle u, u \rangle = 0$ if and only if $u = 0$,   (positivity).

Properties (2) and (3) imply that this inner product is 'linear in the first slot', but because of the conjugation in (2) it is not 'linear in the second slot'. Using properties (1) and (3) it can be shown that the statement $\langle u, (v + w) \rangle = \langle u, v \rangle + \langle u, w \rangle$ is nevertheless true. We say that the inner product is **anti-linear**, or **sesquilinear**.

As before, an important quantity derived from the inner product is the **norm** of u, written $\|u\|$, defined by $\|u\| = \sqrt{\langle u, u \rangle}$. Referring back to Example 4.6.1

$$\|u\| = \sqrt{18}, \qquad \|v\| = \sqrt{91}.$$

In general, the norm $\|u\|$ is calculated from:

$$\|u\|^2 = \langle u, u \rangle$$
$$= \sum_{i=1}^{n} u_i u_i^*$$
$$= \sum_{i=1}^{n} |u_i|^2.$$

Any vector with norm equal to 1 is said to be a **unit vector** and any vector can be **normalised** by dividing by its norm.

Definition 4.11 Orthogonal and orthonormal sets of vectors
*Two vectors, u and v, are said to be **orthogonal** (with respect to the defined inner product) if $\langle u, v \rangle = 0$. A set of vectors the elements of which are both normalised and mutually orthogonal is said to be an **orthonormal** set.*

In the previous discussion our examples referred to row vectors. Precisely the same calculations can be performed on column vectors.

Example 4.6.2

With the previously given definition of the inner product on $\mathbb{C}^n$, show that the vectors u and v are orthogonal, where

$$u = \begin{pmatrix} 1 \\ 3 \\ -5 \end{pmatrix}, \quad v = \begin{pmatrix} -2 \\ 4 \\ 2 \end{pmatrix}.$$

Solution

Note that these vectors have real elements and so

$$\langle u, v \rangle = \sum_{i=1}^{3} u_i v_i^* = \sum_{i=1}^{3} u_i v_i = (1)(-2) + (3)(4) + (-5)(2) = 0$$

and hence u and v are orthogonal.

We now turn to the second definition of the inner product on $\mathbb{C}^n$, the one favoured by many physicists and computer scientists and which lends itself well to bra-ket notation, as we shall see later.

4.6.3   The inner product on $\mathbb{C}^n$: Definition 2

Definition 4.12 The inner product on $\mathbb{C}^n$
Consider two row vectors in $\mathbb{C}^n$. Suppose $u = (u_1, u_2, \ldots, u_n)$ and $v = (v_1, v_2, \ldots, v_n)$ where $u_i, v_i \in \mathbb{C}$. We now define the inner product, written as $\langle u, v \rangle$, to be

$$\langle u, v \rangle = u_1^* v_1 + u_2^* v_2 + \ldots + u_n^* v_n = \sum_{i=1}^{n} u_i^* v_i$$

*where * denotes complex conjugate.*

Observe this definition is very similar to the previous one but now the complex conjugate is taken of the elements of the first vector. Applying this definition to the vectors of Example 4.6.1 you should verify that:

$$\langle u, v \rangle = 16 + 15i$$

$$\langle v, u \rangle = 16 - 15i$$

$$\langle u, u \rangle = 18 \qquad \text{as previously}$$

$$\langle v, v \rangle = 91 \qquad \text{as previously.}$$

Observe that the results of finding $\langle u, v \rangle$ depend on the definition used. We summarise several properties of this inner product below: if $u = (u_1, u_2, \ldots, u_n)$, $v = (v_1, v_2, \ldots, v_n)$, $w = (w_1, w_2, \ldots, w_n)$ where $u_i, v_i, w_i \in \mathbb{C}$, the inner product, $\langle u, v \rangle = \sum_{i=1}^{n} u_i^* v_i$ satisfies:

1. $\langle u, v \rangle = \langle v, u \rangle^*$,
2. $k\langle u, v \rangle = \langle k^* u, v \rangle = \langle u, kv \rangle$,   where $k \in \mathbb{C}$
3. $\langle u, v + w \rangle = \langle u, v \rangle + \langle u, w \rangle$,   (distributivity)
4. $\langle u, u \rangle \geq 0$ and $\langle u, u \rangle = 0$ if and only if $u = 0$,   (positivity).

where * denotes complex conjugate.

Properties (2) and (3) imply that this inner product is 'linear in the second slot', but because of the conjugation in (2) it is not 'linear in the first slot'. Using properties (1) and (3) it can be shown that the statement $\langle (u + v), w \rangle = \langle u, w \rangle + \langle v, w \rangle$ is nevertheless true. As with Definition 1, we say that this inner product is anti-linear, or sesquilinear. The norm of a vector u is found, as previously, from $\|u\|^2 = \langle u, u \rangle$.

Exercises

4.16 Calculate, using both definitions, the inner product, $\langle z, w \rangle$, of $z \in \mathbb{C}^2$ and $w \in \mathbb{C}^2$ where $z = (z_1, z_2)$, $w = (w_1, w_2)$.

4.17 Suppose $z \in \mathbb{C}^2$ and $z = (z_1, z_2)$. Show that $\|z\|^2 = |z_1|^2 + |z_2|^2$.

4.18 Calculate, using both definitions, the inner product, $\langle z, w \rangle$, of $z \in \mathbb{C}$ and $w \in \mathbb{C}$. Show that $\|z\|^2 = zz^* = |z|^2$.

4.19 Consider $x, y \in \mathbb{B}^4$ where $x = (x_1, x_2, x_3, x_4)$, $y = (y_1, y_2, y_3, y_4)$ and where $x_i, y_i \in \mathbb{B}$. Define the inner product $\langle x, y \rangle$ to be $\displaystyle\sum_{i=1}^{4} x_i y_i$ where the multiplication and addition operations are carried out modulo 2. Find $\langle x, y \rangle$ when:

a) $x = (1, 1, 1, 0)$, $y = (0, 1, 1, 1)$,

b) $x = (0, 0, 1, 1)$, $y = (1, 0, 0, 0)$,

c) $x = (0, 0, 0, 0)$, $y = (0, 0, 0, 0)$,

d) $x = (1, 1, 1, 1)$, $y = (1, 1, 1, 1)$.

In all cases show that the inner product is equivalent to calculating

$$(x_1 \wedge y_1) \oplus (x_2 \wedge y_2) \oplus (x_3 \wedge y_3) \oplus (x_4 \wedge y_4).$$

4.20 Let $\psi = \alpha_1 e_1 + \alpha_2 e_2 + \ldots + \alpha_n e_n$, where $\alpha_i \in \mathbb{R}$ and

$$e_1 = \begin{pmatrix} 1 \\ 0 \\ 0 \\ \vdots \\ 0 \end{pmatrix}, \quad e_2 = \begin{pmatrix} 0 \\ 1 \\ 0 \\ \vdots \\ 0 \end{pmatrix}, \quad \ldots, \quad e_n = \begin{pmatrix} 0 \\ 0 \\ 0 \\ \vdots \\ 1 \end{pmatrix}.$$

Show that:

a) $\langle e_1, \psi \rangle = \alpha_1$,   b) $\langle e_2, \psi \rangle = \alpha_2$, $\ldots$,   c) $\langle e_n, \psi \rangle = \alpha_n$.

4.7 End-of-chapter exercises

1. Show that the vectors $\frac{1}{\sqrt{2}} \begin{pmatrix} 1 \\ 1 \end{pmatrix}$ and $\frac{1}{\sqrt{2}} \begin{pmatrix} 1 \\ -1 \end{pmatrix}$ are unit vectors and are mutually orthogonal.

2. Write the qubit $|\psi\rangle = \sqrt{\frac{2}{3}}|0\rangle + \sqrt{\frac{1}{3}}|1\rangle$ as a column vector and find its norm.

3. Given $u = \begin{pmatrix} 1 + i \\ -2 - 3i \end{pmatrix}$ and $v = \begin{pmatrix} 5 + 2i \\ 1 - 2i \end{pmatrix}$, evaluate the linear combination $3u + 7v$.

4. Given $u = \begin{pmatrix} 1+i \\ -2-3i \end{pmatrix}$ and $v = \begin{pmatrix} 5+2i \\ 1-2i \end{pmatrix}$ find the inner products $\langle u, v \rangle$, $\langle v, u \rangle$, $\langle u, u \rangle$, $\langle v, v \rangle$ using both Definitions 1 and 2.

5. Write the vectors $u = \begin{pmatrix} 1 \\ 0 \end{pmatrix}$ and $v = \begin{pmatrix} 0 \\ 1 \end{pmatrix}$ as linear combinations of the vectors $\begin{pmatrix} 1 \\ 1 \end{pmatrix}$ and $\begin{pmatrix} 1 \\ -1 \end{pmatrix}$.

6. Given the vectors $u = \begin{pmatrix} 2 \\ -1 \end{pmatrix}$, $v = \begin{pmatrix} -5 \\ -2 \end{pmatrix}$ and $w = \begin{pmatrix} -1 \\ 1 \end{pmatrix}$, choose one of the vectors and express it as a linear combination of the other two.

7. Given the vectors $u = \begin{pmatrix} 1 \\ 1 \\ 0 \end{pmatrix}$, $v = \begin{pmatrix} 1 \\ 1 \\ 1 \end{pmatrix}$, and $w = \begin{pmatrix} 1 \\ 0 \\ 1 \end{pmatrix}$ show that it is impossible to write any one of the vectors as a linear combination of the other two.

8. Show that

$$\left\{ \frac{1}{\sqrt{2}} \begin{pmatrix} 1 \\ 0 \\ 0 \\ 1 \end{pmatrix}, \frac{1}{\sqrt{2}} \begin{pmatrix} 0 \\ 1 \\ 1 \\ 0 \end{pmatrix}, \frac{1}{\sqrt{2}} \begin{pmatrix} 1 \\ 0 \\ 0 \\ -1 \end{pmatrix}, \frac{1}{\sqrt{2}} \begin{pmatrix} 0 \\ 1 \\ -1 \\ 0 \end{pmatrix} \right\}$$

is an orthonormal set (the so-called Bell basis).

9. Consider $x, a \in \mathbb{B}^n$ where $x = (x_1, x_2, \ldots, x_n)$, $a = (a_1, a_2, \ldots, a_n)$ and where $x_i, a_i \in \mathbb{B}$. Define the inner product $\langle x, a \rangle$ to be $\sum_{i=1}^{n} x_i a_i$ where the multiplication and addition operations are carried out modulo 2. Show that:

    a) if $x = (1, 0, 0, 0, \ldots, 0)$ then $x \cdot a$ yields the value a_1,

    b) if $x = (0, 1, 0, 0, \ldots, 0)$ then $x \cdot a$ yields the value a_2,

    c) if $x_i = 1$, $x_j = 0$ for $j = 1, \ldots, n$, $j \neq i$ then $x \cdot a$ yields the value a_i.

10. Let $|+\rangle = \frac{1}{\sqrt{2}}(|0\rangle + |1\rangle)$ and let $|-\rangle = \frac{1}{\sqrt{2}}(|0\rangle - |1\rangle)$ show that:

    a) $\frac{1}{\sqrt{2}}(|+\rangle + |-\rangle) = |0\rangle$,

    b) $\frac{1}{\sqrt{2}}(|+\rangle - |-\rangle) = |1\rangle$.

11. Consider $u, v \in \mathbb{R}^n$ with inner product $\langle u, v \rangle = \sum_{i=1}^{n} u_i v_i$. Show that if u and v are unit vectors the angle, θ, between them can be found from $\cos \theta = \langle u, v \rangle$.

12. Let $x = (x_1, x_2, x_3) = (x_1 x_2 x_3)$ with $x_i \in \mathbb{B}$. Let $y \in \mathbb{B}^3$. Complete Table 4.1 which evaluates $x \cdot y = \langle x, y \rangle = \sum_{i=1}^{3} x_i y_i$ where the multiplication and addition operations are carried out modulo 2.

13. Let $p, x \in \mathbb{B}^n$, i.e., $p = (p_1 p_2 \ldots p_n)$, $x = (x_1 x_2 \ldots x_n)$ where $x_i, p_i \in \mathbb{B}$. Define the scalar product of p and x to be

$$p \cdot x = (p_1 \wedge x_1) \oplus (p_2 \wedge x_2) \oplus \ldots \oplus (p_n \wedge x_n).$$

    a) Evaluate $p \cdot x$ when $p = (111)$ and $x = (010)$.

    b) Evaluate $p \cdot x$ when $p = (111)$ and $x = (110)$.

TABLE 4.1

Table for question 12

x	y	$x \cdot y$
$x_1 x_2 x_3$	000	0
$x_1 x_2 x_3$	001	x_3
$x_1 x_2 x_3$	010	
$x_1 x_2 x_3$	011	
$x_1 x_2 x_3$	100	
$x_1 x_2 x_3$	101	$x_1 \oplus x_3$
$x_1 x_2 x_3$	110	
$x_1 x_2 x_3$	111	$x_1 \oplus x_2 \oplus x_3$

14. As in Exercise 13, let $p, x \in \mathbb{B}^3 = \mathbb{B} \times \mathbb{B} \times \mathbb{B}$. Consider the sum

$$\sum_{x \in \{0,1\}^3} p \cdot x$$

Note that $\displaystyle\sum_{x \in \{0,1\}^3}$ indicates that the sum is taken over all possible values x in the set $\{0,1\}^3 = \mathbb{B}^3$, i.e.,

$$\{0,1\}^3 = \{000, 001, 010, 011, 100, 101, 110, 111\}.$$

Choose any non-zero value of $p \in \mathbb{B}^3$ and show that $\displaystyle\sum_{x \in \{0,1\}^3} p \cdot x = 4$.

(In Chapter 22 we show that for $p, x \in \mathbb{B}^n$ with $p \neq 0^n$, $\displaystyle\sum_{x=0}^{2^n-1} p \cdot x = 2^{n-1}$, a result used in some quantum algorithms.)

5

Matrices

5.1 Objectives

In quantum computation, matrices 'operate' or act on vectors to move a system from one quantum state to another as it evolves with time. They are also used to model the measurement of quantities of interest. Of particular relevance are the so-called symmetric, Hermitian, orthogonal and unitary matrices which are introduced in this chapter. In subsequent chapters we go on to develop this material and explain the terms 'eigenvalues' and 'eigenvectors' of matrices. When a quantum system is measured, the only possible measured values are the eigenvalues of the measurement matrices, and eigenvectors are used in the representation of the post-measurement state of the system.

The solution of sets of linear simultaneous equations can be accomplished using matrices and we show how this is done. Matrices can also be used to perform transformations of points in two- and three-dimensional space through rotations and reflections. These notions generalise to higher dimensions and are useful tools in some quantum algorithms. The objective of this chapter is to show how calculations with matrices are performed and to explain much of this important terminology.

5.2 Matrices: preliminary definitions

Much of what we have already discussed about row and column vectors can be extended to potentially larger mathematical structures. If we increase the permissible numbers of rows and/or columns, we create rectangular arrays known as **matrices** (singular **matrix**), for example,

$$\begin{pmatrix} 2 & 1 \\ 4 & -3 \\ 7 & 11 \end{pmatrix} \quad \text{and} \quad \begin{pmatrix} 3+2i & -i & 1-i \\ i & 0 & 3-2i \end{pmatrix}.$$

Definition 5.1 Matrix

*An $m \times n$ **matrix** is a rectangular array with m rows and n columns of real or complex numbers. These numbers are called the **elements** of the matrix.*

For example,

$$\begin{pmatrix} 0 & 1 \\ 1 & 0 \end{pmatrix} \quad \text{and} \quad \begin{pmatrix} 0 & -i \\ i & 0 \end{pmatrix}$$

are both 2×2 matrices. We can regard a $1 \times n$ matrix as a row vector, and an $m \times 1$ matrix as a column vector. We will usually use an upper-case letter to denote a matrix.

DOI: 10.1201/9781003264569-5

More generally, if the matrix A has m rows and n columns we can write

$$A = (a_{ij}) = \begin{pmatrix} a_{11} & a_{12} & \cdots & a_{1n} \\ a_{21} & a_{22} & \cdots & a_{2n} \\ \vdots & \vdots & \ddots & \vdots \\ a_{m1} & a_{m2} & \cdots & a_{mn} \end{pmatrix}.$$

Here a_{ij} represents the real or complex element in the ith row and jth column of A.

Matrices which have particular characteristics or properties are given specific names. A **square matrix** has the same number of rows as columns. A **diagonal matrix** is a square matrix with zeros everywhere except possibly on the **leading diagonal**, that is from top-left to bottom-right. An **identity matrix** is a diagonal matrix with 1's on the leading diagonal. For example, the matrices

$$C = \begin{pmatrix} 8 & -3 & 1 \\ -15 & 7 & 2 \\ 0 & 1 & 0 \end{pmatrix}, \quad D = \begin{pmatrix} 4 & 0 & 0 \\ 0 & 2 & 0 \\ 0 & 0 & 11 \end{pmatrix}, \quad \text{and} \quad I = \begin{pmatrix} 1 & 0 \\ 0 & 1 \end{pmatrix}$$

are all square. D and I are diagonal. I is an identity matrix. Generally, we use the symbol I (or I_n or $I_{n \times n}$) to represent an identity matrix. The diagonal matrix $\begin{pmatrix} \lambda_1 & 0 & 0 & 0 \\ 0 & \lambda_2 & 0 & 0 \\ 0 & 0 & \ddots & 0 \\ 0 & 0 & 0 & \lambda_n \end{pmatrix}$ will be denoted $\text{diag}\{\lambda_1, \lambda_2, \ldots, \lambda_n\}$.

Definition 5.2 Transpose of a matrix
*The **transpose** of a matrix A, denoted with a superscript T as in A^T, is found by interchanging its rows and columns. Note that if $A = (a_{ij})$ then $A^T = (a_{ji})$.*

Thus the first row becomes the first column of the transpose, and so on. For example, if

$$M = \begin{pmatrix} 4 + i & 1 - i \\ 5 & -3 - 2i \end{pmatrix} \quad \text{then} \quad M^T = \begin{pmatrix} 4 + i & 5 \\ 1 - i & -3 - 2i \end{pmatrix}.$$

Definition 5.3 Symmetric matrix
*A **symmetric matrix**, A, is a square matrix which is equal to its transpose:*

$$A = A^T.$$

Thus if $A = (a_{ij})$ is symmetric, then $a_{ij} = a_{ji}$.

Thus both

$$A = \begin{pmatrix} 7 & 12 \\ 12 & -5 \end{pmatrix} \quad \text{and} \quad B = \begin{pmatrix} 1 + i & 3 + 5i \\ 3 + 5i & 2 - 2i \end{pmatrix}$$

are symmetric matrices. This is immediately obvious because $A^T = A$ and $B^T = B$, but note also that a symmetric matrix possesses symmetry about its leading diagonal. Of particular interest in quantum computation are real symmetric matrices.

Definition 5.4 Conjugate transpose or Hermitian transpose of a matrix
*The **conjugate transpose**, or **Hermitian transpose** of a matrix A, denoted by $A^\dagger$, is found by taking the complex conjugate of the transpose of A:*

$$A^\dagger = (A^T)^*$$

Note that this is equivalent to $A^\dagger = (A^*)^T$. Here $*$ denotes conjugate, and T denotes transpose.

An important set of matrices used in quantum computation are known as **self-adjoint** or **Hermitian** matrices.

Definition 5.5 Self-adjoint matrix
A **self-adjoint matrix***, H say, is a square matrix which is equal to the complex conjugate of its transpose:*

$$H = (H^T)^* = (H^*)^T = H^\dagger.$$

Self-adjoint matrices play an important role in quantum computation because, as we shall see, they have real eigenvalues which correspond to the measured values of quantum observables.

Example 5.2.1

Suppose

$$H = \begin{pmatrix} 7 & 7+2i \\ 7-2i & -5 \end{pmatrix}.$$

Taking the complex conjugate of each element we find

$$H^* = \begin{pmatrix} 7 & 7-2i \\ 7+2i & -5 \end{pmatrix}.$$

Then, taking the transpose we obtain

$$(H^*)^T = \begin{pmatrix} 7 & 7+2i \\ 7-2i & -5 \end{pmatrix}$$

which is $H^\dagger$. Observe that $H^\dagger = (H^*)^T = H$ so that H is indeed self-adjoint.

Exercises

5.1 Show that the leading diagonal elements of a self-adjoint matrix are necessarily real.

5.2 Show that a real symmetric matrix is self-adjoint.

5.3 The sum of the elements on the leading diagonal of a matrix A is known as its **trace**, and denoted $\text{tr}(A)$. If $A = \begin{pmatrix} 7 & 2 & 1 \\ 8 & 2 & 3 \\ 9 & -1 & -4 \end{pmatrix}$ find $\text{tr}(A)$ and $\text{tr}(A^T)$. What is the trace of an $n \times n$ identity matrix ?

5.3   Matrix arithmetic: addition, subtraction and scalar multiplication

When two matrices have the same size, addition and subtraction are performed, as with vectors, in an element by element fashion, to yield another matrix of the same size. Multiplication by a scalar is achieved by multiplying every element of a matrix by that scalar.

So, if

$$A = \begin{pmatrix} 4 & 1 \\ 5 & -3 \end{pmatrix}, \quad B = \begin{pmatrix} -8 & 2 \\ 0 & -3.5 \end{pmatrix}, \quad C = \begin{pmatrix} 0 & -i \\ i & 0 \end{pmatrix}$$

then

$$A + B = \begin{pmatrix} -4 & 3 \\ 5 & -6.5 \end{pmatrix}, \quad A - B = \begin{pmatrix} 12 & -1 \\ 5 & 0.5 \end{pmatrix}, \quad 7B = \begin{pmatrix} -56 & 14 \\ 0 & -24.5 \end{pmatrix}$$

$$3iC = \begin{pmatrix} 0 & 3 \\ -3 & 0 \end{pmatrix}.$$

Definition 5.6 Matrix addition and scalar multiplication
Let A and B have the same size (i.e., the same number of rows and the same number of columns). If $A = (a_{ij})$, $B = (b_{ij})$ then $A + B = (a_{ij} + b_{ij})$. For a scalar k, $kA = k(a_{ij}) = (ka_{ij})$.

Note that matrix addition is associative, that is

$$(A + B) + C = A + (B + C)$$

so that we can write simply $A + B + C$ and there is no ambiguity. This follows because addition of real or complex numbers is associative.

Example 5.3.1 The set of $m \times n$ matrices forms a group under addition.

Observe that if we add two $m \times n$ matrices the result is another $m \times n$ matrix. So the set is closed under addition. We have noted already that matrix addition is associative. The $m \times n$ matrix with all elements equal to zero is the identity element for this group. (Note that this is not the same as an identity matrix.) The inverse of any $m \times n$ matrix under addition is simply the matrix in which all elements are the negative of those in the original matrix. The remaining group axioms are readily verified. Thus the set of $m \times n$ matrices forms a group under addition.

Exercises

5.4   Show that the sum and difference of two real symmetric $n \times n$ matrices is another real symmetric matrix.

5.5   Show that if an $n \times n$ real symmetric matrix is multiplied by a real scalar the result is another real symmetric matrix.

5.6   If $A = \begin{pmatrix} 9 & 4 \\ 3 & 2 \end{pmatrix}$ find $A + A^T$ and show that this is a symmetric matrix.

5.7   A **skew-symmetric** matrix is one for which $A^T = -A$. If $A = \begin{pmatrix} 9 & 4 \\ 3 & 2 \end{pmatrix}$ find $A - A^T$ and show that the result is a skew-symmetric matrix.

5.8 If A and B are $n \times n$ matrices and k is a scalar show that

    a) $(A + B)^T = A^T + B^T$ b) $(kA)^T = kA^T$.

5.4 The product of two matrices

Matrices of an appropriate size can be multiplied to yield another matrix. Specifically, if A and B are $p \times q$ and $r \times s$ matrices, respectively, we can form the **matrix product** $C = AB$ if and only if $q = r$, that is the number of columns in the first matrix is equal to the number of rows in the second. The result, C, is then a $p \times s$ matrix. The formal definition of the matrix product is as follows:

Definition 5.7 Matrix multiplication
Given A and B of size $p \times q$ and $q \times s$ respectively, the product $C = AB$ is $p \times s$ and given by

$$C = (c_{ij}) \text{ where } c_{ij} = \sum_{k=1}^{q} a_{ik} b_{kj}.$$

The examples which follow demonstrate how this calculation is performed in practice. As will be illustrated, matrix multiplication is, in general, not commutative, that is $AB \neq BA$. However, the following properties hold in general:

Theorem 5.1 *Provided that matrices P, Q and R are of compatible sizes:*

    1. matrix multiplication is associative: $(PQ)R = P(QR)$.

    2. multiplication is distributive over addition: $P(Q + R) = PQ + PR$, and similarly

    3. $(P + Q)R = PR + QR$.

    4. $A(kB) = k(AB)$, where k is a scalar.

Given the expression AB we say that B is **pre-multiplied** by A, or A is **post-multiplied** by B.

Example 5.4.1

(i) Use the formula in the Definition above to calculate c_{21} when $A = \begin{pmatrix} 5 & 3 & -1 \\ 2 & 1 & 8 \end{pmatrix}$ and $B = \begin{pmatrix} 7 \\ 2 \\ 1 \end{pmatrix}$.

(ii) Find the matrix product $C = AB$.

Solution

(i) Note that A has size 2×3, B has size 3×1 and so the product can be formed and will have size 2×1. The calculation of c_{21} is performed as follows:

$$c_{ij} = \sum_{k=1}^{q} a_{ik}b_{kj}$$

$$c_{21} = \sum_{k=1}^{3} a_{2k}b_{k1}$$

$$= a_{21}b_{11} + a_{22}b_{21} + a_{23}b_{31}$$

$$= (2)(7) + (1)(2) + (8)(1)$$

$$= 24.$$

You should try calculating c_{11} in a similar fashion.

(ii) The whole calculation is usually set out as follows:

$$C = AB = \begin{pmatrix} 5 & 3 & -1 \\ 2 & 1 & 8 \end{pmatrix} \begin{pmatrix} 7 \\ 2 \\ 1 \end{pmatrix} = \begin{pmatrix} (5)(7) + (3)(2) + (-1)(1) \\ (2)(7) + (1)(2) + (8)(1) \end{pmatrix} = \begin{pmatrix} 40 \\ 24 \end{pmatrix}.$$

Note from Example 5.4.1 that the product BA cannot be formed because in this order the number of columns in the first matrix is not equal to the number of rows in the second. Even in cases when both AB and BA can be formed, in general $AB \neq BA$ so that matrix multiplication in general is not commutative. However, there are exceptions as we shall see.

Example 5.4.2

Find, if possible, the products AB and BA when

$$A = \begin{pmatrix} 4 \\ -3 \\ 1 \end{pmatrix} \qquad \text{and} \qquad B = \begin{pmatrix} -1 & 2 & 5 \end{pmatrix}.$$

Solution

A has size 3×1 and B has size 1×3. Therefore, in the expression AB the number of columns in the first matrix, 1, is the same as the number of rows in the second. The product can be formed and will have size 3×3:

$$AB = \begin{pmatrix} 4 \\ -3 \\ 1 \end{pmatrix} \begin{pmatrix} -1 & 2 & 5 \end{pmatrix} = \begin{pmatrix} -4 & 8 & 20 \\ 3 & -6 & -15 \\ -1 & 2 & 5 \end{pmatrix}.$$

In the expression BA the number of columns in the first matrix, 3, is the same as the number of rows in the second. The product can be formed and will have size 1×1, that is, a scalar.

$$BA = \begin{pmatrix} -1 & 2 & 5 \end{pmatrix} \begin{pmatrix} 4 \\ -3 \\ 1 \end{pmatrix} = (-1)(4) + (2)(-3) + (5)(1) = -5.$$

We emphasise again that, in general, $AB \neq BA$.

Example 5.4.3

Evaluate the matrix product MP where $M = \begin{pmatrix} 1 & 1 \\ 1 & -1 \end{pmatrix}$ and $P = \begin{pmatrix} e^{i\theta_2} & 0 \\ 0 & e^{i\theta_1} \end{pmatrix}$.

Solution

$$MP = \begin{pmatrix} 1 & 1 \\ 1 & -1 \end{pmatrix} \begin{pmatrix} e^{i\theta_2} & 0 \\ 0 & e^{i\theta_1} \end{pmatrix}$$

$$= \begin{pmatrix} e^{i\theta_2} & e^{i\theta_1} \\ e^{i\theta_2} & -e^{i\theta_1} \end{pmatrix}.$$

Example 5.4.4

Given $H = \frac{1}{\sqrt{2}} \begin{pmatrix} 1 & 1 \\ 1 & -1 \end{pmatrix}$, find a) $H \begin{pmatrix} 1 \\ 0 \end{pmatrix}$, b) $H \begin{pmatrix} 0 \\ 1 \end{pmatrix}$.

Solution

a) $H \begin{pmatrix} 1 \\ 0 \end{pmatrix} = \frac{1}{\sqrt{2}} \begin{pmatrix} 1 & 1 \\ 1 & -1 \end{pmatrix} \begin{pmatrix} 1 \\ 0 \end{pmatrix} = \frac{1}{\sqrt{2}} \begin{pmatrix} 1 \\ 1 \end{pmatrix}$.

b) $H \begin{pmatrix} 0 \\ 1 \end{pmatrix} = \frac{1}{\sqrt{2}} \begin{pmatrix} 1 & 1 \\ 1 & -1 \end{pmatrix} \begin{pmatrix} 0 \\ 1 \end{pmatrix} = \frac{1}{\sqrt{2}} \begin{pmatrix} 1 \\ -1 \end{pmatrix}$.

Recall that in ket notation $\begin{pmatrix} 1 \\ 0 \end{pmatrix} = |0\rangle$ and $\begin{pmatrix} 0 \\ 1 \end{pmatrix} = |1\rangle$. Further, in quantum computation $\frac{1}{\sqrt{2}} \begin{pmatrix} 1 \\ 1 \end{pmatrix}$ is often written $|+\rangle$ and $\frac{1}{\sqrt{2}} \begin{pmatrix} 1 \\ -1 \end{pmatrix}$ as $|-\rangle$ in which case the foregoing results would be written

$$H|0\rangle = |+\rangle, \qquad H|1\rangle = |-\rangle.$$

Example 5.4.5

Given $A = \begin{pmatrix} 7 & -2 \\ 1 & 4 \end{pmatrix}$, find A^2 and show that $A^2 - 11A + 30I = 0$ where 0 here is the 2×2 matrix with all elements zero.

Solution

$$A^2 = \begin{pmatrix} 7 & -2 \\ 1 & 4 \end{pmatrix} \begin{pmatrix} 7 & -2 \\ 1 & 4 \end{pmatrix} = \begin{pmatrix} 47 & -22 \\ 11 & 14 \end{pmatrix}.$$

Then

$$A^2 - 11A + 30I = \begin{pmatrix} 47 & -22 \\ 11 & 14 \end{pmatrix} - 11\begin{pmatrix} 7 & -2 \\ 1 & 4 \end{pmatrix} + 30\begin{pmatrix} 1 & 0 \\ 0 & 1 \end{pmatrix}$$

$$= \begin{pmatrix} 47 & -22 \\ 11 & 14 \end{pmatrix} - \begin{pmatrix} 77 & -22 \\ 11 & 44 \end{pmatrix} + \begin{pmatrix} 30 & 0 \\ 0 & 30 \end{pmatrix}$$

$$= \begin{pmatrix} 0 & 0 \\ 0 & 0 \end{pmatrix}.$$

as required.

Exercises

5.9   Suppose $X = \begin{pmatrix} 0 & 1 \\ 1 & 0 \end{pmatrix}$. Recall that $|0\rangle = \begin{pmatrix} 1 \\ 0 \end{pmatrix}$ and $|1\rangle = \begin{pmatrix} 0 \\ 1 \end{pmatrix}$. Show that

    a) $X|0\rangle = |1\rangle$,   b) $X|1\rangle = |0\rangle$.

5.10   Suppose $H = \frac{1}{\sqrt{2}} \begin{pmatrix} 1 & 1 \\ 1 & -1 \end{pmatrix}$ and $Z = \begin{pmatrix} 1 & 0 \\ 0 & -1 \end{pmatrix}$. Show that $HZH = X$
       where $X = \begin{pmatrix} 0 & 1 \\ 1 & 0 \end{pmatrix}$.

5.11   If x is an $n \times 1$ matrix (column vector) and $I_{n\times n}$ is the $n \times n$ identity matrix, show that $I_{n\times n}x = x$.

5.12   Let $P = \begin{pmatrix} p_{11} & p_{12} & p_{13} \\ p_{21} & p_{22} & p_{23} \\ p_{31} & p_{32} & p_{33} \end{pmatrix}$ and let D be the diagonal matrix $\mathrm{diag}\{\lambda_1, \lambda_2, \lambda_3\} = \begin{pmatrix} \lambda_1 & 0 & 0 \\ 0 & \lambda_2 & 0 \\ 0 & 0 & \lambda_3 \end{pmatrix}$. Show that

$$PD = \begin{pmatrix} \lambda_1 p_{11} & \lambda_2 p_{12} & \lambda_3 p_{13} \\ \lambda_1 p_{21} & \lambda_2 p_{22} & \lambda_3 p_{23} \\ \lambda_1 p_{31} & \lambda_2 p_{32} & \lambda_3 p_{33} \end{pmatrix}.$$

Observe particularly the way that the λ_i on the diagonal of D transfer to the columns of PD.

Definition 5.8 Normal matrix
A square matrix A with real elements for which

$$A A^T = A^T A$$

that is, A commutes with its transpose, is said to be **normal**.

Example 5.4.6

Show that the matrices $A = \begin{pmatrix} 0 & -5 \\ 5 & 0 \end{pmatrix}$ and $B = \begin{pmatrix} 1 & 1 \\ -1 & 1 \end{pmatrix}$ are both normal.

Solution

$$A A^T = \begin{pmatrix} 0 & -5 \\ 5 & 0 \end{pmatrix} \begin{pmatrix} 0 & 5 \\ -5 & 0 \end{pmatrix} = \begin{pmatrix} 25 & 0 \\ 0 & 25 \end{pmatrix}$$

$$A^T A = \begin{pmatrix} 0 & 5 \\ -5 & 0 \end{pmatrix} \begin{pmatrix} 0 & -5 \\ 5 & 0 \end{pmatrix} = \begin{pmatrix} 25 & 0 \\ 0 & 25 \end{pmatrix}$$

$$B B^T = \begin{pmatrix} 1 & 1 \\ -1 & 1 \end{pmatrix} \begin{pmatrix} 1 & -1 \\ 1 & 1 \end{pmatrix} = \begin{pmatrix} 2 & 0 \\ 0 & 2 \end{pmatrix}$$

$$B^T B = \begin{pmatrix} 1 & -1 \\ 1 & 1 \end{pmatrix} \begin{pmatrix} 1 & 1 \\ -1 & 1 \end{pmatrix} = \begin{pmatrix} 2 & 0 \\ 0 & 2 \end{pmatrix}.$$

We see that $A A^T = A^T A$ and $B B^T = B^T B$ as is required for the matrices to be normal.

More generally, for matrices with complex elements, we have the following definition:

Definition 5.9 Normal matrix
A square matrix A with complex elements for which

$$A A^\dagger = A^\dagger A$$

that is, A commutes with its conjugate transpose, is said to be **normal**.

Clearly, in the case of matrices with real elements this is equivalent to $A A^T = A^T A$. All symmetric and self-adjoint matrices are necessarily normal.

Example 5.4.7 Linear functional

Evaluate, if possible, the matrix product AB where

$$A = (\ 3 \quad -5 \quad 2 \ ) \qquad \text{and} \qquad B = \begin{pmatrix} 1 \\ 3 \\ -7 \end{pmatrix}.$$

Solution

A has size 1×3. B has size 3×1. So the product AB can be found and will be of size 1×1, that is, a single number.

$$AB = (\ 3 \quad -5 \quad 2 \ ) \begin{pmatrix} 1 \\ 3 \\ -7 \end{pmatrix} = 3 \times 1 + (-5) \times 3 + 2 \times (-7) = -26.$$

It will become important when we study **linear functionals** (Chapter 9) to observe that we can think of AB as the row vector $A = (3, -5, 2)$ 'operating' on the column vector B which follows it to produce a scalar, i.e., we can think of it, using our knowledge of functions, as $A(B)$, that is as a 'function' A with argument B.

The following theorem will be used when we study linear operators for quantum dynamics.

Theorem 5.2 The transpose of a product
Provided matrices A and B are compatible for multiplication:

$$(AB)^T = B^T A^T.$$

Example 5.4.8 The property $(AB)^T = B^T A^T$.

Suppose $A = \begin{pmatrix} 3 & 1 \\ 2 & 6 \end{pmatrix}$, $B = \begin{pmatrix} -1 & 4 \\ 3 & 8 \end{pmatrix}$. Show that $(AB)^T = B^T A^T$.

Solution

$$AB = \begin{pmatrix} 3 & 1 \\ 2 & 6 \end{pmatrix} \begin{pmatrix} -1 & 4 \\ 3 & 8 \end{pmatrix} = \begin{pmatrix} 0 & 20 \\ 16 & 56 \end{pmatrix}$$

and hence

$$(AB)^T = \begin{pmatrix} 0 & 16 \\ 20 & 56 \end{pmatrix}.$$

Also

$$B^T A^T = \begin{pmatrix} -1 & 3 \\ 4 & 8 \end{pmatrix} \begin{pmatrix} 3 & 2 \\ 1 & 6 \end{pmatrix} = \begin{pmatrix} 0 & 16 \\ 20 & 56 \end{pmatrix}$$

and hence $(AB)^T = B^T A^T$.

The result of the previous example is true more generally, including when the transpose operation is replaced by the conjugate transpose †.

Theorem 5.3 *Provided that the matrices are compatible for multiplication*

$$(AB)^\dagger = B^\dagger A^\dagger.$$

We now introduce a result known as the **completeness relation** or **resolution of the identity**. This result is drawn upon frequently in quantum computations. Consider first the following example.

Example 5.4.9 Matrix multiplication and the resolution of the identity

In this example we start to introduce matrix multiplication using the ket notation.
If $|\alpha_1\rangle = \begin{pmatrix} 1 \\ 0 \end{pmatrix}$ and $|\alpha_2\rangle = \begin{pmatrix} 0 \\ 1 \end{pmatrix}$ evaluate

(a) $|\alpha_1\rangle\langle\alpha_1|$,   (b) $|\alpha_2\rangle\langle\alpha_2|$,   (c) $\sum_{i=1}^{2}|\alpha_i\rangle\langle\alpha_i|$.

Solution

(a) Given $|\alpha_1\rangle = \begin{pmatrix} 1 \\ 0 \end{pmatrix}$ recall that the bra corresponding to this ket is found by taking the conjugate transpose to give $\langle\alpha_1| = \begin{pmatrix} 1 & 0 \end{pmatrix}$. Then

$$|\alpha_1\rangle\langle\alpha_1| = \begin{pmatrix} 1 \\ 0 \end{pmatrix} \begin{pmatrix} 1 & 0 \end{pmatrix} = \begin{pmatrix} 1 & 0 \\ 0 & 0 \end{pmatrix}.$$

(b) Similarly,

$$|\alpha_2\rangle\langle\alpha_2| = \begin{pmatrix} 0 \\ 1 \end{pmatrix} \begin{pmatrix} 0 & 1 \end{pmatrix} = \begin{pmatrix} 0 & 0 \\ 0 & 1 \end{pmatrix}.$$

(c)

$$\sum_{i=1}^{2}|\alpha_i\rangle\langle\alpha_i| = \begin{pmatrix} 1 & 0 \\ 0 & 0 \end{pmatrix} + \begin{pmatrix} 0 & 0 \\ 0 & 1 \end{pmatrix} = \begin{pmatrix} 1 & 0 \\ 0 & 1 \end{pmatrix}.$$

We make an important observation: the final result is an identity matrix.
More generally, given a set of orthonormal n-dimensional vectors $\{|\alpha_i\rangle\}$, $i = 1, \ldots, n$, it can be shown that

$$\sum_{i=1}^{n}|\alpha_i\rangle\langle\alpha_i| = I.$$

This result is the completeness relation or the resolution of the identity.

Example 5.4.10

Evaluate $\sum\limits_{z=0}^{1} z|z\rangle\langle z|$.

Solution

$$\sum_{z=0}^{1} z|z\rangle\langle z| = 0|0\rangle\langle 0| + 1|1\rangle\langle 1|$$

$$= 0\begin{pmatrix} 1 \\ 0 \end{pmatrix}\begin{pmatrix} 1 & 0 \end{pmatrix} + 1\begin{pmatrix} 0 \\ 1 \end{pmatrix}\begin{pmatrix} 0 & 1 \end{pmatrix}$$

$$= 0\begin{pmatrix} 1 & 0 \\ 0 & 0 \end{pmatrix} + 1\begin{pmatrix} 0 & 0 \\ 0 & 1 \end{pmatrix}$$

$$= \begin{pmatrix} 0 & 0 \\ 0 & 1 \end{pmatrix}.$$

Example 5.4.11 Matrix multiplication and permutation matrices

Consider the effect of pre-multiplying the column vector $u = \begin{pmatrix} a \\ b \\ c \end{pmatrix}$ by the matrix $P = \begin{pmatrix} 0 & 0 & 1 \\ 1 & 0 & 0 \\ 0 & 1 & 0 \end{pmatrix}$.

$$Pu = \begin{pmatrix} 0 & 0 & 1 \\ 1 & 0 & 0 \\ 0 & 1 & 0 \end{pmatrix}\begin{pmatrix} a \\ b \\ c \end{pmatrix} = \begin{pmatrix} c \\ a \\ b \end{pmatrix}.$$

Observe that the effect is to re-order or **permute** the elements of u. Thus P is referred to as a **permutation matrix**.

Definition 5.10 Permutation matrix.
A **permutation matrix** *P is a square matrix which has a 1 just once in every row and column and zero everywhere else. Pre-multiplying a column vector u by P has the effect of re-ordering or permuting the elements of u.*

Exercises

5.13   Given $|0\rangle = \begin{pmatrix} 1 \\ 0 \end{pmatrix}$ and $|1\rangle = \begin{pmatrix} 0 \\ 1 \end{pmatrix}$, find

      a) $\sum\limits_{x=0}^{1}|x\rangle\langle x|$,
                      b) $\sum\limits_{x=0}^{1}|\overline{x}\rangle\langle x|$.

5.14   Given $A = \begin{pmatrix} a_{11} & a_{12} \\ a_{21} & a_{22} \end{pmatrix}$, $B = \begin{pmatrix} b_{11} & b_{12} \\ b_{21} & b_{22} \end{pmatrix}$ show that $(AB)^T = B^T A^T$.

5.15   Prove that all symmetric matrices are necessarily normal.

5.16   Prove that all self-adjoint matrices are necessarily normal.

5.17   Given the orthonormal set $\{|\alpha_1\rangle, |\alpha_2\rangle\}$ where $|\alpha_1\rangle = \begin{pmatrix} \frac{1}{\sqrt{2}} \\ \frac{1}{\sqrt{2}} \end{pmatrix}$, $|\alpha_2\rangle = \begin{pmatrix} -\frac{1}{\sqrt{2}} \\ \frac{1}{\sqrt{2}} \end{pmatrix}$,

perform the resolution of the identity to show that $\displaystyle\sum_{i=1}^{2} |\alpha_i\rangle\langle\alpha_i| = I$.

5.5   Block multiplication of matrices

Given two matrices, A and B, which are compatible for multiplication it is possible to partition each matrix into submatrices called **blocks** and then perform multiplication, treating the blocks as though they were simple matrix elements. In order to do this the number of columns of blocks in A must equal the number of rows of blocks in B. Consider the following example which illustrates how this is done.

Example 5.5.1

Suppose we wish to calculate the following product of two 3×3 matrices, P and Q:

$$PQ = \begin{pmatrix} p_{11} & p_{12} & p_{13} \\ p_{21} & p_{22} & p_{23} \\ p_{31} & p_{32} & p_{33} \end{pmatrix} \begin{pmatrix} q_{11} & q_{12} & q_{13} \\ q_{21} & q_{22} & q_{23} \\ q_{31} & q_{32} & q_{33} \end{pmatrix}.$$

Suppose we block P with vertical lines to produce three columns, and block Q with horizontal lines to produce three rows, as shown.

$$PQ = \left(\begin{array}{c|c|c} p_{11} & p_{12} & p_{13} \\ p_{21} & p_{22} & p_{23} \\ p_{31} & p_{32} & p_{33} \end{array} \right) \left(\begin{array}{ccc} q_{11} & q_{12} & q_{13} \\ \hline q_{21} & q_{22} & q_{23} \\ \hline q_{31} & q_{32} & q_{33} \end{array} \right).$$

Now treating each block as a single element we can write

$$PQ = \begin{pmatrix} P_1 & P_2 & P_3 \end{pmatrix} \begin{pmatrix} Q_1 \\ Q_2 \\ Q_3 \end{pmatrix}$$

where $P_1 = \begin{pmatrix} p_{11} \\ p_{21} \\ p_{31} \end{pmatrix}$, $Q_1 = \begin{pmatrix} q_{11} & q_{12} & q_{13} \end{pmatrix}$ and so on. Given the sizes are still compatible for multiplication, that is 1×3 and 3×1, respectively, we find the product, which will be 1×1, is given by

$$PQ = P_1 Q_1 + P_2 Q_2 + P_3 Q_3.$$

It is this form of the product that will be useful when we study eigenvalues and matrix decomposition in Chapter 7. To verify that this form does indeed give the required product observe that

$$PQ = \begin{pmatrix} p_{11} \\ p_{21} \\ p_{31} \end{pmatrix} \begin{pmatrix} q_{11} & q_{12} & q_{13} \end{pmatrix} + \begin{pmatrix} p_{12} \\ p_{22} \\ p_{32} \end{pmatrix} \begin{pmatrix} q_{21} & q_{22} & q_{23} \end{pmatrix} \cdots$$

$$+ \begin{pmatrix} p_{13} \\ p_{23} \\ p_{33} \end{pmatrix} \begin{pmatrix} q_{31} & q_{32} & q_{33} \end{pmatrix}$$

$$= \begin{pmatrix} p_{11}q_{11} & p_{11}q_{12} & p_{11}q_{13} \\ p_{21}q_{11} & p_{21}q_{12} & p_{21}q_{13} \\ p_{31}q_{11} & p_{31}q_{12} & p_{31}q_{13} \end{pmatrix} + \begin{pmatrix} p_{12}q_{21} & p_{12}q_{22} & p_{12}q_{23} \\ p_{22}q_{21} & p_{22}q_{22} & p_{22}q_{23} \\ p_{32}q_{21} & p_{32}q_{22} & p_{32}q_{23} \end{pmatrix} \cdots$$

$$+ \begin{pmatrix} p_{13}q_{31} & p_{13}q_{32} & p_{13}q_{33} \\ p_{23}q_{31} & p_{23}q_{32} & p_{23}q_{33} \\ p_{33}q_{31} & p_{33}q_{32} & p_{33}q_{33} \end{pmatrix}$$

$$= \begin{pmatrix} p_{11}q_{11} + p_{12}q_{21} + p_{13}q_{31} & p_{11}q_{12} + p_{12}q_{22} + p_{13}q_{32} & p_{11}q_{13} + p_{12}q_{23} + p_{13}q_{33} \\ p_{21}q_{11} + p_{22}q_{21} + p_{23}q_{31} & p_{21}q_{12} + p_{22}q_{22} + p_{23}q_{32} & p_{21}q_{13} + p_{22}q_{23} + p_{23}q_{33} \\ p_{31}q_{11} + p_{32}q_{21} + p_{33}q_{31} & p_{31}q_{12} + p_{32}q_{22} + p_{33}q_{32} & p_{31}q_{13} + p_{32}q_{23} + p_{33}q_{33} \end{pmatrix}$$

which is the required matrix product.

Example 5.5.2

Partition the first matrix below into three columns and the second into three rows and perform block multiplication to find PQ.

$$PQ = \begin{pmatrix} 1 & 2 & 3 \\ 4 & 5 & 1 \\ 6 & 0 & 2 \end{pmatrix} \begin{pmatrix} 4 & -1 \\ 2 & 3 \\ 0 & 7 \end{pmatrix}.$$

Solution

We partition the matrices as indicated:

$$PQ = \left(\begin{array}{c|c|c} 1 & 2 & 3 \\ 4 & 5 & 1 \\ 6 & 0 & 2 \end{array} \right) \left(\begin{array}{cc} 4 & -1 \\ \hline 2 & 3 \\ \hline 0 & 7 \end{array} \right)$$

that is,

$$PQ = \begin{pmatrix} P_1 & P_2 & P_3 \end{pmatrix} \begin{pmatrix} Q_1 \\ Q_2 \\ Q_3 \end{pmatrix}.$$

The product is then

$$PQ = P_1 Q_1 + P_2 Q_2 + P_3 Q_3.$$

We can evaluate this as follows:

$$PQ = \begin{pmatrix} 1 \\ 4 \\ 6 \end{pmatrix} (4 \quad -1) + \begin{pmatrix} 2 \\ 5 \\ 0 \end{pmatrix} (2 \quad 3) + \begin{pmatrix} 3 \\ 1 \\ 2 \end{pmatrix} (0 \quad 7)$$

$$= \begin{pmatrix} 4 & -1 \\ 16 & -4 \\ 24 & -6 \end{pmatrix} + \begin{pmatrix} 4 & 6 \\ 10 & 15 \\ 0 & 0 \end{pmatrix} + \begin{pmatrix} 0 & 21 \\ 0 & 7 \\ 0 & 14 \end{pmatrix}$$

$$= \begin{pmatrix} 8 & 26 \\ 26 & 18 \\ 24 & 8 \end{pmatrix}.$$

Example 5.5.3

Consider $P = \begin{pmatrix} p_{11} & p_{12} \\ p_{21} & p_{22} \end{pmatrix}$.

a) State P^T.

b) Evaluate PP^T.

c) Let $p_1 = \begin{pmatrix} p_{11} \\ p_{21} \end{pmatrix}$ and $p_2 = \begin{pmatrix} p_{12} \\ p_{22} \end{pmatrix}$, the first and second columns of P respectively, and use block multiplication to deduce that PP^T can be evaluated as

$$PP^T = p_1 p_1^T + p_2 p_2^T.$$

Solution

a) $P^T = \begin{pmatrix} p_{11} & p_{21} \\ p_{12} & p_{22} \end{pmatrix}$.

b)

$$PP^T = \begin{pmatrix} p_{11} & p_{12} \\ p_{21} & p_{22} \end{pmatrix} \begin{pmatrix} p_{11} & p_{21} \\ p_{12} & p_{22} \end{pmatrix}$$

$$= \begin{pmatrix} p_{11}^2 + p_{12}^2 & p_{11}p_{21} + p_{12}p_{22} \\ p_{21}p_{11} + p_{22}p_{12} & p_{21}^2 + p_{22}^2 \end{pmatrix}.$$

c) Let

$$p_1 = \begin{pmatrix} p_{11} \\ p_{21} \end{pmatrix} \quad \text{and} \quad p_2 = \begin{pmatrix} p_{12} \\ p_{22} \end{pmatrix},$$

so that

$$p_1^T = (p_{11} \quad p_{21}) \quad \text{and} \quad p_2^T = (p_{12} \quad p_{22}).$$

Observe that in block form

$$P = (p_1 \quad p_2) \quad \text{and} \quad P^T = \begin{pmatrix} p_1^T \\ p_2^T \end{pmatrix}.$$

Then

$$PP^T = \begin{pmatrix} p_1 & p_2 \end{pmatrix} \begin{pmatrix} p_1^T \\ p_2^T \end{pmatrix}$$

$$= p_1 p_1^T + p_2 p_2^T$$

$$= \begin{pmatrix} p_{11} \\ p_{21} \end{pmatrix} \begin{pmatrix} p_{11} & p_{21} \end{pmatrix} + \begin{pmatrix} p_{12} \\ p_{22} \end{pmatrix} \begin{pmatrix} p_{12} & p_{22} \end{pmatrix}$$

$$= \begin{pmatrix} p_{11}^2 & p_{11}p_{21} \\ p_{21}p_{11} & p_{21}^2 \end{pmatrix} + \begin{pmatrix} p_{12}^2 & p_{12}p_{22} \\ p_{22}p_{12} & p_{22}^2 \end{pmatrix}$$

$$= \begin{pmatrix} p_{11}^2 + p_{12}^2 & p_{11}p_{21} + p_{12}p_{22} \\ p_{21}p_{11} + p_{22}p_{12} & p_{21}^2 + p_{22}^2 \end{pmatrix}.$$

Comparing this result with the solution to part b) we deduce that

$$PP^T = p_1 p_1^T + p_2 p_2^T$$

where p_1 and p_2 are the first and second columns of P, respectively. This result can be readily generalised to larger $n \times n$ matrices:

$$PP^T = p_1 p_1^T + p_2 p_2^T + \ldots + p_n p_n^T$$

where the p_i, $i = 1, 2, \ldots, n$ are the columns of P.

5.6   Matrices, inner products and ket notation

We shall now consider how the inner product in $\mathbb{C}^n$, is written and applied using ket notation. We have seen that given two column vectors u and v in $\mathbb{C}^n$, their inner product is defined (Definition 2 in Section 4.6.3) as $\langle u, v \rangle = \sum_{i=1}^{n} u_i^* v_i$. So, for example, if $u = \begin{pmatrix} 1 + i \\ 2 - 3i \end{pmatrix}$, and $v = \begin{pmatrix} 4 + 5i \\ 1 + 2i \end{pmatrix}$ then

$$\langle \begin{pmatrix} 1 + i \\ 2 - 3i \end{pmatrix}, \begin{pmatrix} 4 + 5i \\ 1 + 2i \end{pmatrix} \rangle = (1 + i)^*(4 + 5i) + (2 - 3i)^*(1 + 2i)$$

$$= (1 - i)(4 + 5i) + (2 + 3i)(1 + 2i)$$

$$= 9 + i - 4 + 7i$$

$$= 5 + 8i.$$

Observe that the same result can be obtained using matrix multiplication by calculating

$$\begin{pmatrix} 1 - i & 2 + 3i \end{pmatrix} \begin{pmatrix} 4 + 5i \\ 1 + 2i \end{pmatrix}.$$

But the row vector here is simply the bra corresponding to the ket $\begin{pmatrix} 1 + i \\ 2 - 3i \end{pmatrix}$. (Recall that when writing a ket as a bra we take the complex conjugate of each element and then find the transpose.) Thus, sticking with the ket notation, the inner product of $|u\rangle$ and $|v\rangle$ can be written as

$$\langle u | \, | v \rangle \qquad \text{or more concisely} \qquad \langle u | v \rangle.$$

Definition 5.11 The inner product of two kets.
The inner product of two kets (column vectors) $|u\rangle$ and $|v\rangle$ is $\langle u| \, |v\rangle$, also written more concisely as $\langle u|v\rangle$ where the bra $\langle u|$ is found by taking the complex conjugate transpose of the ket $|u\rangle$.

In Section 5.4, we noted that a row vector can be thought of as operating on a column vector to produce a scalar. With this in mind we can think of the bra $\langle u|$ as operating on an arbitrary vector v (of the appropriate size) to produce a complex scalar, that is

$$\langle u| \, (v) : v \to \mathbb{C}.$$

That is, a vector v is mapped by the bra to a scalar in $\mathbb{C}$.

Example 5.6.1

Calculate the inner product $\langle \psi_1 | \psi_2 \rangle$ if $|\psi_1\rangle = \begin{pmatrix} \frac{1+i}{2} \\ \frac{1-i}{2} \end{pmatrix}$ and $|\psi_2\rangle = \begin{pmatrix} \frac{i}{\sqrt{2}} \\ \frac{-i}{\sqrt{2}} \end{pmatrix}$.

Solution

$$\langle \psi_1 | \psi_2 \rangle = \begin{pmatrix} \frac{1-i}{2} & \frac{1+i}{2} \end{pmatrix} \begin{pmatrix} \frac{i}{\sqrt{2}} \\ \frac{-i}{\sqrt{2}} \end{pmatrix}$$

$$= \left(\frac{1-i}{2}\right)\left(\frac{i}{\sqrt{2}}\right) + \left(\frac{1+i}{2}\right)\left(\frac{-i}{\sqrt{2}}\right)$$

$$= \frac{i+1-i+1}{2\sqrt{2}}$$

$$= \frac{1}{\sqrt{2}}.$$

Example 5.6.2

Consider the bra $\langle v|$ and kets $|u\rangle$ and $|w\rangle$, and the combination

$$|u\rangle \langle v| \, |w\rangle.$$

We explore what this might mean. Suppose

$$|u\rangle = \begin{pmatrix} u_1 \\ u_2 \end{pmatrix}, \quad \langle v| = \begin{pmatrix} v_1 & v_2 \end{pmatrix}, \quad |w\rangle = \begin{pmatrix} w_1 \\ w_2 \end{pmatrix}.$$

Then, performing $|u\rangle \langle v|$ first,

$$|u\rangle \langle v| = \begin{pmatrix} u_1 \\ u_2 \end{pmatrix} \begin{pmatrix} v_1 & v_2 \end{pmatrix} = \begin{pmatrix} u_1 v_1 & u_1 v_2 \\ u_2 v_1 & u_2 v_2 \end{pmatrix}$$

so that

$$|u\rangle \langle v| \, |w\rangle = \begin{pmatrix} u_1 v_1 & u_1 v_2 \\ u_2 v_1 & u_2 v_2 \end{pmatrix} \begin{pmatrix} w_1 \\ w_2 \end{pmatrix} = \begin{pmatrix} u_1 v_1 w_1 + u_1 v_2 w_2 \\ u_2 v_1 w_1 + u_2 v_2 w_2 \end{pmatrix}.$$

Alternatively, performing $\langle v| \, |w\rangle$ first,

$$\langle v| \, |w\rangle = \langle v|w\rangle = \begin{pmatrix} v_1 & v_2 \end{pmatrix} \begin{pmatrix} w_1 \\ w_2 \end{pmatrix} = v_1 w_1 + v_2 w_2, \text{ a scalar.}$$

Then, because $\langle v|w\rangle = v_1 w_1 + v_2 w_2$ is a scalar, we perform scalar multiplication:

$$|u\rangle\langle v|\,|w\rangle = \begin{pmatrix} u_1 \\ u_2 \end{pmatrix}(v_1 w_1 + v_2 w_2) = \begin{pmatrix} u_1 v_1 w_1 + u_1 v_2 w_2 \\ u_2 v_1 w_1 + u_2 v_2 w_2 \end{pmatrix}$$

the same result as above.

This example serves to illustrate the associativity of matrix multiplication. Further, it illustrates the following: we see that $|u\rangle\langle v|\,|w\rangle$ can be interpreted as a matrix $|u\rangle\langle v|$ multiplying, or operating on, a vector $|w\rangle$ to produce another vector. Alternatively, it represents the vector $|u\rangle$ being multiplied by the scalar $\langle v|\,|w\rangle$ resulting from an inner product. We shall use this result when discussing projection matrices later in this chapter.

Exercises

5.18   Given $|u\rangle = \begin{pmatrix} 4 \\ -2 \end{pmatrix}$, $\langle v| = (\,5 \quad 7\,)$, $|w\rangle = \begin{pmatrix} 3 \\ 9 \end{pmatrix}$, evaluate

(a) $|u\rangle\langle v|$

(b) $|u\rangle\langle v||w\rangle$

(c) $\langle v||w\rangle$

(d) $(\langle v||w\rangle)\,|u\rangle$.

5.7   The determinant of a square matrix

Definition 5.12 Determinant of a 2×2 matrix

*Consider the square 2×2 matrix $A = \begin{pmatrix} a & b \\ c & d \end{pmatrix}$. The scalar quantity, $ad - bc$, derived from this matrix is called the **determinant**, written det(A) or $|A|$, that is*

$$|A| = \begin{vmatrix} a & b \\ c & d \end{vmatrix} = ad - bc.$$

Example 5.7.1

Given $A = \begin{pmatrix} 1 & 5 \\ -2 & 8 \end{pmatrix}$ its determinant is

$$\begin{vmatrix} 1 & 5 \\ -2 & 8 \end{vmatrix} = (1)(8) - (5)(-2) = 8 + 10 = 18.$$

Given $B = \begin{pmatrix} 4 & 6 \\ 2 & 3 \end{pmatrix}$ its determinant is

$$\begin{vmatrix} 4 & 6 \\ 2 & 3 \end{vmatrix} = (4)(3) - (6)(2) = 0.$$

For larger square matrices, formulae exist for finding determinants. For example, in the case of a 3×3 matrix A, we have

Definition 5.13 Determinant of a 3×3 matrix

$Given\ A = \begin{pmatrix} a_{11} & a_{12} & a_{13} \\ a_{21} & a_{22} & a_{23} \\ a_{31} & a_{32} & a_{33} \end{pmatrix}$, *its determinant is given by*

$$|A| = a_{11} \begin{vmatrix} a_{22} & a_{23} \\ a_{32} & a_{33} \end{vmatrix} - a_{12} \begin{vmatrix} a_{21} & a_{23} \\ a_{31} & a_{33} \end{vmatrix} + a_{13} \begin{vmatrix} a_{21} & a_{22} \\ a_{31} & a_{32} \end{vmatrix}.$$

Note that there are also other, equivalent, formulae for finding the determinant of a 3×3 matrix. The determinant of a 4×4 matrix is found by generalising the previous result in an obvious way and can be found in most Linear Algebra textbooks.

Exercises

5.19   Show that the determinant of the matrix $A = \begin{pmatrix} 2 & -1 & 0 \\ -1 & 2 & -1 \\ 0 & -1 & 2 \end{pmatrix}$ is 4.

5.20   For the matrix A in the previous exercise, find $|A^T|$. Comment upon the result.

5.21   Find $\begin{vmatrix} \cos \omega t & \sin \omega t \\ -\sin \omega t & \cos \omega t \end{vmatrix}$.

5.22   Show that if A is a 2×2 or 3×3 matrix, then $|A| = |A^T|$, a result which is also true for an arbitrary $n \times n$ matrix.

5.8   The inverse of a square matrix

Suppose A is a square matrix. If there exists another matrix B say, such that $AB = BA = I$, an identity matrix, then B is said to be the **inverse** of A. Likewise A is the inverse of B. It is a straightforward matter, left to the reader, to check that

$$\begin{pmatrix} 4 & 1 \\ 3 & 1 \end{pmatrix}\ \text{is the inverse of}\ \begin{pmatrix} 1 & -1 \\ -3 & 4 \end{pmatrix}.$$

It is usual to denote the inverse of A by A^{-1}. We have the following definition:

Definition 5.14 The inverse of a matrix
*Given a square matrix A, its **inverse**, when such exists, is denoted A^{-1} and has the property that*

$$A A^{-1} = A^{-1} A = I$$

*where I is the identity matrix having the same size as A. Matrices which possess an inverse are said to be **non-singular** or **invertible**.*

When it exists, the inverse of a 2×2 matrix $A = \begin{pmatrix} a & b \\ c & d \end{pmatrix}$ is given by

$$A^{-1} = \frac{1}{ad - bc} \begin{pmatrix} d & -b \\ -c & a \end{pmatrix} = \frac{1}{|A|} \begin{pmatrix} d & -b \\ -c & a \end{pmatrix}.$$

Whenever $\det(A) = |A| = ad - bc = 0$ the inverse does not exist, that is, the matrix is **singular**, because $\frac{1}{0}$ is undefined.

Example 5.8.1 The general linear group $GL(n, \mathbb{F})$.

The set of all non-singular $n \times n$ matrices with elements from a field $\mathbb{F}$ (usually $\mathbb{R}$ or $\mathbb{C}$) with the operation of matrix multiplication form a group called the **general linear group**, denoted $GL(n, \mathbb{F})$. Because the elements of this group are non-singular we can be assured that each has a multiplicative inverse. The identity element for the group is the $n \times n$ identity matrix. You should verify that all the group axioms are satisfied. The group $GL(n, \mathbb{F})$ and any subgroups are referred to as **linear groups** or sometimes **matrix groups**.

5.8.1 The inverse of an $n \times n$ matrix

The procedure for calculating the inverse of a general $n \times n$ matrix necessitates the introduction of some new terminology. Suppose $A = (a_{ij})$ is an $n \times n$ matrix. Consider the submatrix obtained from A by removing its i-th row and j-th column. The determinant of this submatrix, written M_{ij}, is a scalar called the **minor** of element a_{ij}. We form the matrix of minors by replacing each element of A with its minor. The **cofactor** of a_{ij}, which we will denote by A_{ij}, is given by $A_{ij} = (-1)^{i+j} M_{ij}$. We can then write down the matrix of cofactors by imposing the **place sign** $(-1)^{i+j}$ on each minor.

$$\begin{pmatrix} A_{11} & A_{12} & A_{13} & \cdots & A_{1n} \\ A_{21} & A_{22} & A_{23} & \cdots & A_{2n} \\ \vdots & \vdots & \vdots & \ddots & \vdots \\ A_{n1} & A_{n2} & A_{n3} & \cdots & A_{nn} \end{pmatrix}.$$

Then the transpose of the matrix of cofactors is called the **classical adjoint** of A denoted $\text{adj}(A)$. The inverse of an $n \times n$ matrix A, when this exists, is given by

$$A^{-1} = \frac{1}{|A|} \text{adj}(A).$$

Example 5.8.2

Find the inverse of $A = \begin{pmatrix} 3 & 1 & 0 \\ 5 & 2 & -1 \\ 1 & 4 & -2 \end{pmatrix}$.

Solution

We begin by calculating the matrix of minors of each element of A. For example, the minor of a_{11} (the element of A which equals 3) is found by removing its row and column and calculating $M_{11} = \begin{vmatrix} 2 & -1 \\ 4 & -2 \end{vmatrix} = -4 - (-4) = 0$. Continuing in this fashion the matrix of minors is

$$M = \begin{pmatrix} 0 & -9 & 18 \\ -2 & -6 & 11 \\ -1 & -3 & 1 \end{pmatrix}.$$

Imposing the place sign $(-1)^{i+j}$ on each minor gives the matrix of cofactors:

$$\begin{pmatrix} 0 & 9 & 18 \\ 2 & -6 & -11 \\ -1 & 3 & 1 \end{pmatrix}.$$

The classical adjoint of A is then the transpose of the matrix of cofactors:

$$\mathrm{adj}(A) = \begin{pmatrix} 0 & 2 & -1 \\ 9 & -6 & 3 \\ 18 & -11 & 1 \end{pmatrix}.$$

The determinant of A is given by the formula in Definition 5.13:

$$|A| = 3 \begin{vmatrix} 2 & -1 \\ 4 & -2 \end{vmatrix} - 1 \begin{vmatrix} 5 & -1 \\ 1 & -2 \end{vmatrix} + 0 \begin{vmatrix} 5 & 2 \\ 1 & 4 \end{vmatrix} = 3(0) - 1(-9) = 9.$$

Then finally, the inverse of A is given by

$$A^{-1} = \frac{1}{|A|} \mathrm{adj}(A) = \frac{1}{9} \begin{pmatrix} 0 & 2 & -1 \\ 9 & -6 & 3 \\ 18 & -11 & 1 \end{pmatrix}.$$

Exercises

5.23 Calculate, if possible, the inverse of $A = \begin{pmatrix} 3 & -2 \\ 4 & -2 \end{pmatrix}$.

5.24 Calculate, if possible, the inverse of $A = \begin{pmatrix} 3 & -1 & 7 \\ 2 & 0 & 1 \\ 5 & -2 & 6 \end{pmatrix}$.

5.25 Calculate the values of λ for which $A = \begin{pmatrix} 2-\lambda & 4 \\ 1.5 & 1-\lambda \end{pmatrix}$ has no inverse.

5.26 Suppose $R = \begin{pmatrix} \cos\theta & -\sin\theta \\ \sin\theta & \cos\theta \end{pmatrix}$.

   a) Find R^T.

   b) Show that $R R^T = R^T R = I$, the 2×2 identity matrix.

   c) Deduce that $R^T = R^{-1}$.

5.9 Similar matrices and diagonalisation

Definition 5.15 Similar matrices
*A square matrix B is said to be **similar** to another square matrix A if there exists an invertible matrix P such that $B = P^{-1}AP$.*

Similar matrices have several properties in common. In particular they share the same special values known as eigenvalues, a property we explore further in Chapter 7.

Example 5.9.1 Similar matrices

Given $P = \begin{pmatrix} 3 & -1 \\ -5 & 2 \end{pmatrix}$, by explicitly evaluating the matrix products show that $B = \begin{pmatrix} -10 & 7 \\ -36 & 22 \end{pmatrix}$ is similar to $A = \begin{pmatrix} 7 & 3 \\ 1 & 5 \end{pmatrix}$.

Solution

Using the formula for the inverse matrix,

$$P^{-1} = \begin{pmatrix} 2 & 1 \\ 5 & 3 \end{pmatrix}.$$

Then

$$\begin{aligned} P^{-1}AP &= \begin{pmatrix} 2 & 1 \\ 5 & 3 \end{pmatrix} \begin{pmatrix} 7 & 3 \\ 1 & 5 \end{pmatrix} \begin{pmatrix} 3 & -1 \\ -5 & 2 \end{pmatrix} \\ &= \begin{pmatrix} 15 & 11 \\ 38 & 30 \end{pmatrix} \begin{pmatrix} 3 & -1 \\ -5 & 2 \end{pmatrix} \\ &= \begin{pmatrix} -10 & 7 \\ -36 & 22 \end{pmatrix}. \end{aligned}$$

Hence, $B = \begin{pmatrix} -10 & 7 \\ -36 & 22 \end{pmatrix}$ is similar to $A = \begin{pmatrix} 7 & 3 \\ 1 & 5 \end{pmatrix}$.

Exercises

5.27   If B is similar to A show that A is similar to B.

Example 5.9.2 Diagonalisation

Given $P = \begin{pmatrix} 1 & 2 \\ 1 & 1 \end{pmatrix}$, by explicitly evaluating the matrix products show that $B = \begin{pmatrix} 5 & 0 \\ 0 & 6 \end{pmatrix}$ is similar to $A = \begin{pmatrix} 7 & -2 \\ 1 & 4 \end{pmatrix}$.

Solution

Using the formula for the inverse matrix, $P^{-1} = \begin{pmatrix} -1 & 2 \\ 1 & -1 \end{pmatrix}$. Then

$$\begin{aligned} P^{-1}AP &= \begin{pmatrix} -1 & 2 \\ 1 & -1 \end{pmatrix} \begin{pmatrix} 7 & -2 \\ 1 & 4 \end{pmatrix} \begin{pmatrix} 1 & 2 \\ 1 & 1 \end{pmatrix} \\ &= \begin{pmatrix} -5 & 10 \\ 6 & -6 \end{pmatrix} \begin{pmatrix} 1 & 2 \\ 1 & 1 \end{pmatrix} \\ &= \begin{pmatrix} 5 & 0 \\ 0 & 6 \end{pmatrix}. \end{aligned}$$

Thus $A = \begin{pmatrix} 7 & -2 \\ 1 & 4 \end{pmatrix}$ is similar to the diagonal matrix $B = \begin{pmatrix} 5 & 0 \\ 0 & 6 \end{pmatrix}$. The process we have observed is known as **diagonalisation** and will be important when we study eigenvalues and eigenvectors used in making quantum measurements.

5.10   Orthogonal matrices

Definition 5.16 Orthogonal matrix

An **orthogonal matrix** A, *is a non-singular square matrix for which its transpose is equal to its inverse:*

$$A^T = A^{-1}.$$

Equivalently, $A^T A = A A^T = I$.

An important property of any orthogonal matrix is that the columns are orthogonal and normalised. The same is true of the rows. Orthogonal matrices are necessarily **normal**, that is they commute with their transpose.

We saw in the previous section that matrices A and B are similar if there exists P such that $B = P^{-1}AP$. We also saw that under some circumstances, which we have yet to explain, the matrix B is diagonal. If in addition, the matrix P is orthogonal, so that $P^{-1} = P^T$, then the condition for similarity becomes $B = P^T AP$. If such a matrix exists, we say that A and B are **orthogonally similar**. If A is orthogonally similar to a diagonal matrix, then A is **orthogonally diagonalisable**. We shall see that all symmetric matrices have this property.

Example 5.10.1

Consider the matrix $P = \begin{pmatrix} -1/\sqrt{2} & 1/\sqrt{2} \\ 1/\sqrt{2} & 1/\sqrt{2} \end{pmatrix}$.

(a) Show that P is an orthogonal matrix.

(b) By evaluating $P^T AP$ when A is the symmetric matrix $\begin{pmatrix} 5 & -1 \\ -1 & 5 \end{pmatrix}$ show that A is orthogonally diagonalisable.

Solution

(a) Consider $P^T P$:

$$P^T P = \begin{pmatrix} -1/\sqrt{2} & 1/\sqrt{2} \\ 1/\sqrt{2} & 1/\sqrt{2} \end{pmatrix} \begin{pmatrix} -1/\sqrt{2} & 1/\sqrt{2} \\ 1/\sqrt{2} & 1/\sqrt{2} \end{pmatrix} = \begin{pmatrix} 1 & 0 \\ 0 & 1 \end{pmatrix} = I.$$

Likewise, $PP^T = I$. Thus P is an orthogonal matrix. Observe that the inner product of the two columns of P is zero. Further, the modulus of each column vector is 1. Thus the columns are orthogonal and normalised. Likewise, the rows.

(b) By explicitly evaluating the matrix products, we find

$$
\begin{aligned}
P^T A P &= \left(\begin{array}{cc} -1/\sqrt{2} & 1/\sqrt{2} \\ 1/\sqrt{2} & 1/\sqrt{2} \end{array} \right) \left(\begin{array}{cc} 5 & -1 \\ -1 & 5 \end{array} \right) \left(\begin{array}{cc} -1/\sqrt{2} & 1/\sqrt{2} \\ 1/\sqrt{2} & 1/\sqrt{2} \end{array} \right) \\
&= \left(\begin{array}{cc} -1/\sqrt{2} & 1/\sqrt{2} \\ 1/\sqrt{2} & 1/\sqrt{2} \end{array} \right) \left(\begin{array}{cc} -6/\sqrt{2} & 4/\sqrt{2} \\ 6/\sqrt{2} & 4/\sqrt{2} \end{array} \right) \\
&= \left(\begin{array}{cc} 6 & 0 \\ 0 & 4 \end{array} \right).
\end{aligned}
$$

Thus the result is a diagonal matrix, and so A is orthogonally diagonalisable.

Example 5.10.2 The orthogonal group $O(2)$ and the special orthogonal group $SO(2)$.

The set of all 2×2 orthogonal matrices, together with matrix multiplication, form a group, denoted $O(2)$. So, given two orthogonal matrices, their product is another orthogonal matrix. The identity element of the group is the identity matrix $I = \left(\begin{array}{cc} 1 & 0 \\ 0 & 1 \end{array} \right)$ which is readily checked to be orthogonal. Every orthogonal matrix possesses an inverse which is also orthogonal. Indeed, this is its transpose. Those matrices in $O(2)$ which have determinant equal to 1, form a subgroup denoted $SO(2)$ referred to as the **special orthogonal group**.

5.11 Unitary matrices

Definition 5.17 Unitary matrix
*A non-singular, complex matrix U such that $U^\dagger = U^{-1}$ (where $U^\dagger$ is the conjugate transpose of U, i.e., $(U^*)^T$) is said to be a **unitary matrix**. Thus*

$$
U U^\dagger = U^\dagger U = I.
$$

The rows of a unitary matrix are pairwise orthogonal so that the inner product of two distinct rows will be zero. Likewise, the columns. The norm of each row or column is equal to 1. Unitary matrices are necessarily normal, that is they commute with their conjugate transpose. These properties are easily verified for the unitary matrices in the examples which follow. Such matrices play a special role in quantum computation because quantum dynamics is via unitary operators acting on the underlying vector space of quantum states.

Example 5.11.1 Pauli matrices.

Consider the following 2×2 matrices known as **Pauli matrices**.

$$
\sigma_0 = \left(\begin{array}{cc} 1 & 0 \\ 0 & 1 \end{array} \right), \quad \sigma_1 = \left(\begin{array}{cc} 0 & 1 \\ 1 & 0 \end{array} \right), \quad \sigma_2 = \left(\begin{array}{cc} 0 & -i \\ i & 0 \end{array} \right), \quad \sigma_3 = \left(\begin{array}{cc} 1 & 0 \\ 0 & -1 \end{array} \right).
$$

It is a straightforward exercise to verify that each of these matrices is unitary, $\sigma_i^{-1} = \sigma_i^\dagger$, $i = 0, \ldots, 3$.

Example 5.11.2

Consider the matrix given by

$$\sigma = \frac{1}{2}(1 + i)(\sigma_0 - i\sigma_1)$$

where σ_0 and σ_1 are the Pauli matrices of Example 5.11.1. Show that this matrix is unitary, i.e., $\sigma^{-1} = \sigma^\dagger$.

Solution

$$\sigma = \frac{1}{2}(1 + i)(\sigma_0 - i\sigma_1)$$

$$= \frac{1}{2}(1 + i)\begin{pmatrix} 1 & -i \\ -i & 1 \end{pmatrix}$$

$$= \begin{pmatrix} \frac{1+i}{2} & \frac{1-i}{2} \\ \frac{1-i}{2} & \frac{1+i}{2} \end{pmatrix}.$$

The conjugate transpose of this matrix is

$$\sigma^\dagger = \begin{pmatrix} \frac{1-i}{2} & \frac{1+i}{2} \\ \frac{1+i}{2} & \frac{1-i}{2} \end{pmatrix}.$$

Evaluating $\sigma \sigma^\dagger$ we find

$$\begin{pmatrix} \frac{1+i}{2} & \frac{1-i}{2} \\ \frac{1-i}{2} & \frac{1+i}{2} \end{pmatrix}\begin{pmatrix} \frac{1-i}{2} & \frac{1+i}{2} \\ \frac{1+i}{2} & \frac{1-i}{2} \end{pmatrix} = \begin{pmatrix} 1 & 0 \\ 0 & 1 \end{pmatrix} = I.$$

Likewise, $\sigma^\dagger \sigma = I$. Thus the given matrix is unitary.

Exercises

5.28   Consider the Pauli matrices of Example 5.11.1. Show that

$$\left[\frac{1}{2}(1 + i)(\sigma_0 - i\sigma_1)\right]^2 = \sigma_1.$$

Deduce that σ_1 has a square root.

We have seen previously that matrices A and B are similar if there exists P such that $B = P^{-1}AP$. We also saw that under some circumstances, which we have yet to explain, the matrix B is diagonal. If in addition, the matrix P is unitary, so that $P^{-1} = P^\dagger$, then the condition for similarity becomes $B = P^\dagger AP$. If such a matrix exists, we say that A and B are **unitarily similar**. If A is unitarily similar to a diagonal matrix, then A is **unitarily diagonalisable**. We shall see that all self-adjoint matrices have this property.

Example 5.11.3 The unitary group $U(2)$ and the special unitary group $SU(2)$.

The set of all 2×2 unitary matrices, together with matrix multiplication, form a group, denoted $U(2)$. So, given two unitary matrices, their product is another unitary matrix. The identity element of the group is the identity matrix $I = \begin{pmatrix} 1 & 0 \\ 0 & 1 \end{pmatrix}$ which is readily checked to be unitary. Every unitary matrix possesses an inverse which is also unitary. Indeed, this is its conjugate transpose. Those matrices in $U(2)$ which have determinant equal to 1, form a subgroup denoted $SU(2)$ referred to as the **special unitary group**. Elements of the unitary group $U(2)$ are used in the representation of 1-qubit gates (Chapter 21).

5.12 Matrices and the solution of linear simultaneous equations

Using our knowledge of matrix multiplication, a set of m linear equations in n unknowns, $x_1, \ldots, x_n$,

$$a_{11}x_1 + a_{12}x_2 + \ldots + a_{1n}x_n = b_1$$
$$a_{21}x_1 + a_{22}x_2 + \ldots + a_{2n}x_n = b_2$$
$$\vdots \qquad \qquad = \vdots$$
$$a_{m1}x_1 + a_{m2}x_2 + \ldots + a_{mn}x_n = b_m$$

can be written in matrix form as

$$\begin{pmatrix} a_{11} & a_{12} & \cdots & a_{1n} \\ a_{21} & a_{22} & \cdots & a_{2n} \\ \vdots & \vdots & \ddots & \vdots \\ a_{m1} & a_{m2} & \cdots & a_{mn} \end{pmatrix} \begin{pmatrix} x_1 \\ x_2 \\ \vdots \\ x_n \end{pmatrix} = \begin{pmatrix} b_1 \\ b_2 \\ \vdots \\ b_m \end{pmatrix}$$

or more succinctly as $Ax = b$ where

$$A = \begin{pmatrix} a_{11} & a_{12} & \cdots & a_{1n} \\ a_{21} & a_{22} & \cdots & a_{2n} \\ \vdots & \vdots & \ddots & \vdots \\ a_{m1} & a_{m2} & \cdots & a_{mn} \end{pmatrix}, \quad x = \begin{pmatrix} x_1 \\ x_2 \\ \vdots \\ x_n \end{pmatrix}, \quad \text{and } b = \begin{pmatrix} b_1 \\ b_2 \\ \vdots \\ b_m \end{pmatrix}.$$

The $m \times n$ matrix A is then called the **coefficient matrix**. It is important to note that when solving any such set of linear simultaneous equations just three possibilities exist:

- there is a unique solution (i.e., unique values for each of $x_1, x_2, \ldots, x_n$).

- no solutions exist whatsoever, and the equations are said to be **inconsistent**.

- there is an infinite number of different solutions.

In the special case when A is square and non-singular, its inverse A^{-1} exists and we can formally write the solution as

$$x = A^{-1}b$$

by pre-multiplying $Ax = b$ by A^{-1} and using the properties that $A^{-1}A = I$ and $Ix = x$, where I is an identity matrix. Thus by finding the inverse matrix, the solution can be determined. Recall A^{-1} exists if $|A| \neq 0$.

In the case when all b_i are zero, i.e., $b = 0$, the equations are said to be **homogeneous**. Note that if the equations are homogeneous, i.e., $Ax = 0$, and if $|A| \neq 0$ then $x = A^{-1}0 = 0$. The only solution is the **trivial** one for which $x_i = 0$ for all i.

An important consequence is that if we seek **non-trivial** solutions of a set of homogeneous equations, $Ax = 0$, then we will require $|A| = 0$ in order for such solutions to exist. Further, if A has rank r (the rank is the number of linearly independent rows of A, see Chapter 6) then the number of linearly independent solutions of the homogeneous system $Ax = 0$ is $n - r$.

Example 5.12.1

Use the matrix inverse, if possible, to solve the equations

(a) $\begin{aligned} x_1 + 2x_2 &= 13 \\ 2x_1 - 5x_2 &= 8 \end{aligned}$

(b) $\begin{aligned} x_1 + 2x_2 &= 0 \\ 2x_1 - 5x_2 &= 0 \end{aligned}$

(c) $\begin{aligned} x_1 + 2x_2 &= 0 \\ 2x_1 + 4x_2 &= 0. \end{aligned}$

Solution

(a) Note that the set of equations is not homogeneous, i.e., the set is inhomogeneous. In matrix form, $Ax = b$, we have

$$\begin{pmatrix} 1 & 2 \\ 2 & -5 \end{pmatrix} \begin{pmatrix} x_1 \\ x_2 \end{pmatrix} = \begin{pmatrix} 13 \\ 8 \end{pmatrix}.$$

Note $|A| = -9 \neq 0$. Using the formula for the matrix inverse, $A^{-1} = -\frac{1}{9}\begin{pmatrix} -5 & -2 \\ -2 & 1 \end{pmatrix}$ from which $x = A^{-1}b$, that is

$$x = \begin{pmatrix} x_1 \\ x_2 \end{pmatrix} = -\frac{1}{9}\begin{pmatrix} -5 & -2 \\ -2 & 1 \end{pmatrix} \begin{pmatrix} 13 \\ 8 \end{pmatrix} = \begin{pmatrix} 9 \\ 2 \end{pmatrix}.$$

It follows that $x_1 = 9$, $x_2 = 2$ is the unique solution.

(b) The equations are homogeneous. In matrix form, $Ax = b$, we have

$$\begin{pmatrix} 1 & 2 \\ 2 & -5 \end{pmatrix} \begin{pmatrix} x_1 \\ x_2 \end{pmatrix} = \begin{pmatrix} 0 \\ 0 \end{pmatrix}.$$

The inverse of A exists (it is the same as in part (a)).

$$x = \begin{pmatrix} x_1 \\ x_2 \end{pmatrix} = -\frac{1}{9}\begin{pmatrix} -5 & -2 \\ -2 & 1 \end{pmatrix} \begin{pmatrix} 0 \\ 0 \end{pmatrix} = \begin{pmatrix} 0 \\ 0 \end{pmatrix}.$$

Hence, only the trivial solution, $x_1 = x_2 = 0$ exists.

(c) The equations are homogeneous. Here, the matrix form, $Ax = b$, of the equations is

$$\begin{pmatrix} 1 & 2 \\ 2 & 4 \end{pmatrix} \begin{pmatrix} x_1 \\ x_2 \end{pmatrix} = \begin{pmatrix} 0 \\ 0 \end{pmatrix}.$$

Observe that $|A| = 0$ and hence A^{-1} does not exist, and this method of solution is inappropriate. Nevertheless, solutions do exist as we shall demonstrate shortly.

In the following section, we present a systematic process for solving simultaneous linear equations called **Gaussian elimination**.

5.12.1 Gaussian elimination

Gaussian elimination is a systematic method for solving sets of linear equations. To solve the linear system $Ax = b$, we form the **augmented matrix** by appending the column vector b to the right-hand side of the coefficient matrix to give:

$$\begin{pmatrix} a_{11} & a_{12} & \cdots & a_{1n} & b_1 \\ a_{21} & a_{22} & \cdots & a_{2n} & b_2 \\ \vdots & \vdots & \ddots & \vdots & \vdots \\ a_{m1} & a_{m2} & \cdots & a_{mn} & b_m \end{pmatrix}.$$

Next we perform any of the following so-called **elementary row operations**:

- interchange any two rows,

- multiply any row by a non-zero constant,

- add or subtract any row from any other.

A little thought will show that these operations are just the same ones as applied to eliminate one of the variables when solving the sort of simultaneous equations met in a school course. The aim is to achieve **row echelon form** in which

- any rows of zeros are at the bottom,

- each successive row contains more leading zeros than the previous one.

Once the augmented matrix is in row echelon form, **back-substitution** leads to the solution as we demonstrate in the examples below. Using this method reveals the behaviour mentioned previously: there may be a unique solution, no solutions at all, or an infinite number of solutions. This behaviour is also illustrated in the examples below. Note that there are multiple ways of achieving echelon form leading to different yet equivalent expressions for the solution.

Example 5.12.2 Gaussian elimination – unique solution

Use Gaussian elimination to show that the following linear system of equations has a unique solution and determine it.

$$2x - y + z = -2$$
$$x + 2y + 3z = -1$$
$$2x + 2y - z = 8.$$

Solution

The augmented matrix is

$$\begin{pmatrix} 2 & -1 & 1 & -2 \\ 1 & 2 & 3 & -1 \\ 2 & 2 & -1 & 8 \end{pmatrix}.$$

Leave row 1 alone. Replace row 2 with the result of subtracting row 1 from twice row 2, denoted $R_2 \rightarrow 2R_2 - R_1$. Then perform the operation $R_3 \rightarrow R_3 - R_1$.

The result is the matrix:

$$\begin{pmatrix} 2 & -1 & 1 & -2 \\ 0 & 5 & 5 & 0 \\ 0 & 3 & -2 & 10 \end{pmatrix}.$$

The aim of these operations is to introduce as many leading zeros as possible.

Continuing, leave row 2 alone and perform $R_3 \rightarrow 5R_3 - 3R_2$ to give

$$\begin{pmatrix} 2 & -1 & 1 & -2 \\ 0 & 5 & 5 & 0 \\ 0 & 0 & -25 & 50 \end{pmatrix}.$$

At this point, the equations are in echelon form. Observe that as we move down the rows each successive row contains more leading zeros than the previous one. In this example there are no rows of zeros, but if there were then they should be at the bottom.

The final row, re-expressed as an equation is $-25z = 50$ from which $z = -2$. The process of back-substitution then begins by substituting this value of z into the row above. This then reads:

$$5y + 5z = 5y + 5(-2) = 0$$
$$5y = 10$$
$$y = 2.$$

Finally, the solutions already obtained for y and z are substituted into the first equation:

$$2x - y + z = 2x - (2) + (-2) = -2$$
$$2x = 2$$
$$x = 1.$$

So the unique solution is

$$x = 1, \quad y = 2, \quad z = -2.$$

Example 5.12.3 Gaussian elimination – inconsistent equations

Use Gaussian elimination to show that the following linear system of equations has no solution, i.e., it is inconsistent.

$$2x + y - z = -3$$
$$3x + 2y + z = 6$$
$$x + y + 2z = 8.$$

Solution

The augmented matrix is

$$\begin{pmatrix} 2 & 1 & -1 & -3 \\ 3 & 2 & 1 & 6 \\ 1 & 1 & 2 & 8 \end{pmatrix}.$$

We reduce this matrix to echelon form. Leave row 1 alone and then perform the operations $R_2 \rightarrow 2R_2 - 3R_1$, and $R_3 \rightarrow 2R_3 - R_1$:

$$\begin{pmatrix} 2 & 1 & -1 & -3 \\ 0 & 1 & 5 & 21 \\ 0 & 1 & 5 & 19 \end{pmatrix}.$$

Continuing, leave row 2 alone and perform $R_3 \rightarrow R_3 - R_2$ to give

$$\begin{pmatrix} 2 & 1 & -1 & -3 \\ 0 & 1 & 5 & 21 \\ 0 & 0 & 0 & -2 \end{pmatrix}.$$

We now have echelon form, but notice that an issue arises. The final line reads $0x + 0y + 0z = -2$, i.e., $0 = -2$ which is clearly impossible. This signals to us that the equations are inconsistent and there are no solutions.

Example 5.12.4 Gaussian elimination – an infinite number of solutions

Use Gaussian elimination to show that the following linear system of equations has an infinite number of solutions and write down a general expression for them.

$$x + y - z = 1$$
$$3x - y + 5z = 3$$
$$7x + 2y + 3z = 7.$$

Solution

The augmented matrix is

$$\begin{pmatrix} 1 & 1 & -1 & 1 \\ 3 & -1 & 5 & 3 \\ 7 & 2 & 3 & 7 \end{pmatrix}.$$

We reduce the augmented matrix to echelon form. Leave row 1 alone and perform the operations $R_2 \rightarrow R_2 - 3R_1$, $R_3 \rightarrow R_3 - 7R_1$. This gives

$$\begin{pmatrix} 1 & 1 & -1 & 1 \\ 0 & -4 & 8 & 0 \\ 0 & -5 & 10 & 0 \end{pmatrix}.$$

Continuing, perform $R_3 \rightarrow 4R_3 - 5R_2$:

$$\begin{pmatrix} 1 & 1 & -1 & 1 \\ 0 & -4 & 8 & 0 \\ 0 & 0 & 0 & 0 \end{pmatrix}.$$

Observe that on this occasion there is a row of zeros and this row is at the bottom of the matrix. This row provides no information; it merely states $0 = 0$, but leads to the concept of a **free variable**. Rows 1 and 2 correspond to the equations

$$x + y - z = 1$$
$$-4y + 8z = 0.$$

In the echelon form, the first row starts off with a non-zero x, the second with a non-zero y. There is now no row that starts off with a non-zero z. It is this variable that we shall choose as a free variable. We can choose the value of z to be anything we please, say $z = \mu$, where μ is our choice. We then continue the back-substitution to obtain

$$z = \mu$$

$$-4y + 8z = -4y + 8\mu = 0 \text{ so that } y = 2\mu.$$

$$x + y - z = x + 2\mu - \mu = 1 \text{ so that } x = 1 - \mu.$$

We can therefore formally write the solution as $x = 1 - \mu$, $y = 2\mu$, $z = \mu$ or often as

$$\begin{pmatrix} x \\ y \\ z \end{pmatrix} = \begin{pmatrix} 1 - \mu \\ 2\mu \\ \mu \end{pmatrix} = \begin{pmatrix} 1 \\ 0 \\ 0 \end{pmatrix} + \mu \begin{pmatrix} -1 \\ 2 \\ 1 \end{pmatrix}.$$

Clearly then, there is an infinite number of different solutions as μ is allowed to vary.

Example 5.12.5 Gaussian elimination – an infinite number of solutions

Use Gaussian elimination to show that the following linear system of equations has an infinite number of solutions and write down a general expression for them.

$$x + 7y - z + w = 1$$
$$y - z \qquad = 3$$

Solution

In this example there are four unknowns and only two equations. The augmented matrix is

$$\begin{pmatrix} 1 & 7 & -1 & 1 & 1 \\ 0 & 1 & -1 & 0 & 3 \\ 0 & 0 & 0 & 0 & 0 \\ 0 & 0 & 0 & 0 & 0 \end{pmatrix} \qquad \text{or more simply} \qquad \begin{pmatrix} 1 & 7 & -1 & 1 & 1 \\ 0 & 1 & -1 & 0 & 3 \end{pmatrix}$$

Observe that the augmented matrix is in row echelon form. The first row starts off with a non-zero x, the second with a non-zero y. There are no rows starting with a non-zero z or a non-zero w. We thus create two free variables, $w = \lambda$, $z = \mu$ say. Then back-substitution gives

$$y = 3 + z$$
$$= 3 + \mu$$

$$x = 1 - 7y + z - w$$
$$= 1 - 7(3 + \mu) + \mu - \lambda$$
$$= -20 - 6\mu - \lambda.$$

We can therefore formally write the solution as

$$\begin{pmatrix} x \\ y \\ z \\ w \end{pmatrix} = \begin{pmatrix} -20 - 6\mu - \lambda \\ 3 + \mu \\ \mu \\ \lambda \end{pmatrix} = \begin{pmatrix} -20 \\ 3 \\ 0 \\ 0 \end{pmatrix} + \mu \begin{pmatrix} -6 \\ 1 \\ 1 \\ 0 \end{pmatrix} + \lambda \begin{pmatrix} -1 \\ 0 \\ 0 \\ 1 \end{pmatrix}$$

Clearly then, there is a doubly infinite number of different solutions as μ and λ are allowed to vary.

Systems of linear equations which have an infinite number of solutions are important in the study of eigenvectors in Chapter 7. In turn, eigenvectors are used in the representation of the states of a quantum system (Chapter 13).

Example 5.12.6 Linear simultaneous equations with arithmetic mod 2

In some quantum algorithms it is necessary to solve linear simultaneous equations whilst performing arithmetic modulo 2. Consider the set of linear simultaneous equations

$$
\begin{aligned}
x_1 + x_2 + x_3 \quad &= 1 \\
x_2 \quad + x_4 &= 1 \\
x_1 \quad + x_3 \quad &= 1 \\
x_2 + x_3 + x_4 &= 1
\end{aligned}
$$

where $x_1, \ldots, x_4$ are binary variables 0 or 1. We seek a solution using arithmetic modulo 2.

Solution

We can apply Gaussian elimination in the way shown previously making use of addition modulo 2 in which:

$$
0 + 0 = 0, \quad 0 + 1 = 1 + 0 = 1, \quad 1 + 1 = 0.
$$

The augmented matrix is

$$
\begin{pmatrix}
1 & 1 & 1 & 0 & 1 \\
0 & 1 & 0 & 1 & 1 \\
1 & 0 & 1 & 0 & 1 \\
0 & 1 & 1 & 1 & 1
\end{pmatrix}.
$$

We apply the row operation $R_3 \to R_3 - R_1$ to obtain:

$$
\begin{pmatrix}
1 & 1 & 1 & 0 & 1 \\
0 & 1 & 0 & 1 & 1 \\
0 & 1 & 0 & 0 & 0 \\
0 & 1 & 1 & 1 & 1
\end{pmatrix}.
$$

where we note that $-1 = 1 \mod 2$. Then applying $R_3 \to R_3 - R_2$, $R_4 \to R_4 - R_2$ gives:

$$
\begin{pmatrix}
1 & 1 & 1 & 0 & 1 \\
0 & 1 & 0 & 1 & 1 \\
0 & 0 & 0 & 1 & 1 \\
0 & 0 & 1 & 0 & 0
\end{pmatrix}.
$$

Interchanging rows 3 and 4:

$$
\begin{pmatrix}
1 & 1 & 1 & 0 & 1 \\
0 & 1 & 0 & 1 & 1 \\
0 & 0 & 1 & 0 & 0 \\
0 & 0 & 0 & 1 & 1
\end{pmatrix}.
$$

which is now in row echelon form. Back-substitution then yields:

$$
x_4 = 1, \ x_3 = 0
$$

$$
x_2 = 1 - x_4 = 1 - 1 = 0
$$

$$x_1 = 1 - x_2 - x_3 = 1 - 0 - 0 = 1.$$

The unique solution is then

$$x_1 = 1, \ x_2 = 0, \ x_3 = 0, \ x_4 = 1.$$

You should verify that this is indeed a solution by substituting into the original equations.

Example 5.12.7 Linear simultaneous equations with arithmetic mod 2 – free variables

Consider the set of linear simultaneous equations

$$
\begin{aligned}
x_1 \quad\ + x_3 \quad\quad\ &= 1 \\
x_2 + x_3 + x_4 &= 0 \\
x_1 + x_2 + x_3 \quad\ &= 0 \\
x_1 + x_2 \quad\ + x_4 &= 1
\end{aligned}
$$

where $x_1, \ldots, x_4$ are binary variables 0 or 1. We seek a solution using arithmetic modulo 2.

Solution

The augmented matrix is

$$
\begin{pmatrix}
1 & 0 & 1 & 0 & 1 \\
0 & 1 & 1 & 1 & 0 \\
1 & 1 & 1 & 0 & 0 \\
1 & 1 & 0 & 1 & 1
\end{pmatrix}.
$$

We apply the following row operations $R_3 \to R_3 - R_1$, $R_4 \to R_4 - R_1$ to obtain:

$$
\begin{pmatrix}
1 & 0 & 1 & 0 & 1 \\
0 & 1 & 1 & 1 & 0 \\
0 & 1 & 0 & 0 & 1 \\
0 & 1 & 1 & 1 & 0
\end{pmatrix}.
$$

Then apply $R_3 \to R_3 - R_2$, $R_4 \to R_4 - R_2$ to obtain

$$
\begin{pmatrix}
1 & 0 & 1 & 0 & 1 \\
0 & 1 & 1 & 1 & 0 \\
0 & 0 & 1 & 1 & 1 \\
0 & 0 & 0 & 0 & 0
\end{pmatrix}.
$$

Observe that in echelon form there is no row starting off with a non-zero x_4. We choose this to be free: $x_4 = \mu$ say where $\mu \in \{0, 1\}$. Then using back-substitution

$$x_3 = 1 - x_4 = 1 - \mu$$

$$x_2 + (1 - \mu) + \mu = 0, \ \text{so that} \ x_2 = -1 = 1 (\mathrm{mod}\ 2)$$

$$x_1 = 1 - x_3 = 1 - (1 - \mu) = \mu.$$

Hence the solution is $x_1 = \mu$, $x_2 = 1$, $x_3 = 1 - \mu \ (= 1 + \mu)$, $x_4 = \mu$, where $\mu \in \{0, 1\}$. This solution can also be expressed as

$$
\begin{pmatrix}
x_1 \\
x_2 \\
x_3 \\
x_4
\end{pmatrix}
=
\begin{pmatrix}
0 \\
1 \\
1 \\
0
\end{pmatrix}
+ \mu
\begin{pmatrix}
1 \\
0 \\
1 \\
1
\end{pmatrix}.
$$

When we study quantum algorithms in Chapters 22 and 23, we will require the solution of a linear system of simultaneous equations with Boolean variables 0 and 1. For these applications, we shall write addition using the exclusive-or operator $\oplus$, and multiplication using the Boolean operator $\wedge$, or 'and'. Consider the following example.

Example 5.12.8

Solve the following equation to find s_1 and s_2 where $s_1, s_2 \in \{0, 1\}$.

$$1 \wedge s_1 \oplus 1 \wedge s_2 = 0.$$

Solution

Observe that with the Gaussian elimination notation introduced previously, this equation is equivalent to the echelon form

$$s_1 + s_2 = 0.$$

So select s_2 to be a free variable, $s_2 = \mu$ where $\mu \in \{0, 1\}$. Then $s_1 = -\mu$. We could write the solution as

$$\begin{pmatrix} s_1 \\ s_2 \end{pmatrix} = \mu \begin{pmatrix} -1 \\ 1 \end{pmatrix} = \mu \begin{pmatrix} 1 \\ 1 \end{pmatrix}$$

because $-1 = 1 \pmod 2$. Later we shall see the solution written as a string of binary values $s_1 s_2$, so in this case there are two distinct solutions 00 (when $\mu = 0$) and 11 (when $\mu = 1$).

Example 5.12.9

Solve the following equation to find s_1 and s_2 where $s_1, s_2 \in \{0, 1\}$.

$$0 \wedge s_1 \oplus 1 \wedge s_2 = 0.$$

Solution

The given equation is equivalent to writing $0s_1 + 1s_2 = 0$. Thus $s_2 = 0$ and s_1, a free variable, can be either 0 or 1. We can write each solution as a string of binary values, $s_1 s_2$, that is 00 and 10.

Exercises

5.29   Solve the homogeneous equations
$$x_1 + 4x_2 = 0$$
$$3x_1 + 12x_2 = 0.$$

5.30   Use Gaussian elimination to solve

$$x + 2y - 3z + 2w = 2$$
$$2x + 5y - 8z + 6w = 5$$
$$3x + 4y - 5z + 2w = 4.$$

5.13   Matrix transformations

We have seen that a point in the xy plane with Cartesian coordinates (x, y) can be represented by a column vector $\begin{pmatrix} x \\ y \end{pmatrix}$. By pre-multiplying this vector by carefully chosen 2×2 matrices, the point can be relocated elsewhere in the plane. For example, the point might be rotated through an angle θ say around the origin; it might be reflected in any chosen line through the origin. Such an operation is referred to as a **transformation** and is achieved using a transformation matrix. In this section we describe transformation matrices which perform rotations and reflections in the plane.

Definition 5.18 Rotation matrix.
The transformation matrix $W(\theta)$ that rotates a point anticlockwise about the origin by an angle θ is

$$W(\theta) = \begin{pmatrix} \cos\theta & -\sin\theta \\ \sin\theta & \cos\theta \end{pmatrix}.$$

We now derive this transformation matrix. Consider the point $P(x, y)$ in Figure 5.1 and the effect of rotation through an angle θ anticlockwise about the origin to the new position given by the point $Q(x', y')$.

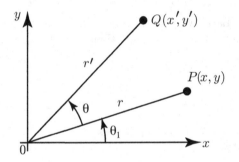

FIGURE 5.1
Rotation in the plane, anticlockwise through angle θ.

The position vectors of P and Q are $r = \begin{pmatrix} x \\ y \end{pmatrix}$ and $r' = \begin{pmatrix} x' \\ y' \end{pmatrix}$, respectively. Clearly rotation around the origin preserves the length of the arm OP and hence $|r| = |r'|$. We now derive a transformation matrix that will achieve such a rotation. Referring to Figure 5.1 observe that

$$\cos\theta_1 = \frac{x}{|r|}, \quad \sin\theta_1 = \frac{y}{|r|}$$

$$\cos(\theta_1 + \theta) = \frac{x'}{|r'|}, \quad \sin(\theta_1 + \theta) = \frac{y'}{|r'|}.$$

Expanding the equations in the previous line using trigonometric identities (see Appendix B), we have

$$\cos\theta_1 \cos\theta - \sin\theta_1 \sin\theta = \frac{x'}{|r'|}, \quad \sin\theta_1 \cos\theta + \cos\theta_1 \sin\theta = \frac{y'}{|r'|}$$

so that

$$x' = |r'|(\cos\theta_1 \cos\theta - \sin\theta_1 \sin\theta), \quad y' = |r'|(\sin\theta_1 \cos\theta + \cos\theta_1 \sin\theta).$$

Noting $|r'| = |r|$, we have

$$x' = |r|\cos\theta_1 \cos\theta - |r|\sin\theta_1 \sin\theta, \quad y' = |r|\sin\theta_1 \cos\theta + |r|\cos\theta_1 \sin\theta.$$

But $x = |r|\cos\theta_1$ and $y = |r|\sin\theta_1$ and hence

$$x' = x\cos\theta - y\sin\theta, \quad y' = x\sin\theta + y\cos\theta.$$

Writing these equations in matrix form, we have

$$\begin{pmatrix} x' \\ y' \end{pmatrix} = \begin{pmatrix} \cos\theta & -\sin\theta \\ \sin\theta & \cos\theta \end{pmatrix} \begin{pmatrix} x \\ y \end{pmatrix}.$$

Let us label the matrix which achieves this rotation through θ by $W(\theta)$, that is

$$W(\theta) = \begin{pmatrix} \cos\theta & -\sin\theta \\ \sin\theta & \cos\theta \end{pmatrix}.$$

It is straightforward to verify that the matrix $W(\theta)$ is an orthogonal matrix and in common with all orthogonal matrices it preserves the length of the vector upon which it operates. This is to be expected because the point P is being rotated about the origin and hence its distance from the origin remains the same.

Definition 5.19 Reflection in the x axis.
The transformation matrix U that reflects a point in the x axis is

$$U = \begin{pmatrix} 1 & 0 \\ 0 & -1 \end{pmatrix}.$$

We now derive this transformation matrix. Consider the reflection of the point $P(x,y)$ in the x axis as shown in Figure 5.2. Clearly, under this operation the x coordinate does not change, but the new y coordinate is $-y$. Thus Q is the point $(x, -y)$. Labelling the matrix

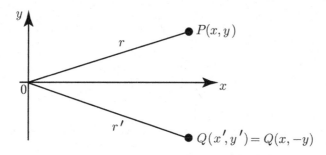

FIGURE 5.2
Reflection in the x axis.

which achieves this transformation by U we have

$$\begin{pmatrix} x' \\ y' \end{pmatrix} = U \begin{pmatrix} x \\ y \end{pmatrix} = \begin{pmatrix} 1 & 0 \\ 0 & -1 \end{pmatrix} \begin{pmatrix} x \\ y \end{pmatrix}.$$

Transformations such as these can be performed one after the other using matrix multiplication. For example

$$W(\phi)\, U\, W(-\phi) \begin{pmatrix} x \\ y \end{pmatrix}$$

first applies $W(-\phi)$ to the point $\begin{pmatrix} x \\ y \end{pmatrix}$ which is a clockwise rotation through ϕ, followed by reflection in the x axis, followed by an anticlockwise rotation through ϕ. This can be shown to be equivalent to reflection in the line through the origin with gradient $\tan\phi$ (Figure 5.3).

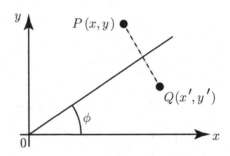

FIGURE 5.3
Reflection in the line through the origin with gradient $\tan\phi$.

5.14   Projections

We have seen that vectors in $\mathbb{R}^2$ and $\mathbb{R}^3$ can be visualised conveniently. Such visualisations are helpful in defining quantities such as the length of a vector and the angle between vectors. These concepts can then be generalised to more complicated vector spaces for which visualisation is not feasible. Consider Figure 5.4 which shows arbitrary vectors a and n in $\mathbb{R}^2$ or $\mathbb{R}^3$. A line has been drawn from P to meet n at a right-angle. The distance OQ is

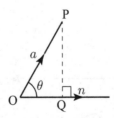

FIGURE 5.4
OQ is the projection of a onto n.

called the **orthogonal projection** of a onto n, or alternatively 'the component of a in the direction of n'. Observe, by simple trigonometry, that the length of OQ is $|a|\cos\theta$. Suppose we take the inner product (here the Euclidean inner product) of a with the unit vector $\hat{n}$:

$$\langle a|\hat{n}\rangle = |a|\,|\hat{n}|\cos\theta$$
$$= |a|\cos\theta.$$

We conclude that the orthogonal projection of a onto n is given by the scalar product $\langle a|\hat{n}\rangle$. If we seek a vector in the same direction as n but with length equal to the projection OQ then this is simply $\langle a|\hat{n}\rangle\hat{n}$. In ket notation this vector is $\langle a|\hat{n}\rangle\,|\hat{n}\rangle$.

This concept of orthogonal projection of one vector onto another is generalised from $\mathbb{R}^2$ and $\mathbb{R}^3$ to vectors of complex numbers in the following example.

Example 5.14.1 Projection matrices

Consider the kets $|\psi\rangle$ and $|\phi\rangle$ in $\mathbb{C}^n$. We might depict them schematically as in Figure 5.5 although of course we are now dealing with vectors of complex numbers. Let $|\psi\rangle$ be a unit

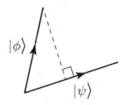

FIGURE 5.5
The projection of $|\phi\rangle$ onto $|\psi\rangle$.

vector. Suppose we wish to project $|\phi\rangle$ onto $|\psi\rangle$. Consider the expression

$$(|\psi\rangle\langle\psi|)\,|\phi\rangle.$$

Observe that $|\psi\rangle\langle\psi|$ is a matrix which multiplies the ket $|\phi\rangle$. But using associativity of matrix multiplication, we can consider first $\langle\psi|\phi\rangle$ which is the inner product of $|\psi\rangle$ and $|\phi\rangle$, and of course, a complex scalar, c say.

Thus

$$(|\psi\rangle\langle\psi|)\,|\phi\rangle = c|\psi\rangle.$$

The effect of multiplying $|\phi\rangle$ by $|\psi\rangle\langle\psi|$ has been to project it onto $|\psi\rangle$. The matrix $|\psi\rangle\langle\psi|$ is referred to as a **projection matrix**.

Definition 5.20 Projection matrix.
Given a unit vector $|\psi\rangle$, the matrix which projects an arbitrary $|\phi\rangle$ onto $|\psi\rangle$ is given by

$$|\psi\rangle\langle\psi|$$

*and is referred to as a **projection matrix**.*

Example 5.14.2

Find the projection matrix which projects an arbitrary $|\phi\rangle$ onto the unit vector $|\psi\rangle = \begin{pmatrix} \sqrt{3}/2 \\ 1/2 \end{pmatrix}$. Find the orthogonal projection of $\begin{pmatrix} 1 \\ 5 \end{pmatrix}$ onto $|\psi\rangle$.

Solution

The projection matrix is given by

$$|\psi\rangle\langle\psi| = \begin{pmatrix} \sqrt{3}/2 \\ 1/2 \end{pmatrix} \begin{pmatrix} \sqrt{3}/2 & 1/2 \end{pmatrix} = \begin{pmatrix} 3/4 & \sqrt{3}/4 \\ \sqrt{3}/4 & 1/4 \end{pmatrix}.$$

Then

$$\begin{pmatrix} 3/4 & \sqrt{3}/4 \\ \sqrt{3}/4 & 1/4 \end{pmatrix} \begin{pmatrix} 1 \\ 5 \end{pmatrix} = \begin{pmatrix} (3+5\sqrt{3})/4 \\ (5+\sqrt{3})/4 \end{pmatrix}.$$

5.15   End-of-chapter exercises

1. Find the values of λ for which the linear system

    $$x + 4y = \lambda x$$
    $$2x + 3y = \lambda y$$

    has non-trivial solutions.

2. Solve the following equations using Gaussian elimination

    $$2x - y + z = 2$$
    $$-2x + y + z = 4$$
    $$6x - 3y - 2z = -9.$$

3. If $A = \begin{pmatrix} 2 & 1 & 3 \\ 4 & 2 & 1 \\ -1 & 3 & 2 \end{pmatrix}$ and $B = \begin{pmatrix} 1 & -7 & 0 \\ 0 & 2 & 5 \\ 3 & 4 & 5 \end{pmatrix}$ find A^T, B^T and $(AB)^T$.

    Deduce that $(AB)^T = B^T A^T$.

4. Show that the Pauli matrices

    $$\sigma_0 = \begin{pmatrix} 1 & 0 \\ 0 & 1 \end{pmatrix}, \quad \sigma_1 = \begin{pmatrix} 0 & 1 \\ 1 & 0 \end{pmatrix}, \quad \sigma_2 = \begin{pmatrix} 0 & -i \\ i & 0 \end{pmatrix}, \quad \sigma_3 = \begin{pmatrix} 1 & 0 \\ 0 & -1 \end{pmatrix}$$

    are self-inverse.

5. Show that the projection matrix $P = \begin{pmatrix} 3/4 & \sqrt{3}/4 \\ \sqrt{3}/4 & 1/4 \end{pmatrix}$ has the property that $P^2 = P$. This is a property possessed by all projection matrices.

6. Given $A = \begin{pmatrix} 3 & 2 \\ -3 & -4 \end{pmatrix}$, find A^2 and show that $A^2 + A - 6I = 0$, the zero matrix.

7. Suppose $H = \begin{pmatrix} 3 & 1+i \\ 1-i & 2 \end{pmatrix}$.

    a) Find the adjoint matrix $H^\dagger$.

    b) Confirm that H is self-adjoint.

    c) Show that H is normal, i.e $HH^\dagger = H^\dagger H$.

    d) Show that if $P = \begin{pmatrix} -\frac{1+i}{2} & 1+i \\ 1 & 1 \end{pmatrix}$, then $P^{-1}HP = D$ where D is a diagonal matrix with 1 and 4 on the leading diagonal. Deduce that H is similar to D.

8. The **commutator** of two matrices A and B is written $[A, B]$ and is defined to be $AB - BA$.

a) Deduce that if A and B commute, then the commutator is zero.

b) For the Pauli matrices given in Q4 above, show that $[\sigma_2, \sigma_3] = 2i\sigma_1$.

9. Given $H = \begin{pmatrix} 1 & 1 \\ 1 & -1 \end{pmatrix}$ and $P = \begin{pmatrix} e^{i\theta_2} & 0 \\ 0 & e^{i\theta_1} \end{pmatrix}$ show that

$$\frac{1}{2}HPH = e^{i\frac{\theta_2+\theta_1}{2}} \begin{pmatrix} \cos\frac{\theta}{2} & -i\sin\frac{\theta}{2} \\ -i\sin\frac{\theta}{2} & \cos\frac{\theta}{2} \end{pmatrix}$$

where $\theta = \theta_1 - \theta_2$.

10. Find a 4×4 permutation matrix which permutes the Boolean strings $00, 01, 10, 11$ to $01, 00, 11, 10$.

11. Show that the matrix $U = \begin{pmatrix} 0 & 0 & 0 & 1 \\ 1 & 0 & 0 & 0 \\ 0 & 1 & 0 & 0 \\ 0 & 0 & 1 & 0 \end{pmatrix}$ is unitary.

12. It can be shown that all permutation matrices are orthogonal. Find a counter-example which demonstrates that the converse is not true.

13. Given $H = \frac{1}{\sqrt{2}} \begin{pmatrix} 1 & 1 \\ 1 & -1 \end{pmatrix}$ show that

a) $H|0\rangle = \frac{1}{\sqrt{2}}(|0\rangle + |1\rangle)$,

b) $H|1\rangle = \frac{1}{\sqrt{2}}(|0\rangle - |1\rangle)$.

14. Given $Z = \begin{pmatrix} 1 & 0 \\ 0 & -1 \end{pmatrix}$, show that

a) $Z|0\rangle = |0\rangle$,   b) $Z|1\rangle = -|1\rangle$,

c) $Z|+\rangle = |-\rangle$,   d) $Z|-\rangle = |+\rangle$,

where $|+\rangle = \frac{1}{\sqrt{2}} \begin{pmatrix} 1 \\ 1 \end{pmatrix}$ and $|-\rangle = \frac{1}{\sqrt{2}} \begin{pmatrix} 1 \\ -1 \end{pmatrix}$.

15. Let $A = \begin{pmatrix} a_{11} & a_{12} \\ a_{21} & a_{22} \end{pmatrix}$, $b = \begin{pmatrix} b_1 \\ b_2 \end{pmatrix}$, $c = \begin{pmatrix} c_1 \\ c_2 \end{pmatrix}$, and let k be a scalar.

a) Evaluate Ab.   b) Evaluate Ac.

c) Evaluate $A(b + c)$.   d) $A(kb)$.

e) Deduce that $A(b + c) = Ab + Ac$.   f) Deduce that $A(kb) = kAb$.

Results e) and f) mean that the matrix A is a 'linear operator', further details of which are given in Chapter 11.

6

Vector spaces

DOI: 10.1201/9781003264569-6

6.1 Objectives

Vector spaces are the mathematical structures in which the state vectors of quantum computation live. When endowed with an inner product, a vector space becomes an inner product space. Of particular relevance are complex inner product spaces. An objective of this chapter is to provide a formal definition of a vector space and give several examples. You will see that, in this context, a 'vector' is a much more general object than vectors such as those in $\mathbb{R}^2$ and $\mathbb{R}^3$. However, the already familiar operations of vector addition and scalar multiplication are essential ingredients of all vector spaces. We demonstrate how vectors in a vector space can be constructed using a set of vectors called a basis and see how vectors can be written in terms of different bases. Of particular importance are orthonormal bases in which the basis vectors are mutually orthogonal and normalised. Finally, we introduce cosets and explain how a quotient space is constructed.

6.2 Definition of a vector space

Definition 6.1 Vector space

Suppose $\mathbb{F}$ is a field (usually the set of real numbers $\mathbb{R}$, or the set of complex numbers $\mathbb{C}$). Let V be a nonempty set $\{u, v, w, \ldots\}$ upon which we can define operations called addition and scalar multiplication such that for any $u, v \in V$ and $k \in \mathbb{F}$, the sum $u + v \in V$ and the product $ku \in V$. Then V is called a **vector space over the field** $\mathbb{F}$ *if the following axioms are satisfied:*

1. *$(u + v) + w = u + (v + w)$ - **associativity of addition***

2. *there exists a **zero element** in V, or **identity under addition**, denoted 0, such that $u + 0 = u$ and $0 + u = u$ for all $u \in V$.*

3. *for each u, there exists an **inverse under addition**, denoted $-u$, also in V, such that $u + (-u) = 0$.*

4. *$u + v = v + u$ - **commutativity of addition***

5. *$k(u + v) = ku + kv$, for $u, v \in V$, $k \in \mathbb{F}$ - **distributivity***

6. *for $a, b \in \mathbb{F}$, $(a + b)u = au + bu$ - **distributivity***

7. *for $a, b \in \mathbb{F}$, $(ab)u = a(bu)$ - **associativity of scalar multiplication***

8. *for $1 \in \mathbb{F}$, $1u = u$.*

The elements of any vector space V are hereafter referred to as **vectors** irrespective of whether they 'look like' the familiar vectors in $\mathbb{R}^2$ and $\mathbb{R}^3$. Because for any $u, v \in V$ and

$k \in \mathbb{F}$, the sum $u + v \in V$ and the product $ku \in V$ we say that the set V is **closed** under the operations of addition and scalar multiplication. Note that axioms 1-4 mean that $(V, +)$ is an Abelian group.

Example 6.2.1 The vector space of n-tuples of complex numbers

The set of n-tuples of complex numbers, denoted $\mathbb{C}^n$, with the usual, natural operations of addition and multiplication of complex numbers is a vector space over $\mathbb{C}$. By way of example consider the specific case when $n = 2$ and the vector space is $\mathbb{C}^2$ over the field $\mathbb{C}$. Observe that the column vectors u and v where

$$u = \begin{pmatrix} 1 + i \\ -7 + 2i \end{pmatrix} \quad \text{and} \quad v = \begin{pmatrix} 2 + 3i \\ 5 - 4i \end{pmatrix}$$

are both elements of $\mathbb{C}^2$. As we have seen previously, their sum is defined in a natural way to be

$$u + v = \begin{pmatrix} 1 + i \\ -7 + 2i \end{pmatrix} + \begin{pmatrix} 2 + 3i \\ 5 - 4i \end{pmatrix} = \begin{pmatrix} 3 + 4i \\ -2 - 2i \end{pmatrix}$$

which is clearly also an element of $\mathbb{C}^2$. The scalar multiple, ku, for $k \in \mathbb{C}$, is given by

$$ku = k \begin{pmatrix} 1 + i \\ -7 + 2i \end{pmatrix} = \begin{pmatrix} k(1 + i) \\ k(-7 + 2i) \end{pmatrix}$$

which is clearly also an element of $\mathbb{C}^2$. So the set $\mathbb{C}^2$ is closed under addition and scalar multiplication. In this example, the zero vector is $\begin{pmatrix} 0 \\ 0 \end{pmatrix} \in \mathbb{C}^2$. You should confirm for yourself that the remaining axioms are satisfied.

We shall write vectors in $\mathbb{C}^n$ as both column vectors and row vectors as the need arises.

Example 6.2.2 The vector space of n-tuples of real numbers

Because all real numbers are complex numbers (with zero imaginary part), the previous example readily interprets when the vectors are real. For example, consider the vector space $\mathbb{R}^3$ over the field $\mathbb{R}$ consisting of 3-tuples of real numbers. Observe that the row vectors u and v, where

$$u = (9, -8, 2) \quad \text{and} \quad v = (-1, 2, 3),$$

are both elements of $\mathbb{R}^3$. Their sum is the vector $(8, -6, 5)$ which is clearly also in $\mathbb{R}^3$. The scalar multiple $5u = 5(9, -8, 2) = (45, -40, 10)$ is also in $\mathbb{R}^3$. So the set $\mathbb{R}^3$ is closed under addition and scalar multiplication. The zero element, or identity under addition is $(0, 0, 0)$. It is straightforward to check that the remaining axioms for a vector space hold.

We generalise the results of the previous two examples by emphasising that for real and complex n-tuples addition and scalar multiplication are defined and performed component-wise, i.e., if $u = (u_1, u_2, \ldots, u_n)$ and $v = (v_1, v_2, \ldots, v_n)$ then

$$u + v = (u_1, u_2, \ldots, u_n) + (v_1, v_2, \ldots, v_n) = (u_1 + v_1, u_2 + v_2, \ldots, u_n + v_n)$$

and, for $k \in \mathbb{F}$,

$$ku = k(u_1, u_2, \ldots, u_n) = (ku_1, ku_2, \ldots, ku_n).$$

We often refer to an n-tuple vector such as $u = (u_1, u_2, \ldots, u_n)$ which has n elements as an n-dimensional vector. This terminology should not be confused with the dimension of a vector space which is not necessarily the same. It is possible, for example, to have a two-dimensional vector space populated with three-dimensional vectors. This will become clear when we discuss subspaces and the basis of a vector space shortly.

Example 6.2.3 The vector space $\mathbb{B} = \{0, 1\}$ with operations $\oplus$ and $\wedge$

Consider the set $\mathbb{B} = \{0, 1\}$. We define addition within this set using $\oplus$, that is the exclusive-or operation or equivalently addition modulo 2. Choosing the underlying field again to be $\{0, 1\}$ (which is the finite field $\mathbb{F}_2$), scalar multiplication can be performed using the Boolean *and*, $\wedge$, or equivalently multiplication modulo 2. It is straightforward to show that $\{0, 1\}$ is closed under the operations $\oplus$ and $\wedge$ and that all the axioms for a vector space are satisfied.

Thus far, the examples of vector spaces that we have seen look like familiar vectors in $\mathbb{R}^2$ and $\mathbb{R}^3$. However, given the stated definition of a vector space, many other mathematical objects can now be thought of as vectors.

Example 6.2.4 The vector space of $m \times n$ matrices over $\mathbb{C}$

We have seen previously that two matrices of the same size can be added. We have also seen how to multiply a matrix by a scalar. In both cases the result is another matrix of the same size. It is straightforward to check that the set of $m \times n$ matrices with the usual addition and scalar multiplication is a vector space. The zero element of the vector space is an $m \times n$ matrix in which all elements are zero. You should verify that the remaining axioms of a vector space are satisfied.

Example 6.2.5 The vector space of functions

Consider the set of all real-valued functions $f : \mathbb{R} \to \mathbb{R}$. For example, the functions

$$f_1(x) = x^2, \ f_2(x) = \sin x, \ \text{and} \ f_3(x) = e^x$$

are all elements of this set. We can define the operation of addition on this set as follows: for any functions f and g

$$(f + g) = (f + g)(x) = f(x) + g(x).$$

Clearly this sum is another real-valued function defined on $\mathbb{R}$ so the set is closed under addition. Given f_1 and f_2 above, then $(f_1 + f_2)(x) = f_1(x) + f_2(x) = x^2 + \sin x$. We define scalar multiplication on a function f, for $k \in \mathbb{R}$, by

$$(kf) = (kf)(x) = k \, f(x).$$

Clearly, the result is another real-valued function, so the set is closed under scalar multiplication. For example, if $k = 5$ and $f_3(x) = e^x$ then

$$(5f_3) = (5f_3)(x) = 5 \, f_3(x) = 5e^x.$$

The zero vector in this vector space is the zero function, $z(x) = 0$ because, for any function $f(x)$,

$$z(x) + f(x) = 0 + f(x) = f(x).$$

The inverse of the function $f(x)$ is $(-f)(x) = -f(x)$ because

$$f(x) + (-f(x)) = 0.$$

For example, the inverse of $f_2(x) = \sin x$ is $(-f_2)(x) = -\sin x$. The remaining vector space axioms are readily verified.

Exercises

6.1   Consider the set of all polynomials of degree 2 or less, with real coefficients. This set includes, for example, the polynomials $p_1(t) = 1 + 2t - t^2$, $p_2(t) = 2 + 3t - 7t^2$, $p_3(t) = 1 + 5t$, $p_4(t) = 6$. Show that this set with the usual operations of addition and multiplication by a real number is a vector space over $\mathbb{R}$. Identify the zero element. Determine the inverse of each of the polynomials given above.

6.3   Inner product spaces

In Chapter 4 we discussed how, given two vectors $u = (u_1, u_2, \ldots, u_n)$, $v = (v_1, v_2, \ldots, v_n)$, we can define their inner product $\langle u, v \rangle$. In the case of vectors in $\mathbb{R}^2$ and $\mathbb{R}^3$, we defined the Euclidean inner product as

$$\langle u, v \rangle = \sum_{i=1}^{2} u_i v_i, \qquad \langle u, v \rangle = \sum_{i=1}^{3} u_i v_i,$$

respectively, and we generalised this for vectors in $\mathbb{R}^n$. We then showed how an inner product can be defined for vectors $u, v \in \mathbb{C}^n$ either as

$$\langle u, v \rangle = \sum_{i=1}^{n} u_i v_i^* \qquad \text{(Definition 1)}$$

where $*$ denotes the complex conjugate, or

$$\langle u, v \rangle = \sum_{i=1}^{n} u_i^* v_i \qquad \text{(Definition 2)}.$$

We noted that the former is the inner product usually preferred by mathematicians whereas the latter is often preferred by physicists and computer scientists. In fact, the latter is better suited to the Dirac bra-ket notation as we shall see.

Definition 6.2 Inner product space
An inner product space is a vector space V over a field $\mathbb{F}$ which is endowed with an inner product. A complex inner product space is a vector space for which the underlying field is $\mathbb{C}$.

The inner product assigns, to each pair of vectors $u, v \in V$, a scalar in the underlying field $\mathbb{F}$. Thus it is a mapping $V \times V \to \mathbb{F}$. To qualify as an inner product, the mapping has to satisfy requirements as listed in Chapter 4. Once an inner product has been defined on a vector space it makes sense to refer to the norm of a vector u, as

$$\|u\| = \sqrt{\langle u, u \rangle}.$$

Further, two vectors $u, v \in V$ are orthogonal if $\langle u, v \rangle = 0$. We shall see in Chapter 13 that the state vectors of quantum computation are elements of a complex inner product space.

Exercises

6.2   Show that if $|\psi\rangle = (\psi_1, \psi_2, \ldots, \psi_n)^T$, where $\psi_i \in \mathbb{C}$, then

$$\||\psi\rangle\| = \sqrt{\sum_{i=1}^{n} |\psi_i|^2}.$$

6.3   Show if $|\psi\rangle = (\psi_1, \psi_2, \ldots, \psi_n)^T$ is a unit vector then $\sum_{i=1}^{n} |\psi_i|^2 = 1$.

6.4   Subspaces

Definition 6.3 Subspace
*A subset U of a vector space V is called a **subspace** if it is a vector space in its own right.*

It follows that the subspace will be closed under vector addition and scalar multiplication. Importantly, this definition requires that the subset U contains the zero element of V.

Example 6.4.1

Consider the vector space $\mathbb{R}^2$. We can represent any (row) vector (x, y) in this space as a directed line segment drawn from the origin to the point with coordinates (x, y). Vectors $(2, 3)$, $(-3, 5)$ and several others are shown in Figure 6.1. Now consider the subset, U, of $\mathbb{R}^2$ comprising any vector of the form $(x, 5x)$. That is,

$$U = \{(x, y) \in \mathbb{R}^2 : y = 5x\}.$$

So this set includes $(1, 5)$, $(2, 10)$, $(3, 15)$ and so on. Importantly, it includes the zero vector $(0, 0)$. Note that if we draw vectors from the origin to the points $(x, 5x)$, all the resulting

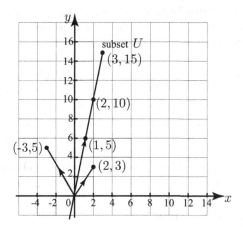

FIGURE 6.1
The subset U of $\mathbb{R}^2$ is a subspace.

points lie on the same straight line passing through the origin. It is straightforward to show that for all the vectors in U, the axioms for a vector space are satisfied, and thus U is a subspace of V. Particularly, adding any two vector in U results in another vector in U. For example, $(1,5) + (2,10) = (3,15)$. Likewise for scalar multiples. Observe that the vector space U is one-dimensional (a straight line) yet is populated by two-dimensional vectors.

Exercises

6.4 Consider the vector space $\mathbb{R}^3$ and row vectors of the form (x, y, z).

  (a) Consider the subset of $\mathbb{R}^3$ comprising vectors of the form $(x, y, 0)$ (i.e., the xy plane). Verify that this is a subspace.

  (b) Consider the subset of $\mathbb{R}^3$ comprising vectors of the form $(0, y, 0)$ (i.e., the y axis). Verify that this is a subspace.

6.5 Consider the subset of $\mathbb{R}^2$ given by

$$U = \{(x, y) \in \mathbb{R}^2 : y = 2x + 1\}.$$

  Explain why this set cannot be a subspace of $\mathbb{R}^2$.

6.6 Show that the set of all 2×2 symmetric matrices, with the usual operations of addition and multiplication by a scalar, is a subspace of the vector space of all 2×2 matrices over $\mathbb{R}$.

6.5   Linear combinations, span and LinSpan

We have seen that, given a vector space V over a field $\mathbb{F}$, we can perform the operations of addition of vectors and multiplication of a vector by a scalar.

Definition 6.4 Linear combination and Span
Given a set of vectors $v_1, v_2, \ldots, v_n \in V$ and a set of scalars $k_1, k_2, \ldots, k_n \in \mathbb{F}$, then the vector formed by scalar multiplication and vector addition

$$v = k_1 v_1 + k_2 v_2 + \ldots + k_n v_n = \sum_{i=1}^{n} k_i v_i$$

*is called a **linear combination** of the vectors $v_1, v_2, \ldots, v_n$. The set of all linear combinations of the vectors $v_1, v_2, \ldots, v_n$ is called the **span** of the set $\{v_1, v_2, \ldots, v_n\}$.*

If we select a subset of V, S say, then the set of all linear combinations of the vectors in S is a subspace referred to as the **subspace spanned by** S or **LinSpan**(S).

Example 6.5.1 Vectors which span $\mathbb{R}^2$

(a) Show that the row vector $(8, 4) \in \mathbb{R}^2$ can be written as a linear combination of the vectors $(1, 1)$ and $(-1, 1)$.

(b) Show that an arbitrary vector $(x, y) \in \mathbb{R}^2$ can be written as a linear combination of the vectors $(1, 1)$ and $(-1, 1)$. Deduce that $(1, 1)$ and $(-1, 1)$ together span $\mathbb{R}^2$.

Solution

(a) We form the linear combination

$$(8, 4) = k_1(1, 1) + k_2(-1, 1)$$

where $k_1, k_2 \in \mathbb{R}$ are yet to be determined. Then, equating components,

$$8 = k_1 - k_2, \qquad 4 = k_1 + k_2.$$

These simultaneous equations have solution $k_1 = 6$, $k_2 = -2$ thus

$$(8, 4) = 6(1, 1) - 2(-1, 1)$$

which is the required linear combination.

(b) We form the linear combination

$$(x, y) = k_1(1, 1) + k_2(-1, 1)$$

where $k_1, k_2 \in \mathbb{R}$. Then, equating components,

$$x = k_1 - k_2, \qquad y = k_1 + k_2.$$

These simultaneous equations have solution $k_1 = \frac{x+y}{2}$, $k_2 = \frac{y-x}{2}$ thus

$$(x, y) = \frac{x + y}{2}(1, 1) + \frac{y - x}{2}(-1, 1)$$

which is the required linear combination. Thus any vector, (x, y), in $\mathbb{R}^2$ can be written in this way and therefore $(1, 1)$ and $(-1, 1)$ span the whole of $\mathbb{R}^2$. As we shall see shortly, there are many other pairs of vectors which also span $\mathbb{R}^2$.

Exercises

6.7 Show that the vectors $(1, 0, 0)$, $(0, 1, 0)$ and $(0, 0, 1)$ span $\mathbb{R}^3$.

6.8 Show that the vectors $(1, 0, 0)$, $(0, 1, 0)$ and $(1, 1, 0)$ do not span $\mathbb{R}^3$. Importantly note that just because we have three vectors in a three-dimensional space, it does not follow that they span the space.

Consider an $m \times n$ matrix A with real elements:

$$A = \begin{pmatrix} a_{11} & a_{12} & a_{13} & \cdots & a_{1n} \\ a_{21} & a_{22} & a_{23} & \cdots & a_{2n} \\ \vdots & \vdots & \vdots & \ddots & \vdots \\ a_{m1} & a_{m2} & a_{m3} & \cdots & a_{mn} \end{pmatrix}.$$

We can regard each row of the matrix as a distinct n-dimensional vector. We can then form linear combinations of these row vectors.

Definition 6.5 Row space and column space of a matrix

*The subspace of $\mathbb{R}^n$ spanned by the rows of an $m \times n$ matrix A is called the **row space** of A. Similarly, we can regard each column as a distinct m-dimensional column vector and the subspace of $\mathbb{R}^m$ spanned by these is called the **column space** of A.*

Example 6.5.2 The row space of matrix A

Consider $A = \begin{pmatrix} 2 & 1 \\ 1 & 1 \end{pmatrix}$. Regarding each row as a vector, we can form linear combinations:

$$k_1(2,1) + k_2(1,1), \qquad k_1, k_2 \in \mathbb{R}.$$

The set of all such linear combinations is the row space of A. Suppose we choose an arbitrary row vector in $\mathbb{R}^2$, (x,y) say. It is of interest to ask whether this is an element of the row space of A. We form

$$(x,y) = k_1(2,1) + k_2(1,1)$$

from which $x = 2k_1 + k_2$, $y = k_1 + k_2$ and solving for k_1, k_2 we find

$$k_1 = x - y, \quad k_2 = 2y - x$$

so that

$$(x,y) = (x-y)(2,1) + (2y-x)(1,1)$$

e.g. $(7,4) = 3(2,1) + 1(1,1)$. Thus any vector in $\mathbb{R}^2$ can be written as a linear combination of the rows of A. Thus the row space of A is in fact $\mathbb{R}^2$. This is not always the case.

Consider the following example.

Example 6.5.3

Consider $B = \begin{pmatrix} 4 & 2 \\ 2 & 1 \end{pmatrix}$. The set of all linear combinations

$$k_1(4,2) + k_2(2,1), \qquad k_1, k_2 \in \mathbb{R}$$

is the row space of B. Observe that this combination can be written

$$2k_1(2,1) + k_2(2,1) = (2k_1 + k_2)(2,1)$$

so that any vector in the row space of B must be a multiple of $(2,1)$. It is not possible to write an arbitrary vector in $\mathbb{R}^2$ as a linear combination of the row vectors. The row space of B does not span the whole of $\mathbb{R}^2$.

Definition 6.6 Null space of a matrix

*Consider an $m \times n$ matrix A and m linear homogeneous equations in n unknowns, $Ax = 0$. The set of solutions is a subspace of $\mathbb{R}^n$ called the **null space** of A.*

We can think of the null space as the set of all vectors that map to the zero vector when multiplied by A.

Consider the following example.

Example 6.5.4 The null space of a matrix A

The set of two linear homogeneous equations in three unknowns

$$x + 3y + z = 0$$
$$y + z = 0$$

has matrix form $AX = 0$ where

$$A = \begin{pmatrix} 1 & 3 & 1 \\ 0 & 1 & 1 \end{pmatrix}, \qquad X = \begin{pmatrix} x \\ y \\ z \end{pmatrix}, \qquad 0 = \begin{pmatrix} 0 \\ 0 \end{pmatrix}.$$

Using Gaussian elimination (see Chapter 5) the full set of solutions can be written in terms of the free variable t as $x = 2t$, $y = -t$, $z = t$, that is

$$X = \begin{pmatrix} 2t \\ -t \\ t \end{pmatrix} = t \begin{pmatrix} 2 \\ -1 \\ 1 \end{pmatrix} \qquad \text{for any } t \in \mathbb{R}.$$

Here, the set of solutions is the subspace spanned by $\begin{pmatrix} 2 \\ -1 \\ 1 \end{pmatrix}$. This subspace is 1-dimensional. Note that this subspace, as expected, includes the zero vector, (obtained by taking $t = 0$). You should verify that any vector in this subspace satisfies the given set of equations and think of this as all vectors in the null space being mapped to the zero vector following multiplication by the matrix A.

Exercises

6.9   Consider the linear homogeneous equation $x + y + z = 0$. Show that the solution space is two-dimensional and is spanned by $\begin{pmatrix} -1 \\ 1 \\ 0 \end{pmatrix}$ and $\begin{pmatrix} -1 \\ 0 \\ 1 \end{pmatrix}$.

6.10   Given $A = \begin{pmatrix} 4 & 1 \\ 2 & 1 \end{pmatrix}$ show that the null space of A contains only the zero vector.

6.11   Find the null space of $A = \begin{pmatrix} 1 & 1 \\ 1 & 1 \end{pmatrix}$.

6.6   Linear independence

Definition 6.7 Linear independence
*Suppose we have a vector space V over a field $\mathbb{F}$. A set of vectors $v_1, v_2, \ldots, v_n \in V$ is said to be **linearly independent** if the only way that the linear combination*

$$k_1 v_1 + k_2 v_2 + \ldots + k_n v_n$$

for $k_1, \ldots, k_n \in \mathbb{F}$, can equal the zero vector, is if all the $k_i, i = 1, \ldots, n$, are zero.

Example 6.6.1

Consider the following linear combination of three vectors in $\mathbb{R}^3$:

$$k_1(1,0,0) + k_2(0,1,0) + k_3(0,0,1)$$

and suppose this sum equals the zero vector $(0,0,0)$. Then we see

$$k_1(1,0,0) + k_2(0,1,0) + k_3(0,0,1) = (k_1,0,0) + (0,k_2,0) + (0,0,k_3)$$
$$= (k_1, k_2, k_3).$$

If this is to equal the zero vector, $(0,0,0)$, then clearly $k_1 = k_2 = k_3 = 0$, i.e., all k_i must be zero. The three given vectors are therefore linearly independent.

Example 6.6.2

Consider the following linear combination of three vectors in $\mathbb{R}^3$:

$$k_1(1,2,3) + k_2(-1,-3,-7) + k_3(3,5,5)$$

and suppose this sum equals the zero vector $(0,0,0)$. Then

$$k_1(1,2,3) + k_2(-1,-3,-7) + k_3(3,5,5) = (k_1, 2k_1, 3k_1) +$$
$$(-k_2, -3k_2, -7k_2) + (3k_3, 5k_3, 5k_3)$$
$$= (k_1 - k_2 + 3k_3, 2k_1 - 3k_2 + 5k_3, 3k_1 - 7k_2 + 5k_3).$$

If this sum equals $(0,0,0)$, then

$$k_1 - k_2 + 3k_3 = 0$$
$$2k_1 - 3k_2 + 5k_3 = 0$$
$$3k_1 - 7k_2 + 5k_3 = 0.$$

Using Gaussian elimination, it is straightforward to verify that there is an infinite number of non-zero solutions of these simultaneous equations e.g. $k_1 = -8$, $k_2 = -2$, $k_3 = 2$. Clearly not all the k_i are zero, and hence, these vectors are not linearly independent. We say that they are **linearly dependent**. Indeed this dependency is apparent when we see that the final vector can be written as a linear combination of the other two:

$$(3,5,5) = 4(1,2,3) + (-1,-3,-7).$$

Exercises

6.12 Show that the row vectors $(1,1)$ and $(-1,1)$ are linearly independent.

6.13 Show that the $\mathbb{R}^3$ vectors $e_1 = (2,1,3)$, $e_2 = (1,0,1)$, and $e_3 = (3,2,1)$ are linearly independent and that they span $\mathbb{R}^3$.

6.14 Consider the vector space of 2×3 matrices and the set of vectors

$$\begin{pmatrix} 1 & 2 & 3 \\ 0 & 1 & 7 \end{pmatrix}, \quad \begin{pmatrix} -2 & -3 & 5 \\ 1 & 0 & 2 \end{pmatrix}, \quad \begin{pmatrix} 2 & 1 & 9 \\ 2 & 2 & 9 \end{pmatrix}, \quad \begin{pmatrix} 3 & 2 & 1 \\ 1 & 1 & 0 \end{pmatrix}.$$

Determine whether these vectors are linearly independent. If they are linearly dependent, express one in terms of the other three.

6.15 Consider the vector space of polynomials of degree 2 or less. Determine whether the set of polynomials $p_1(t) = 2t^2 + 3t - 1$, $p_2(t) = t^2 - t - 1$, $p_3(t) = t^2 + t$ is linearly independent.

6.7 Basis of a vector space

Definition 6.8 Basis

*Given a vector space V over a field $\mathbb{F}$, suppose that there is a set of n linearly independent vectors, $\mathcal{E} = \{e_1, e_2, \ldots, e_n\}$, that span the space. Then this set is said to be a **basis** for V, and the vector space V is said to have **finite dimension**, n. We shall often abbreviate 'dimension' to 'dim'. Note that a basis is not unique.*

Because $\mathcal{E}$ spans the vector space V any vector $v \in V$ can be written as a linear combination of the basis vectors, thus

$$v = v_1 e_1 + v_2 e_2 + \ldots + v_n e_n$$

where $v_i \in \mathbb{F}$, $i = 1, \ldots, n$. We will refer to $v_1, v_2, \ldots, v_n$ as the **coordinates** of v with respect to the basis $\mathcal{E}$, sometimes writing the **coordinate vector** $[v]_{\mathcal{E}}$ as

$$[v]_{\mathcal{E}} = \begin{pmatrix} v_1 \\ v_2 \\ \vdots \\ v_n \end{pmatrix}.$$

The subscript $\mathcal{E}$ here indicates that the coordinate vector on the right depends upon the chosen basis. If the same vector is written using a different basis say $\overline{\mathcal{E}} = \{\overline{e}_1, \overline{e}_2, \ldots, \overline{e}_n\}$ and $v = \overline{v}_1 \overline{e}_1 + \overline{v}_2 \overline{e}_2 + \ldots + \overline{v}_n \overline{e}_n$, where $\overline{v}_i \in \mathbb{F}$, then

$$[v]_{\overline{\mathcal{E}}} = \begin{pmatrix} \overline{v}_1 \\ \overline{v}_2 \\ \vdots \\ \overline{v}_n \end{pmatrix}.$$

It is important to recognise that a basis of a vector space is not unique and that any vector can be written in any suitable basis we choose. Later we shall see that there is a need to write the same vector in different bases and we shall discuss how this is done.

Example 6.7.1 The standard basis of $\mathbb{R}^3$ over the field $\mathbb{R}$

Consider the vector space $\mathbb{R}^3$ of row vectors over the field $\mathbb{R}$. The set of vectors, $\mathcal{E} = \{e_1, e_2, e_3\}$ where

$$e_1 = (1, 0, 0), \ e_2 = (0, 1, 0), \ e_3 = (0, 0, 1)$$

is linearly independent and spans $\mathbb{R}^3$. $\mathcal{E}$ as defined here is referred to as the **standard basis** for $\mathbb{R}^3$. Any vector $v \in \mathbb{R}^3$ can be written as a linear combination of these basis vectors. For example

$$v = (4, -3, 11) = 4(1, 0, 0) - 3(0, 1, 0) + 11(0, 0, 1) = 4e_1 - 3e_2 + 11e_3.$$

Then the coordinate vector of v with respect to this basis is

$$[v]_\mathcal{E} = \begin{pmatrix} 4 \\ -3 \\ 11 \end{pmatrix}.$$

We can regard a basis as the building blocks which can be used to generate any vector in the space. Other sets of three vectors in $\mathbb{R}^3$ can be found which are independent and span the space, and hence, the basis is not unique. Observe that because we have used the standard basis here, the components of the coordinate vector are simply the components of the vector $v = (4, -3, 11)$. This will not be the case when other bases are used, as we shall see.

Example 6.7.2

Consider the vector space $\mathbb{R}^2$ of column vectors over the field $\mathbb{R}$. With the standard basis $\mathcal{E} = \{e_1, e_2\}$ where $e_1 = \begin{pmatrix} 1 \\ 0 \end{pmatrix}$ and $e_2 = \begin{pmatrix} 0 \\ 1 \end{pmatrix}$, the coordinate vector of $v = \begin{pmatrix} 4 \\ -8 \end{pmatrix}$ is $[v]_\mathcal{E} = \begin{pmatrix} 4 \\ -8 \end{pmatrix}$. Express v in terms of the basis $\overline{\mathcal{E}} = \{\overline{e}_1, \overline{e}_2\}$ where $\overline{e}_1 = \begin{pmatrix} 1 \\ 1 \end{pmatrix}$ and $\overline{e}_2 = \begin{pmatrix} -1 \\ 1 \end{pmatrix}$.

Solution

Let

$$v = \begin{pmatrix} 4 \\ -8 \end{pmatrix} = k_1 \overline{e}_1 + k_2 \overline{e}_2 = k_1 \begin{pmatrix} 1 \\ 1 \end{pmatrix} + k_2 \begin{pmatrix} -1 \\ 1 \end{pmatrix}$$

Then

$$k_1 - k_2 = 4$$
$$k_1 + k_2 = -8$$

from which $k_1 = -2$, $k_2 = -6$. Then $v = -2\overline{e}_1 - 6\overline{e}_2$ and we can write

$$[v]_{\overline{\mathcal{E}}} = \begin{pmatrix} -2 \\ -6 \end{pmatrix}.$$

This is the coordinate vector of v with respect to the basis $\overline{\mathcal{E}}$. In the following section we show how change of basis can be achieved using matrices.

Definition 6.9 Orthonormal basis
Consider a vector space V and a basis for it. When the basis vectors are normalised (i.e., they each have norm equal to 1), and when they are pairwise mutually orthogonal, the basis is said to be **orthonormal**.

If the basis vectors are not orthogonal, it is possible to carry out a procedure – the Gram-Schmidt process – in order to produce an orthogonal basis. This is explained in Section 6.10.

We have already seen that if A is the $m \times n$ matrix with real elements:

$$A = \begin{pmatrix} a_{11} & a_{12} & a_{13} & \cdots & a_{1n} \\ a_{21} & a_{22} & a_{23} & \cdots & a_{2n} \\ \vdots & \vdots & \vdots & \ddots & \vdots \\ a_{m1} & a_{m2} & a_{m3} & \cdots & a_{mn} \end{pmatrix}$$

we can regard each row of the matrix as a distinct n-dimensional vector. We can then form linear combinations of these row vectors. The subspace of $\mathbb{R}^n$ spanned by these row vectors is the row space of A.

Definition 6.10 Row rank and column rank of a matrix

The number of linearly independent vectors required to span the row space of a matrix A is referred to as the **row rank** *of A. Similarly, the number of linearly independent vectors required to span the column space is referred to as the* **column rank** *of A.*

It is possible to show that for any matrix A, the row rank and column rank are equal and hence we can simply refer to the **rank** of the matrix.

Exercises

6.16 Show that the standard basis in Example 6.7.1 is orthonormal.

6.17 Show that the set $\{(1, 1), (1, -1)\}$ is a basis for $\mathbb{R}^2$, that the basis is orthogonal but not normalised. How would it be normalised ?

Example 6.7.3 The standard basis of $\mathbb{C}^2$ over the field $\mathbb{C}$

Consider the vector space $\mathbb{C}^2$ of row vectors over the field $\mathbb{C}$. Recall that this space includes vectors such as $(1 + i, -7 + 2i)$. The standard basis of this space is $\mathcal{E} = \{e_1, e_2\}$ where $e_1 = (1, 0)$, $e_2 = (0, 1)$. Any vector $z \in \mathbb{C}^2$, that is $z = (z_1, z_2)$ where $z_1, z_2 \in \mathbb{C}$, can be written in terms of the standard basis as

$$z = (z_1, z_2) = k_1 e_1 + k_2 e_2$$
$$= k_1(1, 0) + k_2(0, 1)$$

where $k_1, k_2 \in \mathbb{C}$. Then, by equating respective components, $k_1 = z_1$ and $k_2 = z_2$, so the coordinate vector in the standard basis is

$$[z]_{\mathcal{E}} = \begin{pmatrix} z_1 \\ z_2 \end{pmatrix}.$$

For example, if

$$z = (1 + i, -7 + 2i) = (1 + i)e_1 + (-7 + 2i)e_2$$

the coordinate vector of z with respect to the standard basis $\mathcal{E}$ is

$$[z]_{\mathcal{E}} = \begin{pmatrix} 1 + i \\ -7 + 2i \end{pmatrix}.$$

Example 6.7.4 The computational basis of $\mathbb{C}^2$ over the field $\mathbb{C}$

The column vectors $|0\rangle = \begin{pmatrix} 1 \\ 0 \end{pmatrix}$ and $|1\rangle = \begin{pmatrix} 0 \\ 1 \end{pmatrix}$ form an orthonormal basis for the vector space $\mathbb{C}^2$ of column vectors over the field $\mathbb{C}$. This basis is often referred to as the **computational basis**. We shall often write this basis as the set $B_{\mathbb{C}^2} = \{|0\rangle, |1\rangle\}$.

Example 6.7.5 An alternative basis of $\mathbb{C}^2$ over the field $\mathbb{C}$

The set $\overline{\mathcal{E}} = \{\overline{e}_1, \overline{e}_2\}$ where $\overline{e}_1 = \frac{1}{\sqrt{2}}(1,1)$, $\overline{e}_2 = \frac{1}{\sqrt{2}}(1,-1)$ is an orthonormal basis for $\mathbb{C}^2$. Any row vector in $\mathbb{C}^2$, $z = (z_1, z_2)$ say, can be expressed in terms of this basis as follows:

$$z = (z_1, z_2) = \overline{z}_1 \overline{e}_1 + \overline{z}_2 \overline{e}_2$$
$$= \overline{z}_1 \frac{1}{\sqrt{2}}(1,1) + \overline{z}_2 \frac{1}{\sqrt{2}}(1,-1)$$

where $\overline{z}_1, \overline{z}_2$ are the coordinates of the given vector in this new basis. Thus

$$[z]_{\overline{\mathcal{E}}} = \begin{pmatrix} \overline{z}_1 \\ \overline{z}_2 \end{pmatrix}.$$

These coordinates can be found by equating respective components:

$$z_1 = \frac{1}{\sqrt{2}}\overline{z}_1 + \frac{1}{\sqrt{2}}\overline{z}_2 \qquad (1)$$

$$z_2 = \frac{1}{\sqrt{2}}\overline{z}_1 - \frac{1}{\sqrt{2}}\overline{z}_2 \qquad (2)$$

Forming $(1) + (2)$ gives

$$z_1 + z_2 = \frac{2}{\sqrt{2}}\overline{z}_1 \qquad \text{so that} \qquad \overline{z}_1 = \frac{\sqrt{2}}{2}(z_1 + z_2) = \frac{1}{\sqrt{2}}(z_1 + z_2).$$

Similarly, subtracting (2) from (1) gives

$$z_1 - z_2 = \frac{2}{\sqrt{2}}\overline{z}_2 \qquad \text{so that} \qquad \overline{z}_2 = \frac{\sqrt{2}}{2}(z_1 - z_2) = \frac{1}{\sqrt{2}}(z_1 - z_2).$$

Then we have

$$(z_1, z_2) = \frac{1}{\sqrt{2}}(z_1 + z_2)\overline{e}_1 + \frac{1}{\sqrt{2}}(z_1 - z_2)\overline{e}_2$$

and thus the coordinate vector of z in the new basis is

$$[z]_{\overline{\mathcal{E}}} = \begin{pmatrix} \frac{1}{\sqrt{2}}(z_1 + z_2) \\ \frac{1}{\sqrt{2}}(z_1 - z_2) \end{pmatrix}.$$

We can also represent this transformation from the old to the new basis in matrix form as

$$\begin{pmatrix} \overline{z}_1 \\ \overline{z}_2 \end{pmatrix} = \frac{1}{\sqrt{2}} \begin{pmatrix} 1 & 1 \\ 1 & -1 \end{pmatrix} \begin{pmatrix} z_1 \\ z_2 \end{pmatrix}.$$

The mapping from (z_1, z_2) to $(\overline{z}_1, \overline{z}_2)$ is the discrete Fourier transform on $\mathbb{C}^2$, also known as the Hadamard transform. This transform is fundamental in quantum computation as will be discussed in Chapter 21.

6.8   Change of basis matrices

Suppose we have a basis $\mathcal{E} = \{e_1, e_2, \ldots, e_n\}$ of an n-dimensional vector space V over a field $\mathbb{F}$. Suppose now we have a second basis of V, $\overline{\mathcal{E}} = \{\overline{e}_1, \overline{e}_2, \ldots, \overline{e}_n\}$. We may wish to convert a vector with coordinates in terms of the original basis into one with coordinates in terms of the new basis. We show how this can be achieved with multiplication by a suitable matrix. Each of the original basis vectors must be written as a linear combination of the new ones:

$$e_1 = p_{11}\overline{e}_1 + p_{21}\overline{e}_2 + \ldots + p_{n1}\overline{e}_n$$
$$e_2 = p_{12}\overline{e}_1 + p_{22}\overline{e}_2 + \ldots + p_{n2}\overline{e}_n$$
$$\vdots = \vdots$$
$$e_n = p_{1n}\overline{e}_1 + p_{2n}\overline{e}_2 + \ldots + p_{nn}\overline{e}_n.$$

Here the $p_{ij} \in \mathbb{F}$. The way that the subscripts in the equations above have been numbered is deliberate; the reason for this choice will become apparent. An arbitrary vector $v \in V$ can be written in terms of the original basis as

$$v = c_1 e_1 + c_2 e_2 + \ldots + c_n e_n \qquad c_i \in \mathbb{F}$$

so that its coordinate vector with respect to the original basis is

$$[v]_{\mathcal{E}} = \begin{pmatrix} c_1 \\ c_2 \\ \vdots \\ c_n \end{pmatrix}$$

We can then express v in terms of the new basis $\overline{\mathcal{E}}$ as follows:

$$\begin{aligned} v &= c_1 e_1 + c_2 e_2 + \ldots + c_n e_n \\ &= c_1(p_{11}\overline{e}_1 + p_{21}\overline{e}_2 + \ldots + p_{n1}\overline{e}_n) \\ &\quad + c_2(p_{12}\overline{e}_1 + p_{22}\overline{e}_2 + \ldots + p_{n2}\overline{e}_n) \\ &\quad \ddots \\ &\quad + c_n(p_{1n}\overline{e}_1 + p_{2n}\overline{e}_2 + \ldots + p_{nn}\overline{e}_n). \end{aligned}$$

Rearranging,

$$\begin{aligned} v = & (c_1 p_{11} + c_2 p_{12} + \ldots + c_n p_{1n})\overline{e}_1 \\ & + (c_1 p_{21} + c_2 p_{22} + \ldots + c_n p_{2n})\overline{e}_2 \\ & \ddots \\ & + (c_1 p_{n1} + c_2 p_{n2} + \ldots + c_n p_{nn})\overline{e}_n. \end{aligned}$$

So that, in terms of coordinates,

$$[v]_{\overline{\mathcal{E}}} = \begin{pmatrix} c_1 p_{11} + c_2 p_{12} + \ldots + c_n p_{1n} \\ c_1 p_{21} + c_2 p_{22} + \ldots + c_n p_{2n} \\ \vdots \\ c_1 p_{n1} + c_2 p_{n2} + \ldots + c_n p_{nn} \end{pmatrix} = \begin{pmatrix} p_{11} & p_{12} & \cdots & p_{1n} \\ p_{21} & p_{22} & \cdots & p_{2n} \\ \vdots & \vdots & \vdots & \vdots \\ p_{n1} & p_{n2} & \cdots & p_{nn} \end{pmatrix} \begin{pmatrix} c_1 \\ c_2 \\ \vdots \\ c_n \end{pmatrix}.$$

The $n \times n$ matrix on the right, P say, effects the transformation from the original basis to the new one. Therefore, in general

$$[v]_{new} = P[v]_{original}$$

where the columns of P are the coordinates of the original basis vectors written in terms of the new ones.

Example 6.8.1

Consider the vector $v \in \mathbb{R}^2$, where in the standard basis $[v]_{\mathcal{E}} = \begin{pmatrix} 4 \\ -8 \end{pmatrix}$. Express v in terms of the basis $\overline{\mathcal{E}} = \{\overline{e}_1, \overline{e}_2\}$ where $\overline{e}_1 = \begin{pmatrix} 1 \\ 1 \end{pmatrix}$ and $\overline{e}_2 = \begin{pmatrix} -1 \\ 1 \end{pmatrix}$.

Solution

We begin by writing the original basis vectors in terms of the new ones:

$$\begin{pmatrix} 1 \\ 0 \end{pmatrix} = k_1 \overline{e}_1 + k_2 \overline{e}_2$$

$$= k_1 \begin{pmatrix} 1 \\ 1 \end{pmatrix} + k_2 \begin{pmatrix} -1 \\ 1 \end{pmatrix}$$

from which $k_1 = \frac{1}{2}$, $k_2 = -\frac{1}{2}$ so that $\begin{pmatrix} 1 \\ 0 \end{pmatrix} = \frac{1}{2}\overline{e}_1 - \frac{1}{2}\overline{e}_2$. Likewise

$$\begin{pmatrix} 0 \\ 1 \end{pmatrix} = k_1 \begin{pmatrix} 1 \\ 1 \end{pmatrix} + k_2 \begin{pmatrix} -1 \\ 1 \end{pmatrix}$$

from which $k_1 = \frac{1}{2}$, $k_2 = \frac{1}{2}$ so that $\begin{pmatrix} 0 \\ 1 \end{pmatrix} = \frac{1}{2}\overline{e}_1 + \frac{1}{2}\overline{e}_2$. Now form P in which the columns are the coordinates of the original basis vectors written in terms of the new ones:

$$P = \begin{pmatrix} \frac{1}{2} & \frac{1}{2} \\ -\frac{1}{2} & \frac{1}{2} \end{pmatrix}.$$

Pre-multiplying a vector written in the original basis by P gives the vector in the new basis:

$$[v]_{\overline{\mathcal{E}}} = P[v]_{\mathcal{E}} = \begin{pmatrix} \frac{1}{2} & \frac{1}{2} \\ -\frac{1}{2} & \frac{1}{2} \end{pmatrix} \begin{pmatrix} 4 \\ -8 \end{pmatrix} = \begin{pmatrix} -2 \\ -6 \end{pmatrix}.$$

Compare this with the method of solution in Example 6.7.2.

Example 6.8.2

Consider the vector $v \in \mathbb{R}^2$ where in the standard basis $[v]_{\mathcal{E}} = \begin{pmatrix} \frac{1}{\sqrt{2}} \\ \frac{1}{\sqrt{2}} \end{pmatrix}$. Now suppose the standard basis vectors are rotated anticlockwise through $\frac{\pi}{6}$. Recall from Section 5.13 that multiplying a column vector in $\mathbb{R}^2$ by the matrix $M = \begin{pmatrix} \cos\theta & -\sin\theta \\ \sin\theta & \cos\theta \end{pmatrix}$ effects a rotation through θ. Thus the new basis vectors are $\overline{\mathcal{E}} = \{\overline{e}_1, \overline{e}_2\}$ where

$$\overline{e}_1 = \left(\begin{array}{cc} \sqrt{3}/2 & -1/2 \\ 1/2 & \sqrt{3}/2 \end{array} \right) \left(\begin{array}{c} 1 \\ 0 \end{array} \right) = \left(\begin{array}{c} \sqrt{3}/2 \\ 1/2 \end{array} \right)$$

and

$$\overline{e}_2 = \left(\begin{array}{cc} \sqrt{3}/2 & -1/2 \\ 1/2 & \sqrt{3}/2 \end{array} \right) \left(\begin{array}{c} 0 \\ 1 \end{array} \right) = \left(\begin{array}{c} -1/2 \\ \sqrt{3}/2 \end{array} \right).$$

Writing the original basis in terms of the new basis vectors:

$$\left(\begin{array}{c} 1 \\ 0 \end{array} \right) = \frac{\sqrt{3}}{2}\overline{e}_1 - \frac{1}{2}\overline{e}_2$$

$$\left(\begin{array}{c} 0 \\ 1 \end{array} \right) = \frac{1}{2}\overline{e}_1 + \frac{\sqrt{3}}{2}\overline{e}_2.$$

Therefore, the matrix which effects the change of basis from the original to the new basis is

$$\left(\begin{array}{cc} \frac{\sqrt{3}}{2} & \frac{1}{2} \\ -\frac{1}{2} & \frac{\sqrt{3}}{2} \end{array} \right).$$

Hence, the coordinates of v where $[v]_{\mathcal{E}} = \left(\begin{array}{c} \frac{1}{\sqrt{2}} \\ \frac{1}{\sqrt{2}} \end{array} \right)$ are given by

$$[v]_{\overline{\mathcal{E}}} = \left(\begin{array}{cc} \frac{\sqrt{3}}{2} & \frac{1}{2} \\ -\frac{1}{2} & \frac{\sqrt{3}}{2} \end{array} \right) \left(\begin{array}{c} \frac{1}{\sqrt{2}} \\ \frac{1}{\sqrt{2}} \end{array} \right) = \left(\begin{array}{c} \frac{\sqrt{3}+1}{2\sqrt{2}} \\ \frac{\sqrt{3}-1}{2\sqrt{2}} \end{array} \right).$$

6.9 Orthogonal projections onto subspaces

Consider the real vector space $\mathbb{R}^3$ with the standard basis $e_1 = (1,0,0), e_2 = (0,1,0), e_3 = (0,0,1)$ as described in Example 6.7.1. The vector (x,y,z) is shown in Figure 6.2. Imagine shining a torch from above along the direction of the z axis. The shadow of the vector (x,y,z) on the xy plane is referred to as the **orthogonal projection** onto this plane. The 'projection line' from the point (x,y,z) to the point $(x,y,0)$ is perpendicular, or orthogonal, to the xy plane.

Now consider the subspace of $\mathbb{R}^3$ spanned by just $e_1 = (1,0,0)$ and $e_2 = (0,1,0)$. This is the xy plane. Observe that the projection onto this subspace has two components, one in each of the directions e_1 and e_2. Recall from Section 5.14 that the component of an arbitrary vector a in the direction of vector n is given by the scalar (inner) product $a \cdot \hat{n}$, or $\langle a, \hat{n} \rangle$. Denoting the projected vector by $P_{xy}(x,y,z)$ we have

$$\begin{aligned} P_{xy}(x,y,z) &= \langle (x,y,z), e_1 \rangle e_1 + \langle (x,y,z), e_2 \rangle e_2 \\ &= xe_1 + ye_2 \\ &= (x,y,0). \end{aligned}$$

More generally, if $\phi_0, \ldots, \phi_n$ is an orthonormal basis for an $(n+1)$-dimensional inner product space $\mathcal{H}$, any subset $\{\chi_1, \ldots, \chi_k\}$ of k elements of $\{\phi_0, \ldots, \phi_n\}$ defines a k-dimensional

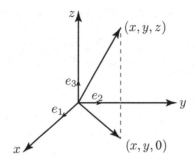

FIGURE 6.2
Orthogonal projections in $\mathbb{R}^3$.

subspace of $\mathcal{H}$ which we shall denote $\mathcal{H}_\chi$. Then, generalising the $\mathbb{R}^3$ case above, for any $\psi \in \mathcal{H}$, the orthogonal projection, $P_\chi \psi$ of ψ onto the subspace $\mathcal{H}_\chi$ is

$$P_\chi \psi = \langle \chi_1, \psi \rangle \chi_1 + \ldots + \langle \chi_k, \psi \rangle \chi_k.$$

Knowledge of such orthogonal projections is essential to an understanding of the measurement of observables (see Chapter 13).

6.10   Construction of an orthogonal basis – the Gram-Schmidt process

We have seen that it is possible to have a basis of a vector space V for which the basis vectors are not orthogonal. The Gram-Schmidt process enables construction of an orthonormal basis. We start by illustrating this process with a simple two-dimensional example and then generalise.

Example 6.10.1 The Gram-Schmidt process

Consider the basis of row vectors in $\mathbb{R}^2$ given by $v_1 = (1, 0)$, $v_2 = (2, 3)$. Firstly, you should verify that these vectors are linearly independent, that they do form a basis for $\mathbb{R}^2$ but that they are not orthogonal.

We begin the process by normalising the given vectors if they are not already normalised: so

$$\hat{v}_1 = (1, 0), \qquad \hat{v}_2 = \frac{1}{\sqrt{13}}(2, 3).$$

We shall let $\hat{u}_1$ and $\hat{u}_2$ stand for the orthonormal basis vectors that we seek. Choose $\hat{u}_1 = \hat{v}_1$.

We then subtract from $\hat{v}_2$ its projection in the direction of $\hat{u}_1$:

$$\hat{v}_2 - \langle \hat{v}_2, \hat{u}_1 \rangle \hat{u}_1 = \frac{1}{\sqrt{13}}(2,3) - \langle \frac{1}{\sqrt{13}}(2,3), (1,0) \rangle (1,0)$$

$$= \frac{1}{\sqrt{13}}(2,3) - \frac{2}{\sqrt{13}}(1,0)$$

$$= (0, \frac{3}{\sqrt{13}}).$$

The result is a vector which is orthogonal to $\hat{u}_1$ but which is not yet normalised. Finally, normalising produces $\hat{u}_2 = (0,1)$. We deduce that an orthonormal basis is $\{(1,0),(0,1)\}$. Observe that the first vector in the new basis is equal to the first vector in the given basis. (Note this new, orthonormal basis is not unique - consider the following example).

Example 6.10.2 The Gram-Schmidt process

Consider the basis of row vectors in $\mathbb{R}^2$ given by $v_1 = (2,3)$ and $v_2 = (1,0)$. We shall construct an orthonormal basis, but starting from $v_1 = (2,3)$.

$$\hat{v}_1 = \frac{1}{\sqrt{13}}(2,3), \qquad \hat{v}_2 = (1,0).$$

As before, we shall let $\hat{u}_1$, and $\hat{u}_2$ stand for the orthonormal basis vectors that we seek. Choose $\hat{u}_1 = \hat{v}_1$. We then subtract from $\hat{v}_2$ its projection in the direction of $\hat{u}_1$:

$$\hat{v}_2 - \langle \hat{v}_2, \hat{u}_1 \rangle \hat{u}_1 = (1,0) - \langle (1,0), \frac{1}{\sqrt{13}}(2,3) \rangle \frac{(2,3)}{\sqrt{13}}$$

$$= (1,0) - \frac{2}{\sqrt{13}} \frac{(2,3)}{\sqrt{13}}$$

$$= (1,0) - \frac{2}{13}(2,3)$$

$$= (\frac{9}{13}, -\frac{6}{13})$$

The result is a vector which is orthogonal to $\hat{u}_1$ but which is not yet normalised. Finally, normalising produces $\hat{u}_2 = (\frac{9}{\sqrt{117}}, -\frac{6}{\sqrt{117}})$. We deduce that an orthonormal basis is $\{\frac{1}{\sqrt{13}}(2,3), (\frac{9}{\sqrt{117}}, -\frac{6}{\sqrt{117}})\}$.

More generally, we have the following procedure.

Given a basis $\{v_1, v_2, \ldots, v_n\}$ of an n-dimensional vector space, an orthonormal basis $\{\hat{u}_1, \hat{u}_2, \ldots, \hat{u}_n\}$ is constructed as follows:

First, normalise each original basis vector to give $\{\hat{v}_1, \hat{v}_2, \ldots, \hat{v}_n\}$.

Let

$$\hat{u}_1 = \hat{v}_1.$$

Subtract from $\hat{v}_2$ its projection in the direction of $\hat{u}_1$. What remains must be orthogonal to $\hat{u}_1$: so

$$u_2 = \hat{v}_2 - \langle \hat{v}_2, \hat{u}_1 \rangle \hat{u}_1, \qquad \hat{u}_2 = \frac{u_2}{\|u_2\|}.$$

Then subtract from $\hat{v}_3$ its projections in the directions of both $\hat{u}_1$ and $\hat{u}_2$. What remains must be orthogonal to both $\hat{u}_1$ and $\hat{u}_2$:

$$u_3 = \hat{v}_3 - \langle \hat{v}_3, \hat{u}_1 \rangle \hat{u}_1 - \langle \hat{v}_3, \hat{u}_2 \rangle \hat{u}_2, \qquad \hat{u}_3 = \frac{u_3}{\|u_3\|}.$$

And continue in like fashion until $\hat{u}_n$ is calculated and the process is complete.

Exercises

6.18   Use the Gram-Schmidt process to produce an orthonormal basis of $\mathbb{R}^3$ from the basis $e_1 = \{1, 1, 0\}$, $e_2 = \{0, 1, 1\}$, $e_3 = \{2, 1, 0\}$.

6.11   The Cartesian product of vector spaces

Suppose we have two vector spaces, V and W, over the same field F. Recall that the Cartesian product $V \times W$ of the sets V and W is defined to be the set of ordered pairs (v, w), with $v \in V$, $w \in W$. For example, the product $\mathbb{C} \times \mathbb{C}$ is the set of ordered pairs of complex numbers (which we have already seen in Example 6.2.1 where $\mathbb{C} \times \mathbb{C}$ is written $\mathbb{C}^2$). Thus, for example,

$$(1 + i, 2 - 5i) \in \mathbb{C} \times \mathbb{C}.$$

We now consider how the structure of a vector space is imposed on the Cartesian product.

Definition 6.11   The Cartesian product of vector spaces
We define the operations of addition and scalar multiplication on the set $V \times W$ in a natural way:

$$(v_1, w_1) + (v_2, w_2) = (v_1 + v_2, w_1 + w_2)$$

$$k(v_1, w_1) = (kv_1, kw_1) \qquad k \in \mathbb{F}.$$

We emphasise here that, for example, $v_1 + v_2$ is the addition of the two vectors v_1 and v_2 in V and not the addition of the components of a vector v. It follows immediately that with these rules, the set $V \times W$ is a vector space over $\mathbb{F}$ as should be verified.

6.12   Equivalence classes defined on vector spaces

Example 6.12.1

Consider a vector space V and a subspace U. Choose any two vectors $v, w \in V$. Define the relation R by

$$wRv \quad \text{if} \quad w - v \in U.$$

So, vectors w and v are related if their difference lies in the subspace U. We show that R is an equivalence relation. Firstly, note that for any $w \in V$, wRw because $w - w = 0 \in U$. This is because U is a subspace and hence contains the zero element (reflexivity). Secondly, note that if wRv then $w - v \in U$. But since U is a subspace, then it contain the inverse $-(w - v) = v - w$. Hence, $v - w \in U$ and so vRw (symmetry). Finally, note that if $w_1 R w_2$ and $w_2 R w_3$ then

$$w_1 - w_2 \in U, \quad w_2 - w_3 \in U.$$

Hence, the sum of these is also in U:

$$(w_1 - w_2) + (w_2 - w_3) = w_1 - w_3 \in U$$

and so $w_1 R w_3$ (transitivity). Thus R defines an equivalence relation on V.

Recall from Chapter 1 that the set of all elements in V equivalent to a fixed element v is called the equivalence class of v, denoted $[v]$ (not to be confused with coordinate vector $[v]_{basis}$ which has a subscript). So using the vector space V, subspace U and the equivalence relation R of Example 6.12.1 we can fix $v \in V$ and calculate its equivalence class $[v]$ by writing down the set of all elements in V that are equivalent to v, that is all w for which $w - v \in U$. By definition $[v] = \{w : wRv\} = \{w : w - v \in U\}$. For any such w, we have $w - v = u \in U$ so $w = v + u$.

6.13　The sum of a vector and a subspace

Definition 6.12 The sum of a vector and a subspace; cosets

　Suppose v is a specific vector in V and that U is a subspace of V. We define the sum of v and U to be the set

$$v + U = \{v + u : u \in U\}.$$

That is, $v + U$ is the set of all vectors obtained by adding to the specific vector v all the vectors in the subspace U.

　The set of points $\{v + u : u \in U\}$ is called an **affine subset** *or* **coset***.*

Example 6.13.1

Referring to Example 6.4.1, suppose we form the set of vectors $v + U$ when $U = \{(x, y) \in \mathbb{R}^2 : y = 5x\}$ and $v = (4, 1)$ say. Thus this set includes the vectors

$$(4, 1), (5, 6), (6, 11), (7, 16), \ldots$$

As before we can depict these vectors by drawing line segments from the origin as shown in Figure 6.3. Observe that the endpoints of the vectors lie on a line parallel to the subspace U. This set of points is referred to as an **affine subset** or **coset** of U. We can repeat this process for other vectors v. The case when $v = (11, 4)$ is illustrated. Note that this second affine subset is a line parallel to the first, and to the subspace U.

Example 6.13.2

As in Example 6.13.1 suppose we form the set of vectors $v + U$ when $U = \{(x, y) \in \mathbb{R}^2 : y = 5x\}$ and $v = (-1, -5)$ say. Observe that in this example $v \in U$. Write down several vectors in this set and deduce that because $v \in U$ then $v + U = U$. Likewise, deduce that $-v + U = U$.

Solution

The subspace U contains vectors of the form $k(1, 5)$ where $k \in \mathbb{R}$. Vectors in the set $v + U$ take the form $(-1, -5) + k(1, 5)$ and hence this set includes, for example,

$$(-1, -5), (0, 0), (1, 5).$$

This set is identical to U. We deduce that when $v \in U$, $v + U = U$. It follows that $-v + U = U$.

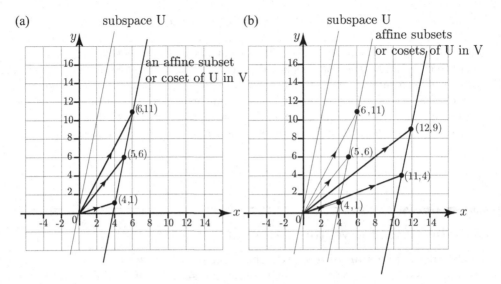

FIGURE 6.3
(a) Subspace U and an affine subset, or coset, of U in V. (b) A second coset.

In general, for any fixed $v \in V$, the set $v + U$ can be thought of as a set of points on a line passing through v and parallel to the line representing the subspace U. As in Example 6.13.1, we refer to such sets as affine subsets or cosets of U in V. Note that these subsets are not subspaces. In yet more general cases when U is any subspace of V, we say that the affine subset $v + U$ is parallel to U. That is, we can make use of the geometric terminology even in more abstract cases.

Returning to the equivalence relation in Example 6.12.1, wRv if $w - v \in U$, and the corresponding equivalence class $[v]$,

$$[v] = \{w : w - v \in U\}, \text{ that is } [v] = \{w : w = v + u \text{ for } u \in U\}$$

observe that the affine subset generated by v is its equivalence class $[v]$.

6.14   The quotient space

Definition 6.13 *Let V be a vector space and U a subspace of V. The set of all affine subsets of V that are parallel to U is called the **quotient space** V/U.*

So we think of fixing U, and let v vary throughout V:

$$V/U = \{v + U : v \in V\}$$

that is, V/U is the set of cosets of U in V. Note the subtle distinction. Here we are fixing the subspace U and varying v throughout V, in contrast to the earlier definition of the sum of a vector and a subspace. It is possible to show that these affine subsets partition V into mutually disjoint subsets. This is equivalent to saying that the set of equivalence classes $[v]$ for $v \in V$ partitions V into mutually disjoint subsets. Every vector v lies in one and only one of these subsets.

Addition and scalar multiplication can be defined on the set V/U in the following way, so that this set is a vector space. Suppose we select a subspace U, and then choose any $v, w \in V$ and $k \in F$. Then

$$v + U \in V/U, \quad w + U \in V/U.$$

Define the sum of these elements as follows:

$$(v + U) + (w + U) = (v + w) + U.$$

Define multiplication by a scalar as

$$k(v + U) = (kv) + U.$$

With these definitions, the quotient space can be shown to satisfy the axioms of a vector space. The zero of the quotient space V/U is the subspace U itself which is the coset corresponding to the zero element of V.

6.15   End-of-chapter exercises

1. Let $|a\rangle$ and $|b\rangle$ be orthogonal unit vectors in a real inner product space. Consider the linear combination $|\psi\rangle = \alpha|a\rangle + \beta|b\rangle$ where $\alpha, \beta \in \mathbb{R}$. Show that $|\psi\rangle$ is a unit vector if $\alpha^2 + \beta^2 = 1$.

2. Let $|a\rangle$ and $|b\rangle$ be orthogonal unit vectors in a complex inner product space. Consider the linear combination $|\psi\rangle = \alpha|a\rangle + \beta|b\rangle$ where $\alpha, \beta \in \mathbb{C}$. Show that $|\psi\rangle$ is a unit vector if $|\alpha|^2 + |\beta|^2 = 1$.

3. Consider the set, P_1, of all polynomials in t of degree 0 or 1.

   (a) Show that this set with the usual operations of addition and multiplication by a scalar is a vector space over $\mathbb{R}$.
   (b) Show that the set $\{e_0, e_1\} = \{1, t\}$ is a basis for this vector space (the standard basis).
   (c) Show that the set $\{\bar{e}_0, \bar{e}_1\} = \{1, 1 - t\}$ is an alternative basis.
   (d) Express the polynomial $p = 5 + 2t$ in terms of the alternative basis in part (c).

4. Consider the vector space over $\mathbb{R}$, P_2 say, of all polynomials in t of degree 2 or less. Show that the set $\{f_0, f_1, f_2\} = \{1, t, t^2\}$ is a basis for this vector space.

5. Consider the following set of vectors in $\mathbb{R}^4$:

$$u = \begin{pmatrix} 1 \\ 2 \\ 0 \\ 1 \end{pmatrix}, \quad v = \begin{pmatrix} 2 \\ 2 \\ 0 \\ 1 \end{pmatrix}, \quad w = \begin{pmatrix} 1 \\ 1 \\ 1 \\ 0 \end{pmatrix}.$$

Determine whether the vector $r = \begin{pmatrix} 8 \\ 11 \\ 1 \\ 5 \end{pmatrix}$ is an element of the space spanned

by u, v and w.

6. Consider the set

$$\left\{\frac{1}{\sqrt{3}}(1,1,1), \frac{1}{\sqrt{3}}(1, e^{2\pi i/3}, e^{4\pi i/3}), \frac{1}{\sqrt{3}}(1, e^{4\pi i/3}, e^{8\pi i/3})\right\}.$$

(a) Show that the set is orthonormal.

(b) Show that any vector in $\mathbb{C}^3$ can be expressed in terms of vectors from this set and deduce that the set is a basis for $\mathbb{C}^3$.

7. Consider $\mathbb{C}^2$, the set of ordered pairs of complex numbers (z_1, z_2) where $z_1, z_2 \in \mathbb{C}$. With addition and scalar multiplication defined by

$$(z_1, z_2) + (w_1, w_2) = (z_1 + w_1, z_2, +w_2), \quad \alpha(z_1, z_2) = (\alpha z_1, \alpha z_2)$$

where $z_1, z_2, w_1, w_2, \alpha \in \mathbb{C}$, show that $\mathbb{C}^2$ satisfies the axioms for a vector space over $\mathbb{C}$.

8. Consider the vector space $\mathbb{C}^2$ over $\mathbb{C}$. Show that the set

$$\{(i, 0), (0, i)\}$$

is a basis and express $z = (z_1, z_2)$ in terms of this basis.

9. Consider $\mathbb{C}^2$, the set of ordered pairs of complex numbers (z_1, z_2) where $z_1, z_2 \in \mathbb{C}$. As a vector space over the set of real numbers $\mathbb{R}$ this is a four-dimensional vector space. Show that the set

$$\{(1, 0), (0, 1), (i, 0), (0, i)\}$$

is a basis, and express the vector $z = (3 + 4i, -7 - 2i)$ in terms of this basis.

7

Eigenvalues and eigenvectors of a matrix

7.1 Objectives

We have already pointed out that in quantum computation vectors are used to represent the state of a system. Matrices are used in the modelling of the measurement of observables and to move a system from one state to another by acting on state vectors. Measured values of observables are permitted to take on only certain special values, and these values are related to the eigenvalues of the measurement matrix. Post-measurement the system is in a state represented by an eigenvector or eigenvectors. The objective of this chapter is to introduce these essential topics and to show how eigenvalues and eigenvectors are calculated. In quantum computation it is relevant to consider cases in which the eigenvalues are all real and the eigenvectors are orthogonal. We discuss the conditions under which this happens.

7.2 Preliminary definitions

Given a square $n \times n$ matrix A with elements from a field $\mathbb{F}$ (real or complex numbers), we have seen that we can find the product of A and an $n \times 1$ matrix (a column vector), v, to obtain Av, which will also be an $n \times 1$ matrix (a column vector). For any given $n \times n$ matrix A, we now consider particular sets of non-zero vectors v which have the special property that $Av = \lambda v$ where λ is a scalar in the field $\mathbb{F}$. That is, the result of multiplying the vector v by A is a scalar multiple of v. (To be of any interest, we will require the vectors v to be non-zero vectors since the equation $Av = \lambda v$ is always true if $v = 0$.)

Definition 7.1 Eigenvalues and eigenvectors of a matrix
*If A is an $n \times n$ matrix, a non-zero vector v is an **eigenvector** of A if*

$$Av = \lambda v$$

*for some scalar λ called an **eigenvalue** of A.*

Geometrically, if we were working in $\mathbb{R}^2$ or $\mathbb{R}^3$ this means that multiplication by A results in a vector which is in the same direction as v if $\lambda > 0$ and is in the opposite direction to v if $\lambda < 0$. Such vectors are called **eigenvectors** of A corresponding to the **eigenvalue** λ. The vector – scalar pairs v, λ are significant in the study of quantum computation because quantum states can be expressed in terms of eigenvectors, and eigenvalues are related to the possible measured values of observables.

DOI: 10.1201/9781003264569-7

7.3   Calculation of eigenvalues

We begin by discussing how to calculate the eigenvalues of a matrix A. Recall that if I is the $n \times n$ identity matrix then $Iv = v$. We can then write $Av = \lambda v$ as $Av = \lambda Iv$ from which

$$Av - \lambda Iv = 0$$

or

$$(A - \lambda I)v = 0.$$

Here 0 represents a column vector of zeros, i.e., the zero vector. This equation represents a set of n linear homogeneous equations. To see this, consider the following example.

Example 7.3.1

If $A = \begin{pmatrix} 1 & 4 \\ 2 & 3 \end{pmatrix}$ and $v = \begin{pmatrix} v_1 \\ v_2 \end{pmatrix}$, write down the set of linear homogeneous equations corresponding to $(A - \lambda I)v = 0$.

Solution

$$
\begin{aligned}
A - \lambda I &= \begin{pmatrix} 1 & 4 \\ 2 & 3 \end{pmatrix} - \lambda \begin{pmatrix} 1 & 0 \\ 0 & 1 \end{pmatrix} \\
&= \begin{pmatrix} 1 & 4 \\ 2 & 3 \end{pmatrix} - \begin{pmatrix} \lambda & 0 \\ 0 & \lambda \end{pmatrix} \\
&= \begin{pmatrix} 1 - \lambda & 4 \\ 2 & 3 - \lambda \end{pmatrix}.
\end{aligned}
$$

Then $(A - \lambda I)v = 0$ becomes

$$\begin{pmatrix} 1 - \lambda & 4 \\ 2 & 3 - \lambda \end{pmatrix} \begin{pmatrix} v_1 \\ v_2 \end{pmatrix} = \begin{pmatrix} 0 \\ 0 \end{pmatrix}$$

and writing out in full

$$
\begin{aligned}
(1 - \lambda)v_1 + 4v_2 &= 0 \\
2v_1 + (3 - \lambda)v_2 &= 0
\end{aligned}
$$

which is the required set of linear homogeneous equations in the unknowns v_1 and v_2.

Returning to the more general equation $(A - \lambda I)v = 0$, non-trivial solutions exist only if v is in the null space of $A - \lambda I$. Recall from Section 6.5 that the null space is the set of all vectors which map to the zero vector. Equivalently, from Section 5.12, $A - \lambda I$ must be singular, that is

$$\det(A - \lambda I) = |A - \lambda I| = 0.$$

Definition 7.2 Eigenvalues and the characteristic equation
The eigenvalues, λ, of a square $n \times n$ matrix A are found by solving the **characteristic equation**

$$\det(A - \lambda I) = |A - \lambda I| = 0.$$

As we shall see in the following examples, this is a polynomial equation of degree n. It has n roots (some of which may be complex) and hence n eigenvalues (some of which may be repeated).

Definition 7.3 Algebraic multiplicity
The number of times a particular eigenvalue occurs in the solution of the characteristic equation is called its algebraic multiplicity.

Definition 7.4 Spectrum
The set of eigenvalues of A, $\{\lambda_i\}$, is referred to as the spectrum of A.

Consider the following examples.

Example 7.3.2

Show that the matrix $A = \begin{pmatrix} 1 & 4 \\ 2 & 3 \end{pmatrix}$ has two real distinct eigenvalues.

Solution

The characteristic equation is

$$|A - \lambda I| = \left| \begin{pmatrix} 1 & 4 \\ 2 & 3 \end{pmatrix} - \lambda \begin{pmatrix} 1 & 0 \\ 0 & 1 \end{pmatrix} \right| = \left| \begin{matrix} 1 - \lambda & 4 \\ 2 & 3 - \lambda \end{matrix} \right| = 0,$$

that is

$$(1 - \lambda)(3 - \lambda) - 8 = \lambda^2 - 4\lambda - 5 = 0.$$

Factorising,

$$(\lambda - 5)(\lambda + 1) = 0$$
$$\lambda = 5, -1.$$

Thus there are two real eigenvalues: 5 and -1. The spectrum of A is $\{5, -1\}$.

Example 7.3.3

Show that the matrix $A = \begin{pmatrix} 1 & -1 \\ 1 & 1 \end{pmatrix}$ has two distinct complex eigenvalues.

Solution

The characteristic equation is

$$|A - \lambda I| = \left| \begin{pmatrix} 1 & -1 \\ 1 & 1 \end{pmatrix} - \lambda \begin{pmatrix} 1 & 0 \\ 0 & 1 \end{pmatrix} \right| = \left| \begin{matrix} 1 - \lambda & -1 \\ 1 & 1 - \lambda \end{matrix} \right| = 0,$$

that is

$$(1 - \lambda)(1 - \lambda) + 1 = \lambda^2 - 2\lambda + 2 = 0.$$

Then
$$\lambda = \frac{-(-2) \pm \sqrt{(-2)^2 - 4(1)(2)}}{2} = \frac{2 \pm \sqrt{-4}}{2} = 1 \pm i.$$

The characteristic equation has complex roots $1 + i$, $1 - i$, and thus the eigenvalues are $1 + i$ and $1 - i$. The observation that the eigenvalues are not real is important. As we shall see, because eigenvalues are related to the measured values of observables, in quantum computation they must be real.

As a further example, it is straightforward to verify that the matrix $A = \begin{pmatrix} 4 & 1 \\ -1 & 2 \end{pmatrix}$ has a repeated real eigenvalue $\lambda = 3$. Thus the algebraic multiplicity of this eigenvalue is 2.

Exercises

7.1   Find the eigenvalues of $\begin{pmatrix} 1 & 0 \\ 0 & 0 \end{pmatrix}$. Write down the spectrum and comment on algebraic multiplicity.

7.2   If A is a real symmetric 2×2 matrix, prove that the eigenvalues of A are real.

7.3   Show that if D is a diagonal matrix, then the elements on the diagonal are the eigenvalues of D.

7.4   An upper triangular matrix U is a square matrix for which all elements below the leading diagonal are zero. Show that for such a 2×2 or 3×3 matrix, the elements on the diagonal are the eigenvalues of U. This result generalises to any $n \times n$ upper triangular matrix.

7.4   Calculation of eigenvectors

For each eigenvalue, λ, of the matrix A we calculate the corresponding eigenvector or eigenvectors, v, using the equation $Av = \lambda v$. We shall see that once we have calculated an eigenvector then any multiple of that vector is also an eigenvector: that is, eigenvectors are only defined up to a constant multiple. Thus we can think of an eigenvector as defining a direction - the **eigendirection**. Often we will **normalise** the eigenvector so that its magnitude is 1. We draw attention to some important characteristics of eigenvalues and eigenvectors which are explored in the examples which follow:

1.   An $n \times n$ matrix may have n distinct eigenvalues and (as a consequence) n linearly independent eigenvectors.

2.   On the other hand, some eigenvalues may be repeated, i.e., an $n \times n$ matrix may not have n distinct eigenvalues. However, it is still possible for the matrix to possess n linearly independent eigenvectors.

3.   When some eigenvalues are repeated, it is possible that a full set of n linearly independent eigenvectors does not exist.

4.   We shall also find that in some cases, the eigenvectors are mutually orthogonal.

Example 7.4.1 Matrix A possesses distinct eigenvalues and a full set of eigenvectors

In Example 7.3.2, we showed that the matrix $A = \begin{pmatrix} 1 & 4 \\ 2 & 3 \end{pmatrix}$ has eigenvalues $\lambda = 5, -1$. Find an eigenvector corresponding to each of these values. Show that these eigenvectors are linearly independent and hence that a full set ($n = 2$) of eigenvectors exists. Show that the eigenvectors are not orthogonal.

Solution

First consider $\lambda = 5$. Writing $v = \begin{pmatrix} v_1 \\ v_2 \end{pmatrix}$ then $Av = 5v$ implies

$$\begin{pmatrix} 1 & 4 \\ 2 & 3 \end{pmatrix} \begin{pmatrix} v_1 \\ v_2 \end{pmatrix} = 5 \begin{pmatrix} v_1 \\ v_2 \end{pmatrix}.$$

Writing out these equations explicitly and rearranging we obtain the homogeneous set

$$-4v_1 + 4v_2 = 0$$
$$2v_1 - 2v_2 = 0.$$

These equations can be solved using Gaussian elimination, but note that both are equivalent to $-v_1 + v_2 = 0$, i.e., $v_2 = v_1$. Thus there is an infinite number of solutions each being a multiple of $\begin{pmatrix} 1 \\ 1 \end{pmatrix}$. Equivalently, there is an infinite number of eigenvectors corresponding to eigenvalue $\lambda = 5$ and each is a multiple of $\begin{pmatrix} 1 \\ 1 \end{pmatrix}$. We can explicitly show this by writing $v = \mu \begin{pmatrix} 1 \\ 1 \end{pmatrix}$ where μ is any real non-zero number we wish, referred to as a free variable. Often we will require the eigenvector to be normalised, in which case $v = \frac{1}{\sqrt{2}} \begin{pmatrix} 1 \\ 1 \end{pmatrix} = \begin{pmatrix} 1/\sqrt{2} \\ 1/\sqrt{2} \end{pmatrix}$.

To find the eigenvectors corresponding to $\lambda = -1$, we proceed in a similar way by solving $Av = -v$. You should verify that a suitably normalised eigenvector is $v = \begin{pmatrix} 2/\sqrt{5} \\ -1/\sqrt{5} \end{pmatrix}$. It is straightforward to verify that the two eigenvectors $\begin{pmatrix} 1/\sqrt{2} \\ 1/\sqrt{2} \end{pmatrix}$ and $\begin{pmatrix} 2/\sqrt{5} \\ -1/\sqrt{5} \end{pmatrix}$ are linearly independent. This means that one cannot be written as a multiple of the other. Finally, note that taking the inner product

$$\left\langle \begin{pmatrix} 1/\sqrt{2} \\ 1/\sqrt{2} \end{pmatrix}, \begin{pmatrix} 2/\sqrt{5} \\ -1/\sqrt{5} \end{pmatrix} \right\rangle = \frac{2}{\sqrt{10}} - \frac{1}{\sqrt{10}} = \frac{1}{\sqrt{10}} \neq 0$$

and so the eigenvectors are not orthogonal.

Example 7.4.2 Matrix A possesses distinct eigenvalues and a full set of orthogonal eigenvectors

Show that the eigenvalues of the matrix $A = \begin{pmatrix} 0 & 0 \\ 0 & 1 \end{pmatrix}$ are real and distinct. Find the corresponding eigenvectors and show that these are linearly independent. Deduce that a full set of eigenvectors exists. Show that eigenvectors corresponding to different eigenvalues are orthogonal.

Solution

As previously, the eigenvalues are found by solving $\det(A - \lambda I) = 0$, that is

$$\begin{vmatrix} -\lambda & 0 \\ 0 & 1 - \lambda \end{vmatrix} = \lambda^2 - \lambda = \lambda(\lambda - 1) = 0.$$

Thus the eigenvalues are $\lambda = 0$ and 1. For each eigenvalue in turn we solve $Av = \lambda v$ to determine the eigenvectors v. For $\lambda = 0$, you should verify that a corresponding eigenvector is $\begin{pmatrix} 1 \\ 0 \end{pmatrix}$. For $\lambda = 1$, a corresponding eigenvector is $\begin{pmatrix} 0 \\ 1 \end{pmatrix}$. The two vectors $\begin{pmatrix} 1 \\ 0 \end{pmatrix}$ and $\begin{pmatrix} 0 \\ 1 \end{pmatrix}$ are linearly independent – clearly, one cannot be written as a multiple of the other. Further, taking the inner product,

$$\left\langle \begin{pmatrix} 1 \\ 0 \end{pmatrix}, \begin{pmatrix} 0 \\ 1 \end{pmatrix} \right\rangle = (1)(0) + (0)(1) = 0$$

and so the eigenvectors are orthogonal.

Note that in the previous two examples, the matrices each had two distinct eigenvalues and their respective eigenvectors were linearly independent though not necessarily orthogonal. This result is generalised in the following theorem which we state without proof.

Theorem 7.1 *If an $n \times n$ matrix A has n distinct eigenvalues, then A has a full set of n linearly independent eigenvectors.*

Now consider the following two examples where the eigenvalues are not distinct.

Example 7.4.3 Matrix A has a repeated eigenvalue and only one independent eigenvector

Show that the matrix $A = \begin{pmatrix} -3 & 4 \\ -4 & 5 \end{pmatrix}$ has a repeated real eigenvalue and only one independent eigenvector.

Solution

The eigenvalues are found from

$$\begin{vmatrix} -3 - \lambda & 4 \\ -4 & 5 - \lambda \end{vmatrix} = \lambda^2 - 2\lambda + 1 = (\lambda - 1)^2 = 0$$

and hence, there is a single repeated real eigenvalue $\lambda = 1$. To find any eigenvectors, we solve

$$\begin{pmatrix} -3 & 4 \\ -4 & 5 \end{pmatrix} \begin{pmatrix} v_1 \\ v_2 \end{pmatrix} = 1 \begin{pmatrix} v_1 \\ v_2 \end{pmatrix}$$

that is

$$-4v_1 + 4v_2 = 0$$
$$-4v_1 + 4v_2 = 0$$

that is $v_1 = v_2$, and so there is just one independent (normalised) eigenvector $\frac{1}{\sqrt{2}} \begin{pmatrix} 1 \\ 1 \end{pmatrix}$.
The matrix A does not have a full set of eigenvectors.

Example 7.4.4 Matrix A has a repeated eigenvalue and two independent eigenvectors

Show that the matrix $A = \begin{pmatrix} 3 & 0 \\ 0 & 3 \end{pmatrix}$ has a repeated real eigenvalue and two independent orthogonal eigenvectors.

Solution

The eigenvalues are found from

$$\begin{vmatrix} 3 - \lambda & 0 \\ 0 & 3 - \lambda \end{vmatrix} = (3 - \lambda)^2 = 0$$

and hence, there is a single repeated real eigenvalue $\lambda = 3$. To find any eigenvectors, we solve

$$\begin{pmatrix} 3 & 0 \\ 0 & 3 \end{pmatrix} \begin{pmatrix} v_1 \\ v_2 \end{pmatrix} = 3 \begin{pmatrix} v_1 \\ v_2 \end{pmatrix}$$

that is

$$3v_1 = 3v_1$$
$$3v_2 = 3v_2.$$

These two equations are satisfied by any v_1 and v_2. Thus there are two independent eigenvectors,

$$\begin{pmatrix} 1 \\ 0 \end{pmatrix}, \quad \text{and} \quad \begin{pmatrix} 0 \\ 1 \end{pmatrix},$$

(for example). It is clear that these two eigenvectors are orthogonal. Further, it is readily verified that any linear combination of the two eigenvectors corresponding to $\lambda = 3$ is also an eigenvector.

Observe that in Example 7.4.3 there was a single eigenvalue and a single corresponding eigenvector. On the other hand in Example 7.4.4 there was a single eigenvalue and two independent eigenvectors. We now draw out some general results.

For an $n \times n$ matrix, for each eigenvalue, λ, we can calculate corresponding eigenvectors. It is possible for there to be up to n independent eigenvectors associated with each eigenvalue.

Definition 7.5 Geometric multiplicity
*Suppose λ is an eigenvalue of matrix A. For each λ, the number of linearly independent eigenvectors is called the **geometric multiplicity** of λ.*

For a specific λ its eigenvectors, together with the zero vector, form a vector space called the **eigenspace** of A corresponding to λ. When a one-dimensional eigenspace is generated by an eigenvalue then that eigenvalue is said to be **non-degenerate** (e.g. both eigenvalues in Example 7.4.1). On the other hand when the eigenspace generated by an eigenvalue has dimension greater than 1 the eigenvalue is said to be **degenerate** (see Example 7.4.4

in which there is a two-dimensional eigenspace generated by $\lambda = 3$). Any set of linearly independent vectors which span an eigenspace is said to be a **basis** for that eigenspace.

Example 7.4.5 Repeated eigenvalues, degeneracy and a full set of eigenvectors

Find the eigenvalues and corresponding eigenvectors of the matrix

$$A = \begin{pmatrix} 0 & 0 & 0 & 0 \\ 0 & 1 & 0 & 0 \\ 0 & 0 & 0 & 0 \\ 0 & 0 & 0 & 1 \end{pmatrix}.$$

Solution

We solve $|A - \lambda I| = 0$:

$$
\begin{aligned}
|A - \lambda I| &= \begin{vmatrix} -\lambda & 0 & 0 & 0 \\ 0 & 1-\lambda & 0 & 0 \\ 0 & 0 & -\lambda & 0 \\ 0 & 0 & 0 & 1-\lambda \end{vmatrix} \\
&= -\lambda \begin{vmatrix} 1-\lambda & 0 & 0 \\ 0 & -\lambda & 0 \\ 0 & 0 & 1-\lambda \end{vmatrix} \\
&= -\lambda(1-\lambda) \begin{vmatrix} -\lambda & 0 \\ 0 & 1-\lambda \end{vmatrix} \\
&= -\lambda(1-\lambda)(-\lambda)(1-\lambda) \\
&= \lambda^2(1-\lambda)^2.
\end{aligned}
$$

Solving $\lambda^2(1-\lambda)^2 = 0$ gives the eigenvalues $\lambda = 0$, $\lambda = 1$, both with algebraic multiplicity 2.

$\underline{\lambda = 0}$: we solve $Av = 0v = 0$. Thus

$$\begin{pmatrix} 0 & 0 & 0 & 0 \\ 0 & 1 & 0 & 0 \\ 0 & 0 & 0 & 0 \\ 0 & 0 & 0 & 1 \end{pmatrix} \begin{pmatrix} v_1 \\ v_2 \\ v_3 \\ v_4 \end{pmatrix} = \begin{pmatrix} 0 \\ 0 \\ 0 \\ 0 \end{pmatrix}.$$

It follows immediately that $v_2 = v_4 = 0$ whilst v_1 and v_3 are free variables. Thus there are two independent (and normalised) eigenvectors corresponding to $\lambda = 0$:

$$\begin{pmatrix} 1 \\ 0 \\ 0 \\ 0 \end{pmatrix}, \quad \begin{pmatrix} 0 \\ 0 \\ 1 \\ 0 \end{pmatrix}$$

and thus the eigenvalue $\lambda = 0$ is degenerate.

$\underline{\lambda = 1}$: we solve $Av = v$.

$$\begin{pmatrix} 0 & 0 & 0 & 0 \\ 0 & 1 & 0 & 0 \\ 0 & 0 & 0 & 0 \\ 0 & 0 & 0 & 1 \end{pmatrix} \begin{pmatrix} v_1 \\ v_2 \\ v_3 \\ v_4 \end{pmatrix} = \begin{pmatrix} v_1 \\ v_2 \\ v_3 \\ v_4 \end{pmatrix}.$$

Simplifying:

$$\begin{pmatrix} -1 & 0 & 0 & 0 \\ 0 & 0 & 0 & 0 \\ 0 & 0 & -1 & 0 \\ 0 & 0 & 0 & 0 \end{pmatrix} \begin{pmatrix} v_1 \\ v_2 \\ v_3 \\ v_4 \end{pmatrix} = \begin{pmatrix} 0 \\ 0 \\ 0 \\ 0 \end{pmatrix}.$$

Then $v_1 = v_3 = 0$ whilst v_2 and v_4 are free variables. Thus there are two independent (and normalised) eigenvectors corresponding to $\lambda = 1$:

$$\begin{pmatrix} 0 \\ 1 \\ 0 \\ 0 \end{pmatrix}, \quad \begin{pmatrix} 0 \\ 0 \\ 0 \\ 1 \end{pmatrix}$$

and thus the eigenvalue $\lambda = 1$ is degenerate. Clearly the geometric multiplicity of each eigenvalue is 2. There are thus four independent eigenvectors. We shall meet these again later in the context of 2-qubit quantum states.

Example 7.4.6 Repeated eigenvalue

Find the eigenvalues and eigenvectors of $A = \begin{pmatrix} 5 & -1 & -1 \\ 2 & 2 & -2 \\ 1 & -1 & 3 \end{pmatrix}$.

Solution

Consider $Av = \lambda v$. The eigenvalues are calculated from

$$|A - \lambda I = \begin{vmatrix} 5 - \lambda & -1 & -1 \\ 2 & 2 - \lambda & -2 \\ 1 & -1 & 3 - \lambda \end{vmatrix}$$

$$= (5 - \lambda) \begin{vmatrix} 2 - \lambda & -2 \\ -1 & 3 - \lambda \end{vmatrix} + 1 \begin{vmatrix} 2 & -2 \\ 1 & 3 - \lambda \end{vmatrix} - 1 \begin{vmatrix} 2 & 2 - \lambda \\ 1 & -1 \end{vmatrix}$$

$$= (5 - \lambda)(\lambda^2 - 5\lambda + 4) + 1(8 - 2\lambda) - 1(-4 + \lambda)$$

$$= -\lambda^3 + 10\lambda^2 - 32\lambda + 32 = 0.$$

The cubic equation can be solved either by drawing a graph and searching for the roots or using online or computer software. We find the three eigenvalues are $\lambda = 2, 4, 4$. Now let

$$v = \begin{pmatrix} v_1 \\ v_2 \\ v_3 \end{pmatrix}$$

and consider again $Av = \lambda v$.

$\underline{\lambda = 2.}$

$$3v_1 - v_2 - v_3 = 0$$
$$2v_1 \quad\quad - 2v_3 = 0$$
$$v_1 - v_2 + v_3 = 0.$$

We solve these equations using Gaussian elimination. The augmented matrix is

$$\begin{pmatrix} 3 & -1 & -1 & 0 \\ 2 & 0 & -2 & 0 \\ 1 & -1 & 1 & 0 \end{pmatrix}.$$

Performing the row operations $R_2 \to 3R_2 - 2R_1$, and $R_3 \to 3R_3 - R_1$ produces

$$\begin{pmatrix} 3 & -1 & -1 & 0 \\ 0 & 2 & -4 & 0 \\ 0 & -2 & 4 & 0 \end{pmatrix}.$$

Then $R_3 \to R_3 + R_2$ gives

$$\begin{pmatrix} 3 & -1 & -1 & 0 \\ 0 & 2 & -4 & 0 \\ 0 & 0 & 0 & 0 \end{pmatrix}.$$

Let $v_3 = \mu$, then by back substitution $v_2 = 2\mu$ and $v_1 = \mu$. Thus the eigenvectors corresponding to $\lambda = 2$ have the form $v = \mu \begin{pmatrix} 1 \\ 2 \\ 1 \end{pmatrix}$. Observe that $\lambda = 2$ is non-degenerate.

$\underline{\lambda = 4.}$ This gives rise to the following linear simultaneous equations:

$$v_1 - v_2 - v_3 = 0$$
$$2v_1 - 2v_2 - 2v_3 = 0$$
$$v_1 - v_2 - v_3 = 0$$

all three of which are equivalent. We need only consider $v_1 - v_2 - v_3 = 0$. Let $v_3 = \mu$ and $v_2 = \nu$ from which $v_1 = \mu + \nu$. We can formally write down the eigenvectors as

$$v = \mu \begin{pmatrix} 1 \\ 0 \\ 1 \end{pmatrix} + \nu \begin{pmatrix} 1 \\ 1 \\ 0 \end{pmatrix}.$$

Observe that $\lambda = 4$ is degenerate because its eigenspace has dimension greater than 1.

When calculating eigenvalues and eigenvectors, the foregoing examples make clear that a variety of scenarios are possible; these are associated with multiplicity, linear independence and orthogonality. In the study of quantum computation, the matrices with which we usually work have properties that enable us to be quite specific about the nature of the eigenvalues and eigenvectors. We explore these properties in the following sections.

Exercises

7.5   Find the eigenvalues and eigenvectors of $A = \begin{pmatrix} 1 & 3 \\ 4 & -1 \end{pmatrix}$.

7.6   Find the eigenvalues and eigenvectors of $B = \begin{pmatrix} 4 & -1 & 1 \\ -2 & 4 & 0 \\ -4 & 3 & 1 \end{pmatrix}$.

7.5   Real symmetric matrices and their eigenvalues and eigenvectors

The eigenvalues and eigenvectors of real symmetric matrices have important properties which are highly relevant to the study of quantum computation. We now state these properties prior to discussing and giving examples. We emphasise that attention here is restricted to symmetric matrices with real elements. Symmetric matrices with complex elements do exist but are beyond the scope of this discussion. Given any real symmetric $n \times n$ matrix A:

1. The eigenvalues of A are always real.

2. Eigenvectors arising from distinct eigenvalues are orthogonal.

3. Eigenvectors arising from the same eigenvalue may not be orthogonal but using them it is possible to generate an orthogonal set.

4. The matrix A possesses an orthonormal set of n linearly independent eigenvectors which can be used as a basis for the vector space $\mathbb{C}^n$ over the field $\mathbb{C}$.

5. It is possible to perform a transformation on A to represent it as a diagonal matrix D in a process called orthogonal diagonalisation. Specifically, $D = P^T A P$ where P is an orthogonal matrix constructed using the n orthonormal eigenvectors.

These important points are illustrated in the examples which follow this theorem.

Theorem 7.2 *The eigenvalues of any 2×2 real symmetric matrix are real.*

Proof

Consider the symmetric matrix $A = \begin{pmatrix} a & b \\ b & d \end{pmatrix}$ where $a, b, d \in \mathbb{R}$. Its eigenvalues are found by solving the quadratic characteristic equation:

$$\lambda^2 - (a + d)\lambda + ad - b^2 = 0.$$

Consider the discriminant $\Delta = (a + d)^2 - 4(ad - b^2)$. Rearranging

$$\Delta = a^2 + d^2 + 2ad - 4ad + 4b^2$$
$$= (a - d)^2 + 4b^2.$$

As the sum of two squared real numbers, $\Delta \geq 0$ and so the roots of the quadratic equation, and hence the eigenvalues, are real. Further, for this 2×2 matrix, the roots and hence the eigenvalues are equal if and only if $\Delta = 0$, that is $a = d$ and $b = 0$. In this case the matrix has the form $A = \begin{pmatrix} a & 0 \\ 0 & a \end{pmatrix}$, that is, a multiple of the identity matrix.

∎

Example 7.5.1 Real and distinct eigenvalues

Show that the eigenvalues of the symmetric matrix $A = \begin{pmatrix} 2 & 3 \\ 3 & -2 \end{pmatrix}$ are real and find the corresponding eigenvectors. Show that these eigenvectors are orthogonal.

Solution

The eigenvalues are found by solving $|A - \lambda I| = 0$, that is

$$\begin{vmatrix} 2 - \lambda & 3 \\ 3 & -2 - \lambda \end{vmatrix} = \lambda^2 - 13 = 0.$$

So there are two distinct real eigenvalues $\lambda = \pm\sqrt{13}$.

For $\lambda = \sqrt{13}$ its eigenvectors are found by solving $Av = \sqrt{13}v$:

$$\begin{pmatrix} 2 & 3 \\ 3 & -2 \end{pmatrix} \begin{pmatrix} v_1 \\ v_2 \end{pmatrix} = \sqrt{13} \begin{pmatrix} v_1 \\ v_2 \end{pmatrix}$$

$$(2 - \sqrt{13})v_1 + 3v_2 = 0$$
$$3v_1 - (2 + \sqrt{13})v_2 = 0.$$

From which we can take an eigenvector, $v^{(\sqrt{13})}$ say, to be any multiple of $\begin{pmatrix} \frac{2+\sqrt{13}}{3} \\ 1 \end{pmatrix}$.

Likewise eigenvectors corresponding to $\lambda = -\sqrt{13}$, $v^{(-\sqrt{13})}$ say, are multiples of $\begin{pmatrix} \frac{2-\sqrt{13}}{3} \\ 1 \end{pmatrix}$.

If we take the inner product (scalar product) of the two eigenvectors we find

$$\langle v^{(\sqrt{13})}, v^{-(\sqrt{13})} \rangle = \begin{pmatrix} \frac{2+\sqrt{13}}{3} \\ 1 \end{pmatrix} \cdot \begin{pmatrix} \frac{2-\sqrt{13}}{3} \\ 1 \end{pmatrix} = \frac{4-13}{9} + 1 = -1 + 1 = 0$$

and thus the eigenvectors corresponding to the two distinct eigenvalues are orthogonal. When necessary these vectors can be normalised. Moreover, these two vectors can be used to construct an orthogonal basis for $\mathbb{C}^2$. This means that they can be used as building blocks for any vector in $\mathbb{C}^2$.

Example 7.5.2 A repeated eigenvalue

a) Show that the eigenvalues of the symmetric matrix $A = \begin{pmatrix} 3 & 0 \\ 0 & 3 \end{pmatrix}$ are real.
b) Write down a pair of independent eigenvectors which are not orthogonal.
c) Write down a pair of independent eigenvectors which are orthogonal.

Solution

a) The eigenvalues are found by solving $|A - \lambda I| = 0$, that is

$$\begin{vmatrix} 3-\lambda & 0 \\ 0 & 3-\lambda \end{vmatrix} = \lambda^2 - 6\lambda + 9 = (\lambda - 3)^2 = 0.$$

So $\lambda = 3$ (twice). That is, the eigenvalue $\lambda = 3$ has algebraic multiplicity 2.
b) The eigenvectors are found by solving $Av = 3v$:

$$\begin{pmatrix} 3 & 0 \\ 0 & 3 \end{pmatrix} \begin{pmatrix} v_1 \\ v_2 \end{pmatrix} = 3 \begin{pmatrix} v_1 \\ v_2 \end{pmatrix}$$

which reduces to simply $0v_1 + 0v_2 = 0$ so that both v_1 and v_2 are free variables μ_1 and μ_2 say. The most general form of an eigenvector is then $v = \begin{pmatrix} \mu_1 \\ \mu_2 \end{pmatrix}$. Suppose $\mu_1 = 1, \mu_2 = 1$, then $v = \begin{pmatrix} 1 \\ 1 \end{pmatrix}$ is an eigenvector. Alternatively, suppose $\mu_1 = 1, \mu_2 = 0$, then $v = \begin{pmatrix} 1 \\ 0 \end{pmatrix}$ is also an eigenvector. It is easily verified that $\begin{pmatrix} 1 \\ 1 \end{pmatrix}$ and $\begin{pmatrix} 1 \\ 0 \end{pmatrix}$ are linearly independent. They are not orthogonal since their inner product is $1 \neq 0$.

c) Suppose $\mu_1 = 1, \mu_2 = 0$, then $v = \begin{pmatrix} 1 \\ 0 \end{pmatrix}$ is an eigenvector. Alternatively, suppose $\mu_1 = 0, \mu_2 = 1$, then $v = \begin{pmatrix} 0 \\ 1 \end{pmatrix}$ is also an eigenvector. These eigenvectors are clearly independent, orthogonal and have norm 1. We observe that even when there is a single (repeated) eigenvalue there are two independent eigenvectors and these can be chosen to be orthonormal.

The behaviour illustrated in this example generalises to larger $n \times n$ symmetric matrices which always have real eigenvalues and possess an orthonormal set of n linearly independent eigenvectors. These can be used as a basis for the vector space $\mathbb{C}^n$ over the field $\mathbb{C}$.

Example 7.5.3 Repeated eigenvalues: generating an orthonormal set of eigenvectors

Find the eigenvalues and corresponding eigenvectors of the real symmetric matrix

$$A = \begin{pmatrix} 5 & 2 & 2 \\ 2 & 5 & 2 \\ 2 & 2 & 5 \end{pmatrix}.$$

Produce an orthonormal set of three eigenvectors.

Solution

The eigenvalues are found by solving the characteristic equation:

$$\begin{vmatrix} 5-\lambda & 2 & 2 \\ 2 & 5-\lambda & 2 \\ 2 & 2 & 5-\lambda \end{vmatrix} = (5-\lambda) \begin{vmatrix} 5-\lambda & 2 \\ 2 & 5-\lambda \end{vmatrix} - 2 \begin{vmatrix} 2 & 2 \\ 2 & 5-\lambda \end{vmatrix} + 2 \begin{vmatrix} 2 & 5-\lambda \\ 2 & 2 \end{vmatrix}$$

$$= (5-\lambda)(\lambda^2 - 10\lambda + 21) - 2(6 - 2\lambda) + 2(-6 + 2\lambda)$$

$$= -\lambda^3 + 15\lambda^2 - 63\lambda + 81 = 0.$$

This cubic equation can be solved, for example by plotting a graph to locate the roots, or by using a computer or an on-line package. It factorises as

$$(-\lambda + 3)(\lambda - 3)(\lambda - 9) = 0$$

so that the eigenvalues are $\lambda = 3, 3$ and 9.

To find the eigenvectors corresponding to $\lambda = 3$ we solve $2v_1 + 2v_2 + 2v_3 = 0$. (Note all three equations are the same). We deduce there are two free variables, so let $v_3 = \mu$ and $v_2 = \nu$ then $v_1 = -\mu - \nu$. The eigenvectors corresponding to $\lambda = 3$ then take the general form

$$\begin{pmatrix} v_1 \\ v_2 \\ v_3 \end{pmatrix} = \mu \begin{pmatrix} -1 \\ 0 \\ 1 \end{pmatrix} + \nu \begin{pmatrix} -1 \\ 1 \\ 0 \end{pmatrix}.$$

Clearly, two independent eigenvectors are $\begin{pmatrix} -1 \\ 0 \\ 1 \end{pmatrix}$ and $\begin{pmatrix} -1 \\ 1 \\ 0 \end{pmatrix}$. With this choice, the inner product is 1, so they are not orthogonal. An eigenvector corresponding to $\lambda = 9$ is $\begin{pmatrix} 1 \\ 1 \\ 1 \end{pmatrix}$ as is readily verified.

We now construct an orthonormal set. Applying the Gram-Schmidt process (Chapter 6) produces an orthogonal pair, for example: $\begin{pmatrix} -1 \\ 0 \\ 1 \end{pmatrix}$ and $\begin{pmatrix} -1 \\ 2 \\ -1 \end{pmatrix}$ which when normalised become $\begin{pmatrix} -1/\sqrt{2} \\ 0 \\ 1/\sqrt{2} \end{pmatrix}$ and $\begin{pmatrix} -1/\sqrt{6} \\ 2\sqrt{6} \\ -1/\sqrt{6} \end{pmatrix}$. It is straightforward to show that a normalised eigenvector corresponding to $\lambda = 9$ is $\begin{pmatrix} 1/\sqrt{3} \\ 1/\sqrt{3} \\ 1/\sqrt{3} \end{pmatrix}$. We now have a full set of three orthonormal eigenvectors. You should check that this is indeed the case.

7.6   Diagonalisation of real symmetric matrices

Suppose A is an $n \times n$ real symmetric matrix. As we have seen, symmetric matrices have a full set of n linearly independent eigenvectors. Given any matrix that has n linearly independent eigenvectors it is possible to perform a transformation in order to represent it as a diagonal matrix. We now illustrate how this is achieved.

Form a new matrix, P say, whose columns are the independent eigenvectors of A. (At this stage there is no requirement that the eigenvectors are orthonormal – just linearly independent). Then evaluation of $P^{-1}AP$ yields a diagonal matrix, D, with the eigenvalues on the diagonal (in the same order as the eigenvectors appeared in the columns of P). Thus

$$P^{-1}AP = D.$$

Pre-multiplying by P and post-multiplying by P^{-1} we can rewrite this equivalently as

$$A = PDP^{-1}$$

so that the matrix A has been **factorised** using its eigenvectors and eigenvalues. The transformation from A to PDP^{-1} is known as a **similarity transformation**.

Definition 7.6 A diagonalisable matrix

*The $n \times n$ symmetric matrix A is **diagonalisable** because there exists an $n \times n$ matrix P with the property that*

$$P^{-1}AP = D, \text{ a diagonal matrix containing the eigenvalues.}$$

Equivalently,

$$A = PDP^{-1} \text{ is a so-called matrix factorisation.}$$

P is the matrix whose columns are the eigenvectors of A.

In the previous development there was no requirement that the eigenvectors be orthonormal – just linearly independent. When we impose this additional requirement the matrix P is an orthogonal matrix. Recall from Section 5.10 that such a matrix, P, has the property that $P^{-1} = P^T$. Consequently the similarity transformation which diagonalises A can be written

$$P^T AP = D \qquad \text{or equivalently} \qquad A = PDP^T.$$

Definition 7.7 Orthogonally diagonalisable

The $n \times n$ symmetric matrix A is **orthogonally diagonalisable** *because there exists an $n \times n$ orthogonal matrix P with the property that*

$$P^T A P = D, \text{ a diagonal matrix containing the eigenvalues.}$$

Equivalently,

$$A = P D P^T.$$

P is the matrix whose columns are the normalised eigenvectors of A.

Example 7.6.1

Find the eigenvalues and corresponding eigenvectors of the real symmetric matrix $A = \begin{pmatrix} 4 & 2 \\ 2 & 4 \end{pmatrix}$ and hence diagonalise A.

Solution

The characteristic equation is $\lambda^2 - 8\lambda + 12 = (\lambda - 6)(\lambda - 2) = 0$ so that $\lambda = 2, 6$. Because these are distinct their eigenvectors must be orthogonal as we now show. Eigenvectors corresponding to $\lambda = 2, 6$, respectively, are readily shown to be:

$$\begin{pmatrix} 1 \\ -1 \end{pmatrix} \quad \text{and} \quad \begin{pmatrix} 1 \\ 1 \end{pmatrix}.$$

These eigenvectors are linearly independent and orthogonal though not normalised. Form P whose columns are the eigenvectors of A:

$$P = \begin{pmatrix} 1 & 1 \\ -1 & 1 \end{pmatrix} \quad \text{from which} \quad P^{-1} = \frac{1}{2} \begin{pmatrix} 1 & -1 \\ 1 & 1 \end{pmatrix}.$$

It is then straightforward to check that

$$P^{-1} A P = \frac{1}{2} \begin{pmatrix} 1 & -1 \\ 1 & 1 \end{pmatrix} \begin{pmatrix} 4 & 2 \\ 2 & 4 \end{pmatrix} \begin{pmatrix} 1 & 1 \\ -1 & 1 \end{pmatrix} = \begin{pmatrix} 2 & 0 \\ 0 & 6 \end{pmatrix}$$

the result being a diagonal matrix, D say, with the eigenvalues on the diagonal.

Now suppose we require that the eigenvectors be orthonormal before we form the matrix P. So let the normalised eigenvectors be

$$\begin{pmatrix} \frac{1}{\sqrt{2}} \\ -\frac{1}{\sqrt{2}} \end{pmatrix} \quad \text{and} \quad \begin{pmatrix} \frac{1}{\sqrt{2}} \\ \frac{1}{\sqrt{2}} \end{pmatrix}.$$

Then take P to be

$$P = \begin{pmatrix} \frac{1}{\sqrt{2}} & \frac{1}{\sqrt{2}} \\ -\frac{1}{\sqrt{2}} & \frac{1}{\sqrt{2}} \end{pmatrix}.$$

It is readily verified that, by construction, this matrix is orthogonal (that is its columns are mutually orthogonal and normalised). You should confirm that this is the case. Thus A is orthogonally diagonalisable. You should verify that $P^T A P = D$.

Example 7.6.2

Consider the matrix $A = \begin{pmatrix} 5 & 2 & 2 \\ 2 & 5 & 2 \\ 2 & 2 & 5 \end{pmatrix}$ of Example 7.5.3.

(a) Demonstrate diagonalisation

(b) Demonstrate orthogonal diagonalisation.

Solution

(a) We have seen that the eigenvalues of A are $\lambda = 9, 3, 3$ with corresponding eigenvectors:

$\begin{pmatrix} 1 \\ 1 \\ 1 \end{pmatrix}, \begin{pmatrix} -1 \\ 0 \\ 1 \end{pmatrix}$ and $\begin{pmatrix} -1 \\ 2 \\ -1 \end{pmatrix}$. So form

$$P = \begin{pmatrix} 1 & -1 & -1 \\ 1 & 0 & 2 \\ 1 & 1 & -1 \end{pmatrix}.$$

You should note that this choice is not unique. We could have placed the eigenvectors in a different order, or used different eigenvectors. You should verify that

$$P^{-1} = \frac{1}{6} \begin{pmatrix} 2 & 2 & 2 \\ -3 & 0 & 3 \\ -1 & 2 & -1 \end{pmatrix}.$$

Then evaluating $P^{-1}AP$:

$$P^{-1}AP = \frac{1}{6} \begin{pmatrix} 2 & 2 & 2 \\ -3 & 0 & 3 \\ -1 & 2 & -1 \end{pmatrix} \begin{pmatrix} 5 & 2 & 2 \\ 2 & 5 & 2 \\ 2 & 2 & 5 \end{pmatrix} \begin{pmatrix} 1 & -1 & -1 \\ 1 & 0 & 2 \\ 1 & 1 & -1 \end{pmatrix}$$

$$= \frac{1}{6} \begin{pmatrix} 2 & 2 & 2 \\ -3 & 0 & 3 \\ -1 & 2 & -1 \end{pmatrix} \begin{pmatrix} 9 & -3 & -3 \\ 9 & 0 & 6 \\ 9 & 3 & -3 \end{pmatrix}$$

$$= \frac{1}{6} \begin{pmatrix} 54 & 0 & 0 \\ 0 & 18 & 0 \\ 0 & 0 & 18 \end{pmatrix}$$

$$= \begin{pmatrix} 9 & 0 & 0 \\ 0 & 3 & 0 \\ 0 & 0 & 3 \end{pmatrix}$$

which is the required diagonal matrix. Note that the eigenvalues of A lie on the diagonal in the same order as we placed the eigenvectors in P.

(b) Now form P using the orthonormal eigenvectors from Example 7.5.3, that is:

$$P = \begin{pmatrix} 1/\sqrt{3} & -1/\sqrt{2} & -1/\sqrt{6} \\ 1/\sqrt{3} & 0 & 2/\sqrt{6} \\ 1/\sqrt{3} & 1/\sqrt{2} & -1/\sqrt{6} \end{pmatrix}.$$

By construction this is an orthogonal matrix. Therefore, its inverse equals its transpose:

$$P^{-1} = P^T = \begin{pmatrix} 1/\sqrt{3} & 1/\sqrt{3} & 1/\sqrt{3} \\ -1/\sqrt{2} & 0 & 1/\sqrt{2} \\ -1/\sqrt{6} & 2/\sqrt{6} & -1/\sqrt{6} \end{pmatrix}.$$

Then

$$P^T A P = \begin{pmatrix} 1/\sqrt{3} & 1/\sqrt{3} & 1/\sqrt{3} \\ -1/\sqrt{2} & 0 & 1/\sqrt{2} \\ -1/\sqrt{6} & 2/\sqrt{6} & -1/\sqrt{6} \end{pmatrix} \begin{pmatrix} 5 & 2 & 2 \\ 2 & 5 & 2 \\ 2 & 2 & 5 \end{pmatrix} \begin{pmatrix} 1/\sqrt{3} & -1/\sqrt{2} & -1/\sqrt{6} \\ 1/\sqrt{3} & 0 & 2/\sqrt{6} \\ 1/\sqrt{3} & 1/\sqrt{2} & -1/\sqrt{6} \end{pmatrix}$$

$$= \begin{pmatrix} 1/\sqrt{3} & 1/\sqrt{3} & 1/\sqrt{3} \\ -1/\sqrt{2} & 0 & 1/\sqrt{2} \\ -1/\sqrt{6} & 2/\sqrt{6} & -1/\sqrt{6} \end{pmatrix} \begin{pmatrix} 9/\sqrt{3} & -3/\sqrt{2} & -3/\sqrt{6} \\ 9/\sqrt{3} & 0 & 6/\sqrt{6} \\ 9/\sqrt{3} & 3/\sqrt{2} & -3/\sqrt{6} \end{pmatrix}$$

$$= \begin{pmatrix} 9 & 0 & 0 \\ 0 & 3 & 0 \\ 0 & 0 & 3 \end{pmatrix}$$

which is the required diagonal matrix. Again note that the eigenvalues of A lie on the diagonal in the same order as we placed the orthonormal eigenvectors in P.

7.7 The spectral theorem for symmetric matrices

We have seen that if A is an $n \times n$ real symmetric matrix, it has an orthonormal set of n eigenvectors and can thus be orthogonally diagonalised by a matrix P say whose columns are the eigenvectors:

$$P^T A P = D.$$

Equivalently, we have the decomposition:

$$A = PDP^T.$$

We now explore a different way in which this can be written leading to the spectral theorem.

Theorem 7.3 Spectral theorem
*Given an $n \times n$ real symmetric matrix A with eigenvalues λ_i and orthonormal eigenvectors p_i the **spectral decomposition** of A is given by*

$$A = PDP^T = \lambda_1 p_1 p_1^T + \lambda_2 p_2 p_2^T + \ldots + \lambda_n p_n p_n^T = \sum_{i=1}^{n} \lambda_i p_i p_i^T.$$

Proof
The proof requires use of block multiplication of matrices and a general result both of which were introduced in Chapter 5. Specifically, given an $n \times n$ matrix P whose columns are the vectors $p_1, p_2, \ldots, p_n$, then

$$P P^T = p_1 p_1^T + p_2 p_2^T + \ldots + p_n p_n^T.$$

(See Example 5.5.3.) We calculate $A = PDP^T$ using block matrix multiplication as follows. First consider PD. Let the columns of P be $p_1, p_2, \ldots, p_n$. That these are in fact column vectors is indicated below by writing $\begin{smallmatrix} p_1 \\ \vdots \end{smallmatrix}$ etc. Similarly, the rows of D are $d_1, d_2, \ldots, d_n$ which we indicate below by $d_1 \cdots$ etc.

$$PD = \begin{pmatrix} p_1 & p_2 & \cdots & p_n \\ \vdots & \vdots & \vdots & \vdots \end{pmatrix} \begin{pmatrix} d_1 & \cdots & \cdots & \cdots \\ d_2 & \cdots & \cdots & \cdots \\ \vdots & \cdots & \cdots & \cdots \\ d_n & \cdots & \cdots & \cdots \end{pmatrix}.$$

Thus

$$PD = \begin{pmatrix} p_1 & p_2 & \cdots & p_n \end{pmatrix} \begin{pmatrix} d_1 \\ d_2 \\ \vdots \\ d_n \end{pmatrix}$$

$$= p_1 d_1 + p_2 d_2 + \ldots + p_n d_n$$

using block matrix multiplication. Because D is diag$\{\lambda_1, \lambda_2, \ldots, \lambda_n\}$ this simplifies to

$$PD = \begin{pmatrix} \lambda_1 p_1 & \lambda_2 p_2 & \cdots & \lambda_n p_n \\ \vdots & \vdots & \vdots & \vdots \end{pmatrix} = \begin{pmatrix} \lambda_1 p_1 & \lambda_2 p_2 & \cdots & \lambda_n p_n \end{pmatrix}.$$

(See, for example, Exercise 5.12.) Then

$$A = PDP^T = \begin{pmatrix} \lambda_1 p_1 & \lambda_2 p_2 & \cdots & \lambda_n p_n \end{pmatrix} \begin{pmatrix} p_1^T \\ p_2^T \\ \vdots \\ p_n^T \end{pmatrix}$$

$$= \lambda_1 p_1 p_1^T + \lambda_2 p_2 p_2^T + \ldots + \lambda_n p_n p_n^T$$

which is often called the spectral theorem or the spectral decomposition of A. This allows A to be expressed as a sum of terms where each term is an eigenvalue multiplied by the matrix which is a projection on to the eigenspace corresponding to that eigenvalue. ∎

Example 7.7.1 Spectral decomposition

In Example 7.6.1, we showed that the matrix $A = \begin{pmatrix} 4 & 2 \\ 2 & 4 \end{pmatrix}$ had eigenvalues $\lambda = 2, 6$ with corresponding eigenvectors $\begin{pmatrix} 1 \\ -1 \end{pmatrix}$ and $\begin{pmatrix} 1 \\ 1 \end{pmatrix}$. Perform the spectral decomposition.

Solution

The given eigenvectors are already orthogonal but we require them to be normalised. So let

$$p_1 = \begin{pmatrix} 1/\sqrt{2} \\ -1/\sqrt{2} \end{pmatrix} \qquad \text{and} \qquad p_2 = \begin{pmatrix} 1/\sqrt{2} \\ 1/\sqrt{2} \end{pmatrix}.$$

Then

$$p_1 p_1^T = \begin{pmatrix} 1/\sqrt{2} \\ -1/\sqrt{2} \end{pmatrix} \begin{pmatrix} 1/\sqrt{2} & -1/\sqrt{2} \end{pmatrix} = \begin{pmatrix} 1/2 & -1/2 \\ -1/2 & 1/2 \end{pmatrix}$$

$$p_2 p_2^T = \begin{pmatrix} 1/\sqrt{2} \\ 1/\sqrt{2} \end{pmatrix} \begin{pmatrix} 1/\sqrt{2} & 1/\sqrt{2} \end{pmatrix} = \begin{pmatrix} 1/2 & 1/2 \\ 1/2 & 1/2 \end{pmatrix}.$$

Then

$$A = \sum_{i=1}^{2} \lambda_i p_i p_i^T = 2 \begin{pmatrix} 1/2 & -1/2 \\ -1/2 & 1/2 \end{pmatrix} + 6 \begin{pmatrix} 1/2 & 1/2 \\ 1/2 & 1/2 \end{pmatrix}.$$

This is the required spectral decomposition, which is readily checked to equal A.

7.8 Self-adjoint matrices and their eigenvalues and eigenvectors

The important results regarding symmetric matrices given in the previous section generalise to the case of complex self-adjoint matrices. In particular, if A is an $n \times n$ complex self-adjoint matrix all the eigenvalues of A are real, the eigenvectors corresponding to different eigenvalues are mutually orthogonal and A possesses an orthonormal set of n eigenvectors. The spectral theorem for self-adjoint matrices is given here without proof.

Theorem 7.4 The spectral theorem for self-adjoint matrices
Given an $n \times n$ complex self-adjoint matrix A with eigenvalues λ_i and orthonormal eigenvectors p_i, the spectral decomposition of A is given by

$$A = PDP^{-1} = PDP^\dagger = \sum_{i=1}^{n} \lambda_i p_i p_i^\dagger.$$

The matrix P whose columns are the orthonormal eigenvectors is a unitary matrix (i.e., $P^{-1} = P^\dagger$). In quantum computing applications the final result above is sometimes written in the equivalent form

$$A = \sum_{i=1}^{n} \lambda_i |p_i\rangle \langle p_i|.$$

7.9 End-of-chapter exercises

1. Show that the characteristic equation of $A = \begin{pmatrix} 4 & -5 \\ 2 & -3 \end{pmatrix}$ is $\lambda^2 - \lambda - 2 = 0$. Show that $A^2 - A - 2I = 0$, where 0 here denotes the zero matrix. Deduce that the matrix A satisfies its characteristic equation (Cayley-Hamilton theorem).

2. Find the eigenvalues and eigenvectors of $A = \begin{pmatrix} 1 & 0 \\ 0 & -1 \end{pmatrix}$.

3. Find a 2×2 matrix $\begin{pmatrix} a & b \\ c & d \end{pmatrix}$ which has eigenvectors $\begin{pmatrix} 1 \\ 0 \end{pmatrix}, \begin{pmatrix} 0 \\ 1 \end{pmatrix}$ and eigenvalues 1 and 0 respectively.

4. Find a 2×2 matrix $\begin{pmatrix} a & b \\ c & d \end{pmatrix}$ which has eigenvectors $\begin{pmatrix} 1 \\ 0 \end{pmatrix}, \begin{pmatrix} 0 \\ 1 \end{pmatrix}$ and eigenvalues 0 and 1, respectively.

5. Find the eigenvalues and eigenvectors of $A = \begin{pmatrix} 0 & 0 \\ 0 & 1 \end{pmatrix}$.

6. Find a 2×2 matrix $\begin{pmatrix} a & b \\ c & d \end{pmatrix}$ which has eigenvectors $\begin{pmatrix} \frac{1}{\sqrt{2}} \\ \frac{1}{\sqrt{2}} \end{pmatrix}, \begin{pmatrix} \frac{1}{\sqrt{2}} \\ -\frac{1}{\sqrt{2}} \end{pmatrix}$ and eigenvalues 1 and -1, respectively.

7. Find the eigenvalues and eigenvectors of

   a) $A = \begin{pmatrix} 0 & 0 & 0 & 0 \\ 0 & 0 & 0 & 0 \\ 0 & 0 & 1 & 0 \\ 0 & 0 & 0 & 1 \end{pmatrix}$

   b) $A = \begin{pmatrix} 0 & 0 & 0 & 0 \\ 0 & 1 & 0 & 0 \\ 0 & 0 & 0 & 0 \\ 0 & 0 & 0 & 1 \end{pmatrix}$.

8. The matrices S_1^2, S_2^2 and S_3^2 defined below arise in the study of so-called Bell states. Find their eigenvalues and eigenvectors:

a) $S_1^2 = \frac{1}{2} \begin{pmatrix} 1 & 0 & 0 & 1 \\ 0 & 1 & 1 & 0 \\ 0 & 1 & 1 & 0 \\ 1 & 0 & 0 & 1 \end{pmatrix}$
   b) $S_2^2 = \frac{1}{2} \begin{pmatrix} 1 & 0 & 0 & -1 \\ 0 & 1 & 1 & 0 \\ 0 & 1 & 1 & 0 \\ -1 & 0 & 0 & 1 \end{pmatrix}$

c) $S_3^2 = \begin{pmatrix} 1 & 0 & 0 & 0 \\ 0 & 0 & 0 & 0 \\ 0 & 0 & 0 & 0 \\ 0 & 0 & 0 & 1 \end{pmatrix}$.

(Hint: see Q13, 14 below).

9. Find the eigenvalues and eigenvectors of

$$\begin{pmatrix} 1 & 0 & 0 & 0 \\ 0 & 0 & 0 & 0 \\ 0 & 0 & 0 & 0 \\ 0 & 0 & 0 & 0 \end{pmatrix}.$$

10. Find the eigenvalues and corresponding eigenvectors of $P = \begin{pmatrix} 1 & 0 \\ 0 & e^{i\phi} \end{pmatrix}$.

11. Using the properties of the transpose and of determinants show that $|A - \lambda I| = |A^T - \lambda I|$. Deduce that A and A^T have the same characteristic equation and hence the same eigenvalues.

12. Suppose v is an eigenvector of B with eigenvalue λ. Suppose B is similar to A, that is, there exists P such that $B = P^{-1}AP$. Show that λ is an eigenvalue of A.

13. If λ is an eigenvalue of A show that $k\lambda$ is an eigenvalue of kA, where k is a scalar.

14. If v is an eigenvector of A with eigenvalue λ show that v is also an eigenvector of kA with eigenvalue $k\lambda$.

8

Group theory

8.1 Objectives

In this chapter, we shall see how a set G that is endowed with a binary operation $\circ$ – a way of combining two set elements to yield another element in the same set – becomes a mathematical structure called a group, denoted $(G, \circ)$. To be a group, the structure must satisfy rules known as group axioms which concern: closure and associativity under the binary operation, and the existence of an identity element and inverse elements within the set G. An objective of this chapter is to explain this terminology and provide several examples of groups.

The evolution of quantum states is modelled through the action of unitary matrices on state vectors. Recall from Chapter 5 that a unitary matrix U satisfies $UU^\dagger = U^\dagger U = I$ where $U^\dagger$ is the conjugate transpose of U. Because quantum computation is reversible, each matrix U must have an inverse and clearly this role is filled by $U^\dagger$. We have already seen that an $n \times n$ identity matrix, I, is unitary and also that matrix multiplication is associative. It is these characteristics that mean the set of $n \times n$ unitary matrices together with the operation of matrix multiplication is a group – the unitary group $U(n)$. These and other relevant aspects of group theory are explored in this chapter.

8.2 Preliminary definitions and the axioms for a group

We noted in Chapter 1 that a **group** is a mathematical structure, denoted $(G, \circ)$, comprising a set G and a binary operation $\circ$, called **composition**, which combines two elements of the set G to produce another element also in G and for which particular rules, or **group axioms**, hold. Specifically, we require closure under the binary operation, associativity of the operation when applied to three (or more) elements, the existence of an identity element within the set G and an inverse for each element of G.

Definition 8.1 Group
*A set G with a binary operation $\circ$ is a **group** $(G, \circ)$ provided:*

    *1. given $a, b \in G$, then $a \circ b \in G$ (**closure**)*

    *2. for any $a, b, c \in G$ then $(a \circ b) \circ c = a \circ (b \circ c)$ (**associativity**)*

    *3. there exists an **identity** or **unit** element $e \in G$ such that $e \circ a = a \circ e = a$ for all $a \in G$.*

    *4. for any $a \in G$, there exists an element also in G, which we denote by a^{-1} (not to be confused with a power or reciprocal), such that $a \circ a^{-1} = a^{-1} \circ a = e$. The element denoted a^{-1} is called the **inverse** of a.*

DOI: 10.1201/9781003264569-8

Here, the binary operation ∘ stands for one of the many possible operations that we shall see in the examples which follow. For example, the binary operation is frequently addition modulo 2 or exclusive-or in which case we write $(G, \oplus)$. On occasions, when the binary operation in $a \circ b$ is clear we omit the operation altogether and write ab. The number of elements in the set G is called the **order** of the group, denoted $|G|$. When the underlying set has finite order (and is small), it can be illustrative to produce a **composition table**, also known as a **Cayley table**, which shows the results of combining all the elements of the group (see Example 8.2.1 below). If, in addition, the operation ∘ is commutative (that is $a \circ b = b \circ a$, for all $a, b \in G$) then $(G, \circ)$ is a **commutative** or **Abelian** group. In Section 1.6 we saw several example of groups from different branches of mathematics, for example the permutation groups P_n, and the symmetric groups S_n. We also saw that if there is a mapping between two groups which is one-to-one, onto and which is structure preserving, we call this correspondence an **isomorphism** and say that the groups are **isomorphic**. Formally, we have the following:

Definition 8.2 Group isomorphism
Given groups $(G, \circ)$ and $(G', *)$, if there exists a one-to-one and onto (i.e., bijective) mapping $\Phi : G \to G'$ such that

$$\Phi(g_1 \circ g_2) = \Phi(g_1) * \Phi(g_2) \qquad \text{where } g_1, g_2 \in G$$

then Φ is an isomorphism and the groups $(G, \circ)$ and $(G', *)$ are said to be isomorphic.

Definition 8.3 Subgroup
*If a subset, H, of G is itself a group with the same operation it is called a **subgroup** of G and we write $H \leq G$.*

Every group has at least two subgroups: the group itself, and the group consisting of just the identity element $\{e\}$, a so-called **trivial** subgroup. Note that the identity element of the group, e, must necessarily be in H in order that $(H, \circ)$ is a subgroup.

Example 8.2.1

Consider the set $G = \{1, -1\}$ and take as the binary operation the usual multiplication of numbers, $\times$. Show that $(G, \times)$ is a group.

Solution

A group composition table is shown in Table 8.1. In general, when reading a composition

TABLE 8.1
Composition
table for $(G, \times)$

$\times$	1	-1
1	1	-1
-1	-1	1

table to find $g_1 \circ g_2$ say, the first element, g_1, is found in the left-most column. The second element, g_2, is found in the top row. Then inspection of the intersection in the table of the corresponding row and column gives the required result. From Table 8.1, it is immediately apparent that $(G, \times)$ is closed: every element that results from a composition is a member of the original set. The associativity of multiplication of integers is obvious. The identity element of the group is 1 since $1 \times 1 = 1$, and $1 \times -1 = -1$. Finally, each element of G has an inverse, an element which multiplies it to give the identity 1. This is also apparent from the table because the identity element occurs in each row and each column of the body of the table. There are two further points to note from this example. Firstly, each element is its own inverse, i.e., is **self-inverse**, and secondly, the composition table is symmetric about the leading diagonal (top-left to bottom-right) which tells us that the group is commutative or Abelian. These are not properties of groups in general.

Example 8.2.2

Consider the set $\mathbb{B} = \{0, 1\}$ and take as the binary operation addition modulo 2, $\oplus$. Show that $(\mathbb{B}, \oplus)$ is a group.

Solution

A group composition table is shown in Table 8.2. Note that working in modulo 2, $1 \oplus 1 = 0$. It is immediately apparent that $(\mathbb{B}, \oplus)$ is closed. Associativity follows because addition is

TABLE 8.2
Composition table for
$(\mathbb{B}, \oplus)$

$\oplus$	0	1
0	0	1
1	1	0

associative. The identity element is 0. Each element has an inverse because the identity element, 0, appears once in each row and column of the body of the table.

Example 8.2.3

For the groups $(G, \times)$ and $(\mathbb{B}, \oplus)$ of Examples 8.2.1 and 8.2.2 find an isomorphism $\Phi : G \to \mathbb{B}$ and deduce that the two groups are isomorphic.

Solution

From an inspection of the group tables, it is apparent that the structures are the same and all that is required is a relabelling of elements. This is done through the mapping Φ. Define

$$\Phi(1_G) = 0_{\mathbb{B}}, \quad \Phi(-1_G) = 1_{\mathbb{B}}$$

where we have inserted subscripts to be explicit about the group to which the elements belong. This mapping is one-to-one and onto. We must also show that for any $g_1, g_2 \in G$,

$$\Phi(g_1 \times g_2) = \Phi(g_1) \oplus \Phi(g_2).$$

Suppose $g_1 = 1_G$, $g_2 = -1_G$. Then

$$\Phi(1_G \times -1_G) = \Phi(-1_G) = 1_{\mathbb{B}}.$$

Also

$$\Phi(1_G) \oplus \Phi(-1_G) = 0_{\mathbb{B}} \oplus 1_{\mathbb{B}} = 1_{\mathbb{B}}.$$

Hence,
$$\Phi(1_G \times -1_G) = \Phi(1_G) \oplus \Phi(-1_G).$$

Likewise, for any other choice of two elements in G, as you should verify. Hence, the two groups are isomorphic.

Example 8.2.4

Verify that $(\mathbb{R}, +)$ is a commutative group.

Solution

Here we are concerned with the set of real numbers, $\mathbb{R}$, and the operation of addition, $+$. It is immediately clear that the operation of adding real numbers is closed: if $a, b \in \mathbb{R}$ then $a + b \in \mathbb{R}$. Further, addition of real numbers is associative: $(a + b) + c = a + (b + c)$. The identity element of this group is $0 \in \mathbb{R}$ because $0 + a = a + 0 = a$ for any $a \in \mathbb{R}$. The negative of any real number is another real number and so the inverse of $a \in \mathbb{R}$ is $-a$, particularly because $a + (-a) = 0$. Thus $(\mathbb{R}, +)$ satisfies the axioms for a group. Because addition of real numbers is commutative, it follows that $(\mathbb{R}, +)$ is a commutative group.

Example 8.2.5 The groups $(\mathbb{B}^n = \{0, 1\}^n, \oplus)$

Consider the set $\mathbb{B}^n = \{0, 1\}^n$ with operation $\oplus$, which is bit-wise addition modulo 2. This is a group. Consider the case $n = 2$, for which

$$\mathbb{B}^2 = \{(0,0), (0,1), (1,0), (1,1)\}$$

which we abbreviate to

$$\mathbb{B}^2 = \{00, 01, 10, 11\}.$$

Bit-wise addition modulo 2 is performed by adding corresponding binary digits and working in modulo 2, thus, for example

$$\begin{array}{cc} 1 & 1 \\ 1 & 0 \\ \hline 0 & 1 \end{array}.$$

The composition table for the group $(\mathbb{B}^2, \oplus)$ is given in Table 8.3.

TABLE 8.3
Composition table for
$(\mathbb{B}^2, \oplus)$

$\oplus$	00	01	10	11
00	00	01	10	11
01	01	00	11	10
10	10	11	00	01
11	11	10	01	00

It is straightforward to verify that the group axioms are satisfied. The identity element is 00 and every element is its own inverse. Further, observe that this group is Abelian.

You should convince yourself that the composition table for the group $(\mathbb{B}^3, \oplus)$ is as given in Table 8.4. Further, note that the group is Abelian and every element is its own inverse.

TABLE 8.4
Composition table for $(\mathbb{B}^3, \oplus)$

$\oplus$	000	001	010	011	100	101	110	111
000	000	001	010	011	100	101	110	111
001	001	000	011	010	101	100	111	110
010	010	011	000	001	110	111	100	101
011	011	010	001	000	111	110	101	100
100	100	101	110	111	000	001	010	011
101	101	100	111	110	001	000	011	010
110	110	111	100	101	010	011	000	001
111	111	110	101	100	011	010	001	000

Example 8.2.6 The group of invertible Boolean functions on $\mathbb{B}^2$

In Section 2.5.2, we discussed several examples of Boolean functions $f : \mathbb{B}^2 \to \mathbb{B}^2$, including the Feynman *cnot* gate $f(x, y) = (x, x \oplus y)$. We noted that whilst there are 4^4 such functions in total, only $4! = 24$ of these are invertible. We also discussed how such functions can be composed. It can be shown that the set of all invertible functions $f : \mathbb{B}^2 \to \mathbb{B}^2$ determines a group under functional composition.

Exercises

8.1   Explain why $(\mathbb{R}^+, +)$ is not a group.

8.2   Explain why $(\mathbb{R} \backslash 0, +)$ is not a group.

8.3   Explain why $(\mathbb{R}, \times)$ is not a group.

8.4   Show that $(\{00, 01\}, \oplus)$ is a two element subgroup of $(\mathbb{B}^2, \oplus)$. How many two element subgroups of $(\mathbb{B}^2, \oplus)$ are there ?

8.5   Show that $(\{000, 011\}, \oplus)$ is a two element subgroup of $(\mathbb{B}^3, \oplus)$. How many two element subgroups of $(\mathbb{B}^3, \oplus)$ are there ?

8.6   Consider the rectangle $ABDC$ shown in Figure 8.1. Observe that L_1 and L_2 are axes of symmetry and that the rectangle possesses rotational symmetry of $180°$. Through a suitable labelling of the symmetry operations draw up a composition table for the symmetry group of the rectangle. This group is called the **Klein-four group**.

8.7   Show that the group $(\mathbb{B}^2, \oplus)$ is isomorphic to the Klein-four group.

8.3   Permutation groups and symmetric groups

We saw in Chapter 2 that a rearrangement or permutation of the elements of a set can be thought of as a bijective function. Therefore, to a collection of permutations we can apply the operation of function composition, $\circ$, which means applying one permutation followed by another.

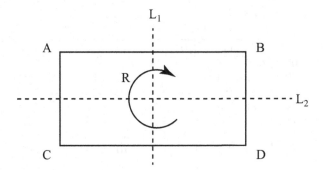

FIGURE 8.1
The symmetries of a rectangle.

Definition 8.4 Permutation group
A collection of permutations, with the operation of function composition ∘, which obey the group axioms is called a **permutation group**.

In this section we shall use cycle notation (see Chapter 2) to represent a permutation. Consider the following example.

Example 8.3.1

Consider the set $A = \{1, 2, 3\}$ and the following set of three permutations, $P = \{\sigma_0, \sigma_1, \sigma_2\}$ where

$$\sigma_0 = \text{ the identity}, \qquad \sigma_1 = (123), \qquad \sigma_2 = (132).$$

It may be helpful to picture the three permutations as shown in Figure 8.2. Note that there are other permutations of the set $\{1, 2, 3\}$ which are not discussed in this example.

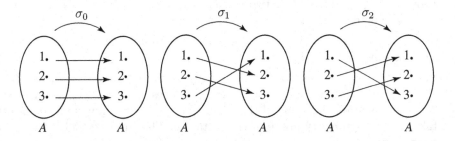

FIGURE 8.2
Some permutations of the set $A = \{1, 2, 3\}$.

Now consider the function composition $\sigma_1 \circ \sigma_2$:

$$(\sigma_1 \circ \sigma_2)(1) = \sigma_1(\sigma_2(1)) = \sigma_1(3) = 1$$

$$(\sigma_1 \circ \sigma_2)(2) = \sigma_1(\sigma_2(2)) = \sigma_1(1) = 2$$

$$(\sigma_1 \circ \sigma_2)(3) = \sigma_1(\sigma_2(3)) = \sigma_1(2) = 3.$$

Observe that $(\sigma_1 \circ \sigma_2)(x) = \sigma_0(x)$, the identity element. Further compositions can be calculated in a similar way. This gives rise to the composition table in Table 8.5 which reveals

TABLE 8.5

Composition table for the permutation group

∘	σ_0	σ_1	σ_2
σ_0	σ_0	σ_1	σ_2
σ_1	σ_1	σ_2	σ_0
σ_2	σ_2	σ_0	σ_1

that the set is closed under composition. It is straightforward to verify associativity. The identity element is σ_0. Every element has an inverse (this is apparent because the identity element appears in every row and every column of the table, and thus, for example, $\sigma_2 \circ \sigma_1 = \sigma_0$, so the inverse of σ_1 is σ_2). Thus the given set of permutations $\{\sigma_0, \sigma_1, \sigma_2\}$ with the binary operation $\circ$ is a permutation group, in fact an Abelian permutation group.

The previous example illustrated a group whose elements were some permutations of $\{1, 2, 3\}$. In addition to the permutations stated there, there are three further possible ones. The six permutations together with the operation $\circ$ form the group of *all* permutations of the three numbers which we call the **symmetric group** S_3. Symmetric groups of higher order are defined in an obvious way.

Definition 8.5 Symmetric group

*The **symmetric group** S_n is the group of all permutations of n elements. The group operation is function composition. Note $|S_n| = n!$.*

For example, the group S_5 is the group of all permutations of five elements, of which there are $5! = 120$.

Example 8.3.2

Consider again Example 8.3.1. Each of the permutations can be represented in an alternative form using a so-called **permutation matrix** and matrix multiplication. Observe that

$$\begin{pmatrix} 1 & 0 & 0 \\ 0 & 1 & 0 \\ 0 & 0 & 1 \end{pmatrix} \begin{pmatrix} 1 \\ 2 \\ 3 \end{pmatrix} = \begin{pmatrix} 1 \\ 2 \\ 3 \end{pmatrix},$$

$$\begin{pmatrix} 0 & 1 & 0 \\ 0 & 0 & 1 \\ 1 & 0 & 0 \end{pmatrix} \begin{pmatrix} 1 \\ 2 \\ 3 \end{pmatrix} = \begin{pmatrix} 2 \\ 3 \\ 1 \end{pmatrix}, \qquad \begin{pmatrix} 0 & 0 & 1 \\ 1 & 0 & 0 \\ 0 & 1 & 0 \end{pmatrix} \begin{pmatrix} 1 \\ 2 \\ 3 \end{pmatrix} = \begin{pmatrix} 3 \\ 1 \\ 2 \end{pmatrix}.$$

Labelling the three permutation matrices as $M_0 = I$, M_1 and M_2, respectively, we can produce the composition table in Table 8.6. It is straightforward to check that the group axioms are satisfied.

8.4 Unitary groups

In quantum computation, unitary matrices play an important role in determining how the state of a computation evolves in time. Recall that a square $(n \times n)$ matrix, U, is unitary if

$$U U^\dagger = U^\dagger U = I, \qquad \text{the } n \times n \text{ identity matrix,}$$

TABLE 8.6

Composition table for the
group of permutation
matrices

	$M_0 = I$	M_1	M_2
M_0	M_0	M_1	M_2
M_1	M_1	M_2	M_0
M_2	M_2	M_0	M_1

where $U^\dagger$ is the conjugate transpose of U. It can be shown that the set of all $n \times n$ unitary matrices with matrix multiplication form a group.

Definition 8.6 Unitary group
The set of all $n \times n$ unitary matrices form a group labelled $U(n)$, called the **unitary group** *of degree n, where the group operation is matrix multiplication.*

$U(n)$ is a subgroup of the **general linear group**, $GL(n, \mathbb{C})$, that is the set of all $n \times n$ invertible matrices with complex elements. The reason why unitary matrices are particularly important is because (a) when they act on a state vector they preserve its norm, and (b) they are invertible. These are two essential properties of the evolution of a quantum state.

Theorem 8.1 *The set of 2×2 unitary matrices $U(2)$ form a group under matrix multiplication.*

Proof
We first prove closure. Suppose $U_1, U_2 \in U(2)$. Then

$$U_1 U_1^\dagger = I, \qquad U_2 U_2^\dagger = I.$$

Then consider the product $(U_1 U_2)(U_1 U_2)^\dagger$:

$$
\begin{aligned}
(U_1 U_2)(U_1 U_2)^\dagger &= (U_1 U_2)(U_2^\dagger U_1^\dagger) \quad &&\text{(since } (U_1 U_2)^\dagger = U_2^\dagger U_1^\dagger\text{)} \\
&= U_1 (U_2\, U_2^\dagger) U_1^\dagger \quad &&\text{(associativity)} \\
&= U_1 (I) U_1^\dagger \\
&= U_1\, U_1^\dagger \\
&= I.
\end{aligned}
$$

Likewise $(U_1 U_2)^\dagger (U_1 U_2) = I$. Hence, $U_1 U_2$ is unitary. We have shown that the product of two unitary matrices is also a unitary matrix of the same size. The group identity element is the 2×2 identity matrix, which is readily shown to be unitary. Associativity follows immediately because matrix multiplication is associative. Finally, each $U \in U(2)$ must have an inverse element in the group, and this role is played by $U^\dagger$ since $U U^\dagger = U^\dagger U = I$. ∎

Example 8.4.1 The Pauli matrices

Consider the following set of 2×2 matrices known as **Pauli matrices**:

$$
\sigma_0 = \begin{pmatrix} 1 & 0 \\ 0 & 1 \end{pmatrix}, \quad
\sigma_1 = \begin{pmatrix} 0 & 1 \\ 1 & 0 \end{pmatrix}, \quad
\sigma_2 = \begin{pmatrix} 0 & -i \\ i & 0 \end{pmatrix}, \quad
\sigma_3 = \begin{pmatrix} 1 & 0 \\ 0 & -1 \end{pmatrix}.
$$

Show that whilst these are all unitary matrices, they do not form a group (e.g. calculate $\sigma_2\sigma_1$). (We note that by extending the set to form the 16 element set

$$\mathcal{P} = \{\pm\sigma_0, \pm i\sigma_0, \pm\sigma_1, \pm i\sigma_1, \pm\sigma_2, \pm i\sigma_2, \pm\sigma_3, \pm i\sigma_3\}$$

it is possible to show that $(\mathcal{P}, \circ)$ is a group – the **Pauli group** – where the binary operation $\circ$ represents matrix multiplication.)

Solution

Note that σ_0 is the identity matrix. That each of these is a unitary matrix is easy to verify. For example, consider σ_2. First check carefully that $\sigma_2^\dagger = \sigma_2$, then

$$\sigma_2\sigma_2^\dagger = \begin{pmatrix} 0 & -i \\ i & 0 \end{pmatrix} \begin{pmatrix} 0 & -i \\ i & 0 \end{pmatrix} = \begin{pmatrix} 1 & 0 \\ 0 & 1 \end{pmatrix}$$

and likewise for $\sigma_2^\dagger\sigma_2$. Hence, σ_2 is a unitary matrix, as are the others. Now form, for example,

$$\sigma_2\sigma_1 = \begin{pmatrix} 0 & -i \\ i & 0 \end{pmatrix} \begin{pmatrix} 0 & 1 \\ 1 & 0 \end{pmatrix} = \begin{pmatrix} -i & 0 \\ 0 & i \end{pmatrix}.$$

The resulting matrix is clearly not a member of the original set, and hence, the closure axiom is not satisfied. The four given Pauli matrices do not form a group.

8.5   Cosets, partitions and equivalence classes

In this section we demonstrate how, given a group $(G, \circ)$ and a subgroup, $(H, \circ)$, we can use the subgroup to divide up or **partition** G into distinct sets known as **cosets**. Elements in these distinct sets share certain properties and as such can be regarded as equivalent in some sense, as we shall explain. We begin by defining cosets.

Definition 8.7 Cosets
Suppose $(G, \circ)$ is a group and $(H, \circ)$ is a subgroup. Now choose any element $a \in G$ that is not in H. We can form the set consisting of all elements of the form $a \circ h$, for $h \in H$. This set is called a **left coset** *of H in G, written $a \circ H$ or aH:*

$$aH = \{a \circ h : h \in H\}.$$

In a similar fashion, a **right coset** *of H in G, written $H \circ a$ or Ha, is given by:*

$$Ha = \{h \circ a : h \in H\}.$$

By choosing a different element $b \in G$, say, we can form its coset in a similar way. We shall find shortly that any element of G is in one and only one coset. Note that, apart from H itself, the cosets are not subgroups. The only subgroup in this discussion is $(H, \circ)$. Further, the identity element of G, e say, must be in H because $(H, \circ)$ is the only subgroup.

Finally, note that the set of all the left cosets of H in G is written G/H. The set of all right cosets is denoted $H\backslash G$.

Example 8.5.1

Consider the sets $G = \{0, 1, 2, 3, 4, 5, 6, 7, 8, 9, 10, 11\}$ and $H = \{0, 3, 6, 9\}$ with the operation of addition modulo 12. Observe $H \subset G$. It is easily verified by inspection of the composition table in Table 8.7 that $(G, + \bmod 12)$ is a group. Further, from Table 8.8, we see that the

TABLE 8.7

Composition table for $(G, + \bmod 12)$

+	0	1	2	3	4	5	6	7	8	9	10	11
0	0	1	2	3	4	5	6	7	8	9	10	11
1	1	2	3	4	5	6	7	8	9	10	11	0
2	2	3	4	5	6	7	8	9	10	11	0	1
3	3	4	5	6	7	8	9	10	11	0	1	2
4	4	5	6	7	8	9	10	11	0	1	2	3
5	5	6	7	8	9	10	11	0	1	2	3	4
6	6	7	8	9	10	11	0	1	2	3	4	5
7	7	8	9	10	11	0	1	2	3	4	5	6
8	8	9	10	11	0	1	2	3	4	5	6	7
9	9	10	11	0	1	2	3	4	5	6	7	8
10	10	11	0	1	2	3	4	5	6	7	8	9
11	11	0	1	2	3	4	5	6	7	8	9	10

set $H = \{0, 3, 6, 9\}$ with the same operation is a subgroup of $(G, + \bmod 12)$, i.e $H \leq G$. We

TABLE 8.8

Composition table
for $(H, + \bmod 12)$

+	0	3	6	9
0	0	3	6	9
3	3	6	9	0
6	6	9	0	3
9	9	0	3	6

now form the left cosets of H in G. We choose any element of G that is not in H (note that were we to choose an element that is in H, its composition with any other element in H will remain in H because H is a subgroup and is therefore closed). Choosing $1 \in G, 1 \notin H$,

$$H = \{0, 3, 6, 9\}$$
$$1 + H = \{1 + h : h \in H\}$$
$$= \{1, 4, 7, 10\}.$$

This is the left coset of H in G generated by the element 1. Moving on to another element in G that is not in H, say 2:

$$2 + H = \{2 + h : h \in H\}$$
$$= \{2, 5, 8, 11\}.$$

This is the left coset of H in G generated by the element 2. Every element of G has now been considered: observe that it is either in H or one of the two additional cosets. (The subgroup H can be regarded as the coset $0 + H$.) Thus a consequence of generating the cosets of H in G is to **partition** the set G into distinct (not overlapping) subsets. Every element of G is in one and only one coset. Within each coset, the elements share particular properties. For example, all elements in H are divisible by 3. All elements in $1 + H$ have remainder 1 when divided by 3. All elements in $2 + H$ have remainder 2 when divided by 3. Thus, in this sense, the elements within each coset are equivalent. We say they are equivalent with respect to the left coset.

Example 8.5.2 Cosets as equivalence classes

Referring to the previous example, we can label the cosets $[0]$, $[1]$, $[2]$ where we now regard 0,1 and 2 as representing all the elements in their coset. The coset $[1]$ is thus a class of equivalent elements – an **equivalence class**. Likewise $[0]$ and $[2]$. That is,

$$[0] = \{h : h \in H\} = H$$

$$[1] = \{1 + h : h \in H\}, \qquad [2] = \{2 + h : h \in H\}.$$

Explicitly:

$$[0] = \{0, 3, 6, 9\}, \quad [1] = \{1, 4, 7, 10\}, \quad [2] = \{2, 5, 8, 11\}.$$

If we define the relation $\sim$ such that

$$b \sim a \text{ if and only if } b = a + h \text{ for some } h \in H$$

then $\sim$ is an equivalence relation. For example, we see that $11 \sim 2$ because $11 = 2 + h$ for some $h \in H$, specifically, $11 = 2 + 9$ and $9 \in H$, and observe that both 2 and 11 are elements of the equivalence class $[2]$.

More generally, for the group $(G, \circ)$, suppose $H \leq G$ and that we have partitioned G into left (or right) cosets. Then we write

$$b \sim a \text{ if there exists } h \in H \text{ such that } b = a \circ h.$$

This means that a and b lie in the same left coset, that is, in the same equivalence class.

Theorem 8.2 *Suppose $H \leq G$ and that we have partitioned G into left (or right) cosets. For $a, b \in G$, if a and b are in the same equivalence class then*

$$a^{-1} \circ b = h \text{ for some } h \in H.$$

Proof

Let $[a]$ be the equivalence class containing a, i.e.,

$$[a] = \{a \circ h : h \in H\}.$$

If $b \in [a]$, that is, $b \sim a$, then $b = a \circ h$ for some $h \in H$. By pre-applying the inverse of a to $b = a \circ h$:

$$a^{-1} \circ b = a^{-1} \circ a \circ h = e \circ h = h \qquad \text{where } e \text{ is the group identity element}$$

we deduce that

$$b \sim a \text{ if } a^{-1} \circ b = h \text{ for some } h \in H.$$

∎

Suppose the left and right cosets of a subgroup are found to be identical. This leads to the following definition.

Definition 8.8 Normal subgroup, $H \triangleleft G$

*Given $H \leq G$, H is said to be a **normal subgroup**, written $H \triangleleft G$, if for every element in G, the left and right cosets are identical.*

It can be shown that any subgroups of a commutative group are normal.

Example 8.5.3 Cosets arising from subgroups of $(\mathbb{B}^3, \oplus)$

Consider the group $(\mathbb{B}^3, \oplus)$ of Example 8.2.5 which we noted was Abelian. For $s \in \mathbb{B}^3$ let K_s be the two element set $\{000, s\}$. It is straightforward to verify that $(K_s, \oplus)$ is a normal subgroup.

Suppose we choose $s = 101$ and consider the subgroup $(K_{101}, \oplus)$ where $K_{101} = \{000, 101\}$. The (left) cosets are then

$$[x] = x \oplus K_{101} = \{x \oplus k : k \in K_{101}\}.$$

It follows that

$$[000] = \{000, 101\}$$
$$[001] = \{001, 100\}$$
$$[010] = \{010, 111\}$$
$$[011] = \{011, 110\}.$$

We see that a consequence of generating the cosets of K_{101} in $\mathbb{B}^3$ has been the partition of $\mathbb{B}^3$ into distinct equivalence classes. You should verify that we could have carried out the same process for any other $s \in \mathbb{B}^3$.

Example 8.5.4

Consider the group $(\mathbb{B}^3, \oplus)$ of Example 8.2.5 which we noted was Abelian.

a) Show that $S = \{000, 010, 100, 110\}$ is a four element subgroup.

b) Find the (left) cosets of S and hence partition $\mathbb{B}^3$.

Solution

a) The composition table for $(S, \oplus)$ is given in Table 8.9. This shows that the set S is closed under $\oplus$. Associativity is inherited from the main group. The identity element is 000. Each element has an inverse because the identity appears in each row and column. Hence $(S, \oplus)$ is a subgroup.

TABLE 8.9

$\oplus$	000	010	100	110
000	000	010	100	110
010	010	000	110	100
100	100	110	000	010
110	110	100	010	000

b) For $x \in \mathbb{B}^3$ the cosets are given by:

$$[x] = x \oplus S = \{x \oplus s : s \in S\}.$$

Then

$$[000] = \{000, 010, 100, 110\} = S$$
$$[001] = \{001, 011, 101, 111\}.$$

This is the required partition of $\mathbb{B}^3$.

Exercises

8.8   Show that for $(H, \circ)$ to be a normal subgroup of $(G, \circ)$, for all $a \in G$ and all $h \in H$, $a^{-1} \circ h \circ a \in H$.

8.9   For $s \in \mathbb{B}^3$ let K_s be the two element set $\{000, s\}$. Show that the cosets of K_{001} in $\mathbb{B}^3$ are the equivalence classes $[000] = \{000, 001\}$, $[010] = \{010, 011\}$, $[100] = \{100, 101\}$, $[110] = \{110, 111\}$.

8.6   Quotient groups

Suppose $(H, \circ)$ is a normal subgroup of $(G, \circ)$. We let G/H be the set of all left cosets of H in G. Because H is a normal subgroup, the set of left cosets is identical to the set of right cosets and we can ignore the distinction. The set G/H and a suitably defined binary operation form a new group called the quotient group, or factor group of G relative to H.

Definition 8.9 Quotient group or factor group

Given $H \triangleleft G$, the set of all (left) cosets of H in G, written G/H, together with the binary operation $\circ$ as defined below and inherited from the group $(G, \circ)$ is a group $(G/H, \circ)$ called the **quotient group** *or* **factor group***.*

Theorem 8.3   *Given $H \triangleleft G$ and cosets $[a], [b], [c] \in G/H$. Define the binary operation $*$ on the cosets as*

$$[a] * [b] = [a \circ b],$$

that is, two cosets are composed by applying the group operation to the representatives of the cosets and determining the coset in which the result lies. Then $(G/H, *)$ is a group.

Proof
First, consider

$$
\begin{aligned}
([a] * [b]) * [c] &= [a \circ b] * [c] \\
&= [(a \circ b) \circ c] \\
&= [a \circ (b \circ c)] \qquad \text{because of associativity of } (G, \circ) \\
&= [a] * [b \circ c] \\
&= [a] * ([b] * [c])
\end{aligned}
$$

and thus the operation $*$ is associative. The identity element in the new group is the coset $[e]$, where e is the identity element of $(G, \circ)$, that is the subgroup H itself, because

$$
\begin{aligned}
[e] * [a] &= [e \circ a] \\
&= [a] \qquad \text{because } e \text{ is the identity in } G.
\end{aligned}
$$

Likewise $[a] * [e] = [a]$ and so $[e]$ is the identity element in $(G/H, *)$. Finally, the inverse of $[a] \in G/H$ is the coset $[a^{-1}] \in G/H$ because

$$
\begin{aligned}
[a] * [a^{-1}] &= [a \circ a^{-1}] \\
&= [e].
\end{aligned}
$$

■

Because we end up using the original group operation ∘ in performing calculations with cosets, it is usual practice to write both operations, ∘ and ∗, with the same symbol ∘ say. Then $(G/H, \circ)$ is known as a quotient group or factor group.

Example 8.6.1

Consider the commutative group $(\mathbb{Z}, +)$ and the subgroup $(5\mathbb{Z}, +)$ where

$$5\mathbb{Z} = \{0, \pm 5, \pm 10, \pm 15, \ldots\}$$

that is, all integer multiples of 5. That $(5\mathbb{Z}, +)$ is a subgroup is readily verified.

Suppose we choose the element $1 \in \mathbb{Z}$, noting $1 \notin 5\mathbb{Z}$. Forming the set consisting of all elements of the form $1 + h$ for $h \in 5\mathbb{Z}$ we obtain

$$\{\ldots, -9, -4, 1, 6, 11, 16, \ldots\} \quad \text{which we denote} \quad 1 + 5\mathbb{Z} \quad \text{or } [1].$$

Likewise, choosing the element $2 \in \mathbb{Z}$, noting $2 \notin 5\mathbb{Z}$ we obtain

$$\{\ldots, -8, -3, 2, 7, 12, 17, \ldots\} \quad \text{which we denote} \quad 2 + 5\mathbb{Z} \quad \text{or } [2].$$

Continuing in this fashion, we define

$$\{\ldots, -7, -2, 3, 8, 13, 18, \ldots\} \quad \text{which we denote} \quad 3 + 5\mathbb{Z} \quad \text{or } [3].$$

$$\{\ldots, -6, -1, 4, 9, 14, 19, \ldots\} \quad \text{which we denote} \quad 4 + 5\mathbb{Z} \quad \text{or } [4].$$

At this stage, every element of $\mathbb{Z}$ is in one and only one of the cosets (with $[0] = 5\mathbb{Z}$). Thus we have partitioned $\mathbb{Z}$ into distinct sets. Every element of any particular coset is considered to be equivalent to any other in its coset and can therefore be regarded as representative of that coset. We now have a set, $\mathbb{Z}/5\mathbb{Z}$, with 5 elements $\{[0], [1], [2], [3], [4]\}$. The set $\mathbb{Z}/5\mathbb{Z}$ with addition is the quotient group, $(\mathbb{Z}/5\mathbb{Z}, +)$.

8.7 End-of-chapter exercises

1. Show that the set of complex numbers $\mathbb{C}$ with the usual addition of complex numbers, $+$, is a group – the additive group of complex numbers $(\mathbb{C}, +)$.

2. Let $\mathbb{Q}\backslash 0$ be the set of rational numbers without 0. Show that with the usual multiplication, $\cdot$, of rational numbers $(\mathbb{Q}\backslash 0, \cdot)$ is a group.

3. Show that $(\mathbb{C}\backslash 0, \cdot)$ is a group, where $\cdot$ is the usual multiplication of complex numbers.

4. Show that the group $(\mathbb{B}^2, \oplus)$ is isomorphic to $(G, \times (\mathrm{mod}\, 8))$ where $G = \{1, 3, 5, 7\}$ and multiplication is performed modulo 8.

5. The matrix $W(\theta) = \begin{pmatrix} \cos\theta & -\sin\theta \\ \sin\theta & \cos\theta \end{pmatrix}$ effects the rotation of a point in the xy plane anticlockwise through an angle θ.

   a) Show that the set of all such matrices forms a group (the special orthogonal group $SO(2)$) under matrix multiplication.

   b) State the identity element of the group.

   c) How many elements are there in this group?

6. Suppose $(G, \circ)$ is a group and suppose H is a subset of G. Show that if $H \neq \emptyset$ and if whenever $a, b \in H$ then $a \circ b^{-1} \in H$ then H is a subgroup.

7. We have seen that $U(2)$ is the unitary group – the group of 2×2 unitary matrices. The group $SU(2)$, the special unitary group is the subgroup of $U(2)$ comprising those unitary matrices with determinant equal to 1. Show that $\begin{pmatrix} \frac{1}{\sqrt{2}} & \frac{1}{\sqrt{2}} \\ -\frac{1}{\sqrt{2}} & \frac{1}{\sqrt{2}} \end{pmatrix}$ is an element of $SU(2)$.

8. We have noted that the four Pauli matrices $\{\sigma_0, \sigma_1, \sigma_2, \sigma_3\}$ do not form a group. Show that the 16 element set

$$\mathcal{P} = \{\pm\sigma_0, \pm i\sigma_0, \pm\sigma_1, \pm i\sigma_1, \pm\sigma_2, \pm i\sigma_2, \pm\sigma_3, \pm i\sigma_3\}$$

is a group – the **Pauli group** – with binary operation the usual matrix multiplication.

9. Consider the additive group of integers modulo 5, $(\mathbb{Z}/5\mathbb{Z}, +)$.

   a) Produce a Cayley table for this group.

   b) Show that by repeatedly adding the element 1, every other element in the group can be generated.

   c) Likewise, show that by repeatedly adding the elements 2, 3 and 4 every other element in the group can be generated.

   If every element of a group can be generated in this way by a single element of the group, then that group is said to be **cyclic**.

   d) Consider the set $\{1, -1, i, -i\}$ with the usual multiplication of complex numbers. Show that this is a cyclic group and determine which elements can be used to generate the group.

10. Suppose $(G, \circ)$ and $(H, *)$ are groups and $f : G \to H$ is an arbitrary function. Suppose further that we define the function $\hat{f} : G \times H \to G \times H$ as

$$\hat{f}(g, h) = (g, h * f(g)), \qquad \text{for } g \in G, h \in H.$$

Suppose $g_1, g_2 \in G$, $h_1, h_2 \in H$. Use the group properties to show that if

$$\hat{f}(g_1, h_1) = \hat{f}(g_2, h_2)$$

then $g_1 = g_2$ and $h_1 = h_2$, and deduce that $\hat{f}$ is a one-to-one function.

11. Consider the groups and functions defined in Q10. Show that for any (a, b) in the co-domain of $\hat{f}$, there exists $g \in G, h \in H$ such that $\hat{f}(g, h) = (a, b)$. Deduce that $\hat{f}$ is an onto function.

12. Prove that if a Cayley table is symmetric about its leading diagonal then the corresponding group is Abelian.

9

Linear transformations

9.1 Objectives

Given two arbitrary vector spaces V and W over the same field $\mathbb{F}$, a transformation is a function which maps each vector in V to a vector in W. Linear transformations are functions which have additional properties which mean that the vector space operations of vector addition and scalar multiplication are preserved. The objective of this chapter is to introduce general linear transformations and shows how they can be represented by matrices. It introduces the terms 'kernel' and 'image' and a result known as the 'dimension theorem'.

These preliminaries pave the way for the study of a particular type of linear transformation known as a linear operator (see Chapter 11). Linear operators are essential to the study of quantum computation because they represent physical observables and determine the time evolution of a quantum system.

9.2 Preliminary information

Suppose we have two vector spaces, V and W say, over the same field, $\mathbb{F}$. In quantum computation the field $\mathbb{F}$ will usually be $\mathbb{R}$ or $\mathbb{C}$. A **function**, or **mapping**, f, assigns to each $v \in V$ a unique element of W. We write $f : V \to W$. We now consider particular functions known as linear transformations.

Definition 9.1 Linear transformation or homomorphism
A mapping, f, is a **linear transformation** *or* **homomorphism** *if it satisfies*

$$f(u + v) = f(u) + f(v)$$
$$f(ku) = kf(u)$$

for any $u, v \in V$ and $k \in \mathbb{F}$.

This requirement means that f preserves the vector space structure: addition of vectors and multiplication of vectors by a scalar (Note that in $f(u + v) = f(u) + f(v)$ the addition on the left-hand side takes place in V, whereas that on the right takes place in W. These additions need not be the same, although it is conventional to use the same symbol $+$).

Definition 9.2 Isomorphism
If a linear transformation f is one-to-one and onto, it is said to be an **isomorphism**.

DOI: 10.1201/9781003264569-9

In this chapter we explore linear transformations in general. In Chapter 11 we consider the specific and important case when the linear transformation maps the vector space V to itself, i.e., $f : V \to V$. Such a transformation is called a **linear operator**.

Example 9.2.1

Consider the vector spaces over $\mathbb{R}$ of row vectors in $\mathbb{R}^3$ and $\mathbb{R}^2$. Let $f : \mathbb{R}^3 \to \mathbb{R}^2$ be defined by

$$f(u_1, u_2, u_3) = (u_1 + u_3, 4u_2 - u_3).$$

a) Find $f(0,0,0)$.
b) Find $f(4,5,-2)$.
c) Show that f is a linear transformation.

Solution

a) $f(0,0,0) = (0,0)$.
b) $f(4,5,-2) = (4 + (-2), 4(5) - (-2)) = (2,22)$.
Observe that in each case a vector in $\mathbb{R}^3$ is mapped to one in $\mathbb{R}^2$. Further, observe that the zero vector in $\mathbb{R}^3$ is mapped to the zero vector in $\mathbb{R}^2$. This is generally true, as we prove below.
(c) To show that f is a linear transformation, choose two arbitrary vectors in $\mathbb{R}^3$, $u = (u_1, u_2, u_3)$, $v = (v_1, v_2, v_3)$. Their sum is $u + v = (u_1 + v_1, u_2 + v_2, u_3 + v_3)$. The images under f are given by

$$f(u) = f(u_1, u_2, u_3) = (u_1 + u_3, 4u_2 - u_3).$$
$$f(v) = f(v_1, v_2, v_3) = (v_1 + v_3, 4v_2 - v_3).$$

$$\begin{aligned}
f(u + v) &= f(u_1 + v_1, u_2 + v_2, u_3 + v_3) \\
&= ((u_1 + v_1) + (u_3 + v_3), 4(u_2 + v_2) - (u_3 + v_3)) \\
&= ((u_1 + u_3) + (v_1 + v_3), (4u_2 - u_3) + (4v_2 - v_3)) \\
&= (u_1 + u_3, 4u_2 - u_3) \quad + \quad (v_1 + v_3, 4v_2 - v_3) \\
&= f(u) + f(v).
\end{aligned}$$

This proves the first of the requirements for f to be a linear transformation. Now consider scalar multiplication:

$$\begin{aligned}
f(ku) = f(k(u_1, u_2, u_3)) &= f(ku_1, ku_2, ku_3) \\
&= (ku_1 + ku_3, 4ku_2 - ku_3) \\
&= k(u_1 + u_3, 4u_2 - u_3) \\
&= kf(u).
\end{aligned}$$

which proves the second requirement. Hence, f is a linear transformation.

Theorem 9.1 *If $f : V \to W$ is a linear transformation then $f(0) = 0$. That is f maps the zero vector in V to the zero vector in W.*

Proof

If f is a linear transformation, we know that $f(ku) = kf(u)$. The underlying field $\mathbb{F}$ must have a zero element, so choose $k = 0$. Then

$$f(0u) = 0f(u), \qquad \text{and hence} \qquad f(0) = 0$$

as required.   ∎

Example 9.2.2

Consider now column vectors in the vector spaces $V = W = \mathbb{R}^2$ over the field $\mathbb{R}$. Let $f : \mathbb{R}^2 \to \mathbb{R}^2$ be given by

$$f\begin{pmatrix} u_1 \\ u_2 \end{pmatrix} = \begin{pmatrix} 3u_1 - 2u_2 \\ u_1 + 4u_2 \end{pmatrix}.$$

(a) Evaluate $f\begin{pmatrix} 4 \\ 0 \end{pmatrix}$, $f\begin{pmatrix} -2 \\ 3 \end{pmatrix}$, $f\begin{pmatrix} 0 \\ 0 \end{pmatrix}$.

(b) Show that f is a linear mapping.

Solution

(a) $f\begin{pmatrix} 4 \\ 0 \end{pmatrix} = \begin{pmatrix} 12 \\ 4 \end{pmatrix}$, $\quad f\begin{pmatrix} -2 \\ 3 \end{pmatrix} = \begin{pmatrix} -12 \\ 10 \end{pmatrix}$, $\quad f\begin{pmatrix} 0 \\ 0 \end{pmatrix} = \begin{pmatrix} 0 \\ 0 \end{pmatrix}$.

(b) We must show that $f(u + v) = f(u) + f(v)$, $f(ku) = kf(u)$. Let

$$u = \begin{pmatrix} u_1 \\ u_2 \end{pmatrix}, \quad v = \begin{pmatrix} v_1 \\ v_2 \end{pmatrix},$$

then

$$u + v = \begin{pmatrix} u_1 \\ u_2 \end{pmatrix} + \begin{pmatrix} v_1 \\ v_2 \end{pmatrix} = \begin{pmatrix} u_1 + v_1 \\ u_2 + v_2 \end{pmatrix}.$$

Note that this is vector addition in the domain. Mapping this sum we find

$$f(u + v) = \begin{pmatrix} 3(u_1 + v_1) - 2(u_2 + v_2) \\ u_1 + v_1 + 4(u_2 + v_2) \end{pmatrix}$$

$$= \begin{pmatrix} 3u_1 + 3v_1 - 2u_2 - 2v_2 \\ u_1 + v_1 + 4u_2 + 4v_2 \end{pmatrix}$$

$$= \begin{pmatrix} (3u_1 - 2u_2) + (3v_1 - 2v_2) \\ (u_1 + 4u_2) + (v_1 + 4v_2) \end{pmatrix}$$

$$= \begin{pmatrix} 3u_1 - 2u_2 \\ u_1 + 4u_2 \end{pmatrix} + \begin{pmatrix} 3v_1 - 2v_2 \\ v_1 + 4v_2 \end{pmatrix}$$

$$= f(u) + f(v) \qquad \text{as required.}$$

Note that now this is vector addition in the co-domain. You should use a similar approach to check the second requirement.

Example 9.2.3 A matrix as a linear transformation.

Consider the case when $V = \mathbb{R}^n$ and $W = \mathbb{R}^m$, over the field $\mathbb{R}$, and the elements of these vector spaces are column vectors. An $m \times n$ matrix A is a linear transformation such that

$$A : \mathbb{R}^n \to \mathbb{R}^m.$$

Take, for example, $A : \mathbb{R}^2 \to \mathbb{R}^3$, that is $m = 3$, $n = 2$, and $A = \begin{pmatrix} 7 & 1 \\ 2 & -3 \\ 4 & 2 \end{pmatrix}$, then

$$A\begin{pmatrix} x \\ y \end{pmatrix} = \begin{pmatrix} 7 & 1 \\ 2 & -3 \\ 4 & 2 \end{pmatrix} \begin{pmatrix} x \\ y \end{pmatrix} = \begin{pmatrix} 7x + y \\ 2x - 3y \\ 4x + 2y \end{pmatrix}.$$

Observe that the matrix A maps a column vector in $\mathbb{R}^2$ to a column vector in $\mathbb{R}^3$. The transformation is linear because, in general, for matrix multiplication,

$$A(u + v) = Au + Av, \qquad A(ku) = kA(u)$$

for $u, v \in \mathbb{R}^n$, $k \in \mathbb{R}$. (See Theorem 5.1.)

9.3   The kernel and image of a linear transformation

Definition 9.3 Kernel of a linear transformation
Given a linear transformation $f : V \to W$ the **kernel** *of f, written ker(f), is the set of vectors in V which map to the zero vector in W, that is*

$$\ker(f) = \{v \in V : f(v) = 0\}.$$

Definition 9.4 Image of a linear transformation
The **image** *of f, written Im(f), is the set of vectors in W which are mapped to by some element in V, that is*

$$\mathrm{Im}(f) = \{w \in W : w = f(v), \text{ for some } v \in V\}.$$

It is possible to show that both the kernel and the image of a linear transformation are subspaces. If ker(f) contains only the zero vector then by definition it has dimension zero.

Example 9.3.1

Consider the linear transformation $f : \mathbb{R}^3 \to \mathbb{R}^2$, $f(x, y, z) = (x + y + z, 2x - y + 3z)$. Find a basis for (a) its kernel and (b) its image.

Solution

(a) The kernel consists of all element of $\mathbb{R}^3$ which map to zero. If $(a, b, c) \in \ker(f)$ then

$$f(a, b, c) = (a + b + c, 2a - b + 3c) = (0, 0).$$

Thus

$$a + b + c = 0$$
$$2a - b + 3c = 0.$$

Solving these equations by reduction to row echelon form gives $(a, b, c) = \mu(-4, 1, 3)$ where $\mu \in \mathbb{R}$ is a free variable. Thus any multiple of $(-4, 1, 3)$ is in the kernel of f. A basis for the kernel is $(-4, 1, 3)$ and thus the dimension of the kernel, also referred to as the **nullity** of f, is 1.

(b) Every vector in V can be written as a linear combination of its basis vectors. To determine the image we consider how each basis vector in V is mapped to W. We shall use the standard basis of $\mathbb{R}^3$. Then

$$f(1, 0, 0) = (1, 2), \quad f(0, 1, 0) = (1, -1), \quad f(0, 0, 1) = (1, 3).$$

Any vector in the image must be a combination of these vectors. Clearly this set is linearly dependent. It is easily verified that the first two are linearly independent and can be used as a basis for $\text{Im}(f)$. The image is therefore the whole of $\mathbb{R}^2$, that is, it has dimension 2. The dimension of the image of f is also known as the **rank** of f.

Note from the previous example in which $f : \mathbb{R}^3 \to \mathbb{R}^2$,

$$\dim \mathbb{R}^3 = \dim \ker(f) + \dim \text{Im}(f).$$

This result is true more generally and is known as the **dimension theorem** which we state without proof:

Theorem 9.2 The dimension theorem

For any linear transformation, $f : V \to W$,

$$dim\, V = dim\, ker(f) + dim\, Im(f).$$

9.4 Linear functionals

We begin by noting that every field is a vector space over itself. Here we consider linear transformations from vector spaces to their underlying fields.

Definition 9.5 Linear functional
A linear transformation from the vector space V to the underlying field $\mathbb{F}$, i.e., $f : V \to \mathbb{F}$ is called a **linear functional** *or* **linear form**.

Example 9.4.1

Consider the vector space over $\mathbb{C}$ of column vectors in $\mathbb{C}^3$, that is vectors of the form

$$\begin{pmatrix} c_1 \\ c_2 \\ c_3 \end{pmatrix}$$

where c_1, c_2 and c_3 are complex numbers. The function $f : \mathbb{C}^3 \to \mathbb{C}$ defined by

$$f \begin{pmatrix} c_1 \\ c_2 \\ c_3 \end{pmatrix} = 3c_1 + 7c_2 - 2c_3$$

is a linear functional. For example, if $u = \begin{pmatrix} 1 + i \\ 2 - 3i \\ 8 \end{pmatrix}$ then

$$f(u) = 3(1 + i) + 7(2 - 3i) - 2(8) = 1 - 18i.$$

It is clear that the functional maps vectors in $\mathbb{C}^3$ to the underlying field $\mathbb{C}$.

The linear functional can be represented as a row vector $\begin{pmatrix} 3 & 7 & -2 \end{pmatrix}$. Then, via matrix multiplication we have

$$\begin{pmatrix} 3 & 7 & -2 \end{pmatrix} \begin{pmatrix} 1 + i \\ 2 - 3i \\ 8 \end{pmatrix} = 1 - 18i.$$

Exercises

9.1   Show that $f : \mathbb{R}^3 \to \mathbb{R}$, $f(x, y, z) = 4x + 2y - 7z$ is a linear functional.

9.2   Determine whether or not $f : \mathbb{R}^3 \to \mathbb{R}$, $f(x, y, z) = 4x + 2y - 7z + 3$ is a linear functional.

9.3   Determine whether or not $f : \mathbb{R}^3 \to \mathbb{R}$, $f(x, y, z) = xyz$ is a linear functional.

9.5   Matrix representations of linear transformations

If V and W are finite dimensional vector spaces, over the same field $\mathbb{F}$, with dimensions n and m respectively then any linear transformation $f : V \to W$ can be represented by an $m \times n$ matrix, A. The image of $v \in V$ under f is then found by premultiplying v by A. In this section we show how this is achieved.

Let V have basis $\{v_1, v_2, \ldots, v_n\}$ and let W have basis $\{w_1, w_2, \ldots, w_m\}$. Then for each basis vector v_i in V we calculate its image under f:

$$f(v_i) = a_{1i}w_1 + a_{2i}w_2 + \ldots + a_{mi}w_m, \qquad i = 1, 2, \ldots, n.$$

Here, the $a_{ji} \in \mathbb{F}$, $j = 1, 2, \ldots, m$, $i = 1, 2, \ldots, n$, are the components of $f(v_i)$ with respect to the chosen basis of W. An arbitrary vector in $v \in V$ is

$$v = c_1 v_1 + c_2 v_2 + \ldots + c_n v_n, \qquad c_i \in \mathbb{F}, \ i = 1, 2, \ldots, n$$

and so its coordinate vector in this basis is

$$[v] = \begin{pmatrix} c_1 \\ c_2 \\ \vdots \\ c_n \end{pmatrix}.$$

Then, because f is linear,

$$\begin{aligned}
f(v) &= f(c_1 v_1 + c_2 v_2 + \ldots + c_n v_n) \\
&= c_1 f(v_1) + c_2 f(v_2) + \ldots + c_n f(v_n) \\
&= c_1(a_{11}w_1 + a_{21}w_2 + \ldots + a_{m1}w_m) \\
&\quad + c_2(a_{12}w_1 + a_{22}w_2 + \ldots + a_{m2}w_m) \\
&\qquad \ddots \\
&\quad + c_n(a_{1n}w_1 + a_{2n}w_2 + \ldots + a_{mn}w_m).
\end{aligned}$$

Rearranging the right-hand side:

$$\begin{aligned}
f(v) &= (c_1 a_{11} + c_2 a_{12} + \ldots + c_n a_{1n})w_1 \\
&\quad + (c_1 a_{21} + c_2 a_{22} + \ldots + c_n a_{2n})w_2 \\
&\qquad \ddots \\
&\quad + (c_1 a_{m1} + c_2 a_{m2} + \ldots + c_n a_{mn})w_m
\end{aligned}$$

and so the coordinate vector of the image of v under f is

$$[f(v)] = \begin{pmatrix} c_1 a_{11} + c_2 a_{12} + \ldots + c_n a_{1n} \\ c_1 a_{21} + c_2 a_{22} + \ldots + c_n a_{2n} \\ \vdots \\ c_1 a_{m1} + c_2 a_{m2} + \ldots + c_n a_{mn} \end{pmatrix} = \begin{pmatrix} a_{11} & a_{12} & \cdots & a_{1n} \\ a_{21} & a_{22} & \cdots & a_{2n} \\ \vdots & \vdots & \ddots & \vdots \\ a_{m1} & a_{m2} & \cdots & a_{mn} \end{pmatrix} \begin{pmatrix} c_1 \\ c_2 \\ \vdots \\ c_n \end{pmatrix}.$$

We see that the image of $v \in V$ under f is then found by premultiplying the coordinate vector of v by A where

$$A = \begin{pmatrix} a_{11} & a_{12} & \cdots & a_{1n} \\ a_{21} & a_{22} & \cdots & a_{2n} \\ \vdots & \vdots & \ddots & \vdots \\ a_{m1} & a_{m2} & \cdots & a_{mn} \end{pmatrix}.$$

Thus the *columns* of the matrix A representing the linear transformation f are precisely the coordinates of the images of the domain's basis vectors. Consider the following example.

Example 9.5.1

Consider the following linear transformation $f : \mathbb{R}^2 \to \mathbb{R}^3$:

$$f(x, y) = (3x + 4y, 2x - y, 5x + y).$$

Assume we are working with the standard bases in both vector spaces. We calculate the image of the domain's basis vectors:

$$f(1, 0) = (3, 2, 5) = 3(1, 0, 0) + 2(0, 1, 0) + 5(0, 0, 1).$$
$$f(0, 1) = (4, -1, 1) = 4(1, 0, 0) - 1(0, 1, 0) + 1(0, 0, 1).$$

Now form a matrix, A, whose columns are the coefficients in the rows of the above expressions:

$$A = \begin{pmatrix} 3 & 4 \\ 2 & -1 \\ 5 & 1 \end{pmatrix}.$$

This is the required matrix representation of the linear transformation f with respect to the stated bases. If we now choose any vector in $\mathbb{R}^2$, $\begin{pmatrix} a \\ b \end{pmatrix}$ say, premultiplication by A will give the image. That is

$$\begin{pmatrix} 3 & 4 \\ 2 & -1 \\ 5 & 1 \end{pmatrix} \begin{pmatrix} a \\ b \end{pmatrix} = \begin{pmatrix} 3a + 4b \\ 2a - b \\ 5a + b \end{pmatrix}.$$

In summary, to find the matrix representation of a linear transformation $f : V \to W$, we have the following:

Theorem 9.3

1. Let V have basis $\{v_1, v_2, \ldots, v_n\}$ and let W have basis $\{w_1, w_2, \ldots, w_m\}$.
2. Calculate $f(v_i)$, $i = 1, 2, \ldots, n$, expressing each in terms of the co-domain basis vectors.

3. *Form an $m \times n$ matrix, A, whose columns are the coefficients of each co-domain basis vector obtained in 2. This is the required matrix representation.*

Importantly, note that the representation depends upon the basis chosen. Different bases will give different matrix representations.

Example 9.5.2

Consider $V = \mathbb{R}^3$ with basis $e_1 = \begin{pmatrix} 1 \\ 1 \\ 0 \end{pmatrix}$, $e_2 = \begin{pmatrix} 0 \\ 1 \\ 1 \end{pmatrix}$, $e_3 = \begin{pmatrix} 0 \\ 0 \\ 1 \end{pmatrix}$. Consider $W = \mathbb{R}^2$

with basis $f_1 = \begin{pmatrix} 1 \\ 1 \end{pmatrix}$, $f_2 = \begin{pmatrix} 1 \\ -1 \end{pmatrix}$. Find the matrix representation of $f : \mathbb{R}^3 \to \mathbb{R}^2$,

$f \begin{pmatrix} x \\ y \\ z \end{pmatrix} = \begin{pmatrix} x + y + z \\ x - y - z \end{pmatrix}$ with respect to the given bases.

Solution

First find the image of the basis vectors:

$$f(e_1) = f \begin{pmatrix} 1 \\ 1 \\ 0 \end{pmatrix} = \begin{pmatrix} 2 \\ 0 \end{pmatrix}, \quad f(e_2) = f \begin{pmatrix} 0 \\ 1 \\ 1 \end{pmatrix} = \begin{pmatrix} 2 \\ -2 \end{pmatrix},$$

$$f(e_3) = f \begin{pmatrix} 0 \\ 0 \\ 1 \end{pmatrix} = \begin{pmatrix} 1 \\ -1 \end{pmatrix},$$

and in terms of the given co-domain basis:

$$f(e_1) = 1 \begin{pmatrix} 1 \\ 1 \end{pmatrix} + 1 \begin{pmatrix} 1 \\ -1 \end{pmatrix}, \quad f(e_2) = 0 \begin{pmatrix} 1 \\ 1 \end{pmatrix} + 2 \begin{pmatrix} 1 \\ -1 \end{pmatrix},$$

$$f(e_3) = 0 \begin{pmatrix} 1 \\ 1 \end{pmatrix} + 1 \begin{pmatrix} 1 \\ -1 \end{pmatrix}.$$

Finally, forming the matrix whose columns are the coefficients in the rows above:

$$A = \begin{pmatrix} 1 & 0 & 0 \\ 1 & 2 & 1 \end{pmatrix}.$$

You should check for yourself that this matrix does indeed represent the given transforma-

tion. For example, use it to find $f \begin{pmatrix} 3 \\ 4 \\ 2 \end{pmatrix}$ (remember to work with the correct bases).

9.6   Bilinear maps

Definition 9.6 Bilinear map
*Consider vector spaces U, V, W over the same field $\mathbb{F}$. Let $u, u_1, u_2 \in U$, $v, v_1, v_2 \in V$. Then a **bilinear map**, f, takes vectors from the Cartesian product $U \times V$ to W and satisfies*

$$f : U \times V \to W$$

$$f(u_1 + u_2, v) = f(u_1, v) + f(u_2, v), \qquad f(u, v_1 + v_2) = f(u, v_1) + f(u, v_2),$$
$$kf(u, v) = f(ku, v) = f(u, kv).$$

Example 9.6.1

Let $u = \begin{pmatrix} u^{(1)} \\ u^{(2)} \end{pmatrix}$, $v = \begin{pmatrix} v^{(1)} \\ v^{(2)} \end{pmatrix}$ be elements of $\mathbb{R}^2$. Consider the map $f : \mathbb{R}^2 \times \mathbb{R}^2 \to \mathbb{R}^2$

defined by $f(u, v) = f\left(\begin{pmatrix} u^{(1)} \\ u^{(2)} \end{pmatrix}, \begin{pmatrix} v^{(1)} \\ v^{(2)} \end{pmatrix} \right) = \begin{pmatrix} u^{(2)} v^{(2)} \\ u^{(1)} v^{(1)} \end{pmatrix}$.

a) If $u = \begin{pmatrix} 3 \\ 4 \end{pmatrix}$, $v = \begin{pmatrix} 2 \\ -1 \end{pmatrix}$ find $f(u, v)$.

b) Show that f is a bilinear map.

Solution

a) $f(u, v) = f\left(\begin{pmatrix} 3 \\ 4 \end{pmatrix}, \begin{pmatrix} 2 \\ -1 \end{pmatrix} \right) = \begin{pmatrix} (4)(-1) \\ (3)(2) \end{pmatrix} = \begin{pmatrix} -4 \\ 6 \end{pmatrix}$.

b) Let $u_1 = \begin{pmatrix} u_1^{(1)} \\ u_1^{(2)} \end{pmatrix}$, $u_2 = \begin{pmatrix} u_2^{(1)} \\ u_2^{(2)} \end{pmatrix}$. Let $v = \begin{pmatrix} v^{(1)} \\ v^{(2)} \end{pmatrix}$.

Then

$$u_1 + u_2 = \begin{pmatrix} u_1^{(1)} + u_2^{(1)} \\ u_1^{(2)} + u_2^{(2)} \end{pmatrix}.$$

Applying the bilinear map:

$$f(u_1 + u_2, v) = f\left(\begin{pmatrix} u_1^{(1)} + u_2^{(1)} \\ u_1^{(2)} + u_2^{(2)} \end{pmatrix}, \begin{pmatrix} v^{(1)} \\ v^{(2)} \end{pmatrix} \right) = \begin{pmatrix} (u_1^{(2)} + u_2^{(2)}) v^{(2)} \\ (u_1^{(1)} + u_2^{(1)}) v^{(1)} \end{pmatrix}$$

$$= \begin{pmatrix} u_1^{(2)} v^{(2)} \\ u_1^{(1)} v^{(1)} \end{pmatrix} + \begin{pmatrix} u_2^{(2)} v^{(2)} \\ u_2^{(1)} v^{(1)} \end{pmatrix}$$

$$= f(u_1, v) + f(u_2, v).$$

In a similar way it is straightforward to verify the linearity in the second argument. Further,

$$kf(u, v) = kf\left(\begin{pmatrix} u^{(1)} \\ u^{(2)} \end{pmatrix}, \begin{pmatrix} v^{(1)} \\ v^{(2)} \end{pmatrix} \right) = k \begin{pmatrix} u^{(2)} v^{(2)} \\ u^{(1)} v^{(1)} \end{pmatrix}$$

$$= \begin{pmatrix} ku^{(2)} v^{(2)} \\ ku^{(1)} v^{(1)} \end{pmatrix}$$

$$= f(ku, v) \text{ or alternatively } f(u, kv)$$

as required. Hence, f is a bilinear map.

Exercises

9.4   Let $u = (x_1, x_2)$, $v = (y_1, y_2)$ be elements of $\mathbb{R}^2$. Consider the inner product on $\mathbb{R}^2$ defined by $f : \mathbb{R}^2 \times \mathbb{R}^2 \to \mathbb{R}$, $f(u, v) = \langle u, v \rangle = \sum_{i=1}^{2} x_i y_i$. Show that f is a bilinear map.

9.7   End-of-chapter exercises

1. Consider the linear transformation $f : \mathbb{R}^3 \to \mathbb{R}^3$,

$$f(x, y, z) = (x, y, 0).$$

   This is a projection mapping of a point in three-dimensional space onto the xy plane.

   a) Verify that f is indeed a linear transformation.

   b) Find its matrix representation, P, in the standard basis $(1, 0, 0)$, $(0, 1, 0)$, $(0, 0, 1)$.

   c) A projector, P, is a linear transformation with the property that $P^2 = P$. Verify that P is indeed a projector.

2. Consider the linear transformation $f : \mathbb{C}^2 \to \mathbb{C}^2$,

$$f(z_1, z_2) = (\tfrac{1}{2}z_1 + \tfrac{1}{2}z_2, \tfrac{1}{2}z_1 + \tfrac{1}{2}z_2).$$

   a) Verify that f is indeed a linear transformation.

   b) Find its matrix representation, P, in the standard basis $(1, 0)$, $(0, 1)$.

   c) Show that $P^2 = P$ and hence verify that P is indeed a projector.

3. The transformation $W(\theta) : \mathbb{R}^2 \to \mathbb{R}^2$, which rotates a point in the xy plane anticlockwise through an angle θ is given by the matrix

$$W(\theta) = \begin{pmatrix} \cos\theta & -\sin\theta \\ \sin\theta & \cos\theta \end{pmatrix}.$$

   Show that W is a linear transformation.

4. Let $u = (x_1, x_2, x_3)$, $v = (y_1, y_2, y_3)$ be elements of $\mathbb{R}^3$. Consider the inner product on $\mathbb{R}^3$ defined by $f : \mathbb{R}^3 \times \mathbb{R}^3 \to \mathbb{R}$, $f(u, v) = \langle u, v \rangle = \sum_{i=1}^{3} x_i y_i$. Show that f is a bilinear map.

5. Suppose $f : U \to V$, $g : U \to V$ are linear transformations over a field $\mathbb{F}$. The sum $f + g$ is defined to be the mapping $(f + g) : U \to V$ such that

$$(f + g)(u) = f(u) + g(u).$$

   Scalar multiplication kf, for $k \in \mathbb{F}$, is defined such that

$$(kf)(u) = k\,f(u).$$

   Show that $f + g$ and kf are also linear transformations.

6. Suppose $f : U \to V$ is a linear transformation. Show that $\ker(f)$ and $\mathrm{Im}(f)$ are subspaces of U and V, respectively.

7. Let $f : \mathbb{R}^3 \to \mathbb{R}^3$, $f(x, y, z) = (2x + 3y + z, x + z, y - z)$.

   a) Find a basis for the kernel of f.

   b) Find a basis for the image of f.

   c) Verify the dimension theorem.

8. Consider the transformation $f : \mathbb{C} \to \mathbb{C}$, $f(z) = z^*$, the complex conjugate of z.

   a) Show that if $\mathbb{C}$ is a vector space over $\mathbb{R}$, then f is a linear transformation.

   b) Show that if $\mathbb{C}$ is a vector space over $\mathbb{C}$, then f is not a linear transformation.

   This example highlights the importance of specifying the underlying field clearly.

9. Find the kernel of $A - \lambda I$ where $A = \begin{pmatrix} 3 & 8 \\ 0 & -1 \end{pmatrix}$ and $\lambda = 3$.

10. Consider the linear operator $\mathcal{A} : \mathbb{R}^n \to \mathbb{R}^n$ with matrix representation A. Show that the eigenspace corresponding to an eigenvalue λ is the kernel of $A - \lambda I$.

10

Tensor product spaces

10.1 Objectives

We have already described how a qubit is represented by a vector in the two-dimensional vector space $\mathbb{C}^2$, that is

$$|\psi\rangle = \alpha_0|0\rangle + \alpha_1|1\rangle \qquad \text{where } |\alpha_0|^2 + |\alpha_1|^2 = 1$$

and $\{|0\rangle, |1\rangle\}$ is an orthonormal basis for $\mathbb{C}^2$. The study of quantum systems involving two or more qubits necessitates the development of mathematical tools which capture the interactions between the qubits. This leads to the construction of objects called tensors which are elements of a vector space called a tensor product space. This material is essential since tensors are required for defining the means of data representation in quantum computation.

The objective of this chapter is to define the tensor product $\otimes$ of two vectors and show how this product can be calculated. This leads to the concept of an entangled state. We show how tensor products reside in a tensor product space and define an inner product and norm on that space.

10.2 Preliminary discussion

The **tensor product** is the result of combining two (or more) vectors in a particular way designed to capture the interaction between the various parts of the individual vectors. The results of finding tensor products reside in a vector space called the **tensor product space**, the elements of which are called **tensors**.

The formal construction of a tensor product space is somewhat abstract and can be difficult for the student to understand. So, initially, we develop the tools necessary for performing calculations with tensors. Later in the chapter we give a more formal treatment once some familiarity has been established.

Bilinear maps were defined in Chapter 9. Recall that if U, V, W are vector spaces over the same field $\mathbb{F}$, a bilinear map, f, takes vectors from the Cartesian product $U \times V$ to W, $f: U \times V \to W$, and satisfies

$$f(u_1 + u_2, v) = f(u_1, v) + f(u_2, v), \qquad f(u, v_1 + v_2) = f(u, v_1) + f(u, v_2),$$

$$kf(u, v) = f(ku, v) = f(u, kv).$$

Observe that these properties mean that f behaves linearly in each of the two slots separately (to see this, fix v and then fix u). We have already seen one way in which vectors can be combined: the inner, scalar or dot product, $\cdot$ or $\langle \ , \ \rangle$, in $\mathbb{R}^n$. The inner product is a function which takes two vectors, $u, v \in \mathbb{R}^n$ say, and produces a scalar $c \in \mathbb{R}$. It is straightforward to

DOI: 10.1201/9781003264569-10

show that the (real) inner product possesses the properties required of a bilinear map from $\mathbb{R}^n \times \mathbb{R}^n$ to $\mathbb{R}$. (We note in passing that the inner product on vectors in $\mathbb{C}^n$ is not a bilinear map). The tensor product, which we give the symbol $\otimes$, is constructed in such a way that it satisfies the properties required of a bilinear map.

Definition 10.1 The tensor product as a bilinear map
Consider vector spaces U and V over the same field, $\mathbb{F}$ (usually $\mathbb{C}$). For $u \in U$ and $v \in V$ their tensor product, written $u \otimes v$, is a bilinear map which satisfies the distributive properties given in Table 10.1. In Table 10.1 u, u_1, $u_2 \in U$, v, v_1, $v_2 \in V$, $\lambda \in \mathbb{F}$.

We use the same symbol for the tensor product space $U \otimes V$ in which the tensor products $u \otimes v$ reside. We shall draw upon these properties extensively in our calculations. The following section introduces the way in which tensor product calculations are performed.

10.3   Calculation of tensor products

Given vector spaces U and V of dimensions n and m, respectively, we may select vectors $u \in U$, $v \in V$ to form a tensor product $u \otimes v$. This new vector is an element of the tensor product space $U \otimes V$, a vector space of dimension nm, and is referred to as a **separable tensor**.

Example 10.3.1

Given that the vectors $\begin{pmatrix} 1 \\ 0 \end{pmatrix}$ and $\begin{pmatrix} 0 \\ 1 \end{pmatrix}$ are elements of $\mathbb{C}^2$, then

$$\begin{pmatrix} 1 \\ 0 \end{pmatrix} \otimes \begin{pmatrix} 0 \\ 1 \end{pmatrix}$$

is a tensor in the tensor product space $\mathbb{C}^2 \otimes \mathbb{C}^2$. Recall that using ket notation, we write $\begin{pmatrix} 1 \\ 0 \end{pmatrix} = |0\rangle$ and $\begin{pmatrix} 0 \\ 1 \end{pmatrix} = |1\rangle$, so that the tensor can be written $|0\rangle \otimes |1\rangle$. It is common practice to omit the symbol $\otimes$ and write this as $|0\rangle|1\rangle$, or even more briefly as $|01\rangle$. Similarly, $|1\rangle \otimes |0\rangle$ may be written $|1\rangle|0\rangle$ or $|10\rangle$, and so on.

The tensor product space $U \otimes V$ is a vector space. Within it, tensors can be added and multiplied by scalars to yield new tensors. Thus elements of $U \otimes V$ are linear combinations of separable tensors.

Example 10.3.2

Consider the following tensors in $\mathbb{C}^2 \otimes \mathbb{C}^2$:

$$|00\rangle, \quad |01\rangle, \quad |10\rangle, \quad |11\rangle.$$

Because they are elements of a vector space, we can perform addition and multiplication by a scalar from the field $\mathbb{C}$. Thus the following is an element of $\mathbb{C}^2 \otimes \mathbb{C}^2$:

$$\alpha_{00}|00\rangle + \alpha_{01}|01\rangle + \alpha_{10}|10\rangle + \alpha_{11}|11\rangle$$

where $\alpha_{00}, \alpha_{01}, \alpha_{10}, \alpha_{11} \in \mathbb{C}$.

The properties stated in Table 10.1 can be used to rewrite expressions involving tensors.

TABLE 10.1

Required properties of $\otimes$

(i)	$(u_1 + u_2) \otimes v = u_1 \otimes v + u_2 \otimes v$
(ii)	$u \otimes (v_1 + v_2) = u \otimes v_1 + u \otimes v_2$
(iii)	$\lambda(u \otimes v) = (\lambda u) \otimes v = u \otimes (\lambda v)$

Example 10.3.3

Suppose $u = \begin{pmatrix} 1 \\ 0 \end{pmatrix}$, $v = \begin{pmatrix} 0 \\ 1 \end{pmatrix}$. The tensor $(au + bv) \otimes cu$, where $a, b, c \in \mathbb{C}$ is

$$\left(a \begin{pmatrix} 1 \\ 0 \end{pmatrix} + b \begin{pmatrix} 0 \\ 1 \end{pmatrix} \right) \otimes \left(c \begin{pmatrix} 1 \\ 0 \end{pmatrix} \right)$$

and can be written using the distributive rules as

$$ac \begin{pmatrix} 1 \\ 0 \end{pmatrix} \otimes \begin{pmatrix} 1 \\ 0 \end{pmatrix} + bc \begin{pmatrix} 0 \\ 1 \end{pmatrix} \otimes \begin{pmatrix} 1 \\ 0 \end{pmatrix}.$$

Gaining fluency in manipulating expressions such as these is important for the study of quantum algorithms. In ket notation we would write

$$(a|0\rangle + b|1\rangle) \otimes c|0\rangle = ac|0\rangle \otimes |0\rangle + bc|1\rangle \otimes |0\rangle = ac|00\rangle + bc|10\rangle.$$

Some tensors formed by taking linear combinations within the tensor product space are not separable and are referred to as **entangled**. Consider the following Example.

Example 10.3.4

Consider the following element of the tensor product space $\mathbb{C}^2 \otimes \mathbb{C}^2$ formed as a linear combination of separable tensors:

$$\begin{pmatrix} 1 \\ 0 \end{pmatrix} \otimes \begin{pmatrix} 1 \\ 0 \end{pmatrix} + \begin{pmatrix} 0 \\ 1 \end{pmatrix} \otimes \begin{pmatrix} 0 \\ 1 \end{pmatrix}. \tag{10.1}$$

Show that it is impossible to write this tensor as $u \otimes v$ where $u, v \in \mathbb{C}^2$ and hence that it represents an entangled state.

Solution

Any vectors $u, v \in \mathbb{C}^2$ can be written as linear combinations of basis vectors, that is

$$u = a \begin{pmatrix} 1 \\ 0 \end{pmatrix} + b \begin{pmatrix} 0 \\ 1 \end{pmatrix} \qquad v = c \begin{pmatrix} 1 \\ 0 \end{pmatrix} + d \begin{pmatrix} 0 \\ 1 \end{pmatrix}$$

where $a, b, c, d \in \mathbb{C}$. Hence,

$$u \otimes v = \left(a \begin{pmatrix} 1 \\ 0 \end{pmatrix} + b \begin{pmatrix} 0 \\ 1 \end{pmatrix} \right) \otimes \left(c \begin{pmatrix} 1 \\ 0 \end{pmatrix} + d \begin{pmatrix} 0 \\ 1 \end{pmatrix} \right)$$

which, using the distributivity properties, can be written

$$u \otimes v = ac \begin{pmatrix} 1 \\ 0 \end{pmatrix} \otimes \begin{pmatrix} 1 \\ 0 \end{pmatrix} + ad \begin{pmatrix} 1 \\ 0 \end{pmatrix} \otimes \begin{pmatrix} 0 \\ 1 \end{pmatrix}$$
$$+ bc \begin{pmatrix} 0 \\ 1 \end{pmatrix} \otimes \begin{pmatrix} 1 \\ 0 \end{pmatrix} + bd \begin{pmatrix} 0 \\ 1 \end{pmatrix} \otimes \begin{pmatrix} 0 \\ 1 \end{pmatrix}.$$

Hence, if Equation 10.1 is to be expressed as a separable tensor, we require

$$ac = 1, \ ad = 0, \ bc = 0, \ bd = 1.$$

From $ad = 0$ it follows that either $a = 0$ or $d = 0$ but either of these options makes $ac = 1$ or $bd = 1$ impossible. Thus the given tensor (10.1) cannot be expressed as a separable tensor. It represents an entangled state. This particular state, written briefly as $|00\rangle + |11\rangle$, when normalised is known as a **Bell state** and is prominent in quantum computation.

Exercises

10.1 The tensors $|00\rangle - |11\rangle$, $|01\rangle + |10\rangle$ and $|01\rangle - |10\rangle$, once normalised, are so-called Bell states. Show that each state is entangled.

10.2 Show that $|00\rangle + |10\rangle$ is not an entangled state.

10.3 Show that $|10\rangle + |11\rangle$ is not an entangled state.

10.4 Consider the state

$$|0\rangle|f(0)\rangle + |1\rangle|f(1)\rangle$$

where $f(x) \in \{0, 1\}$. Show that if $f(0) = f(1)$ the state is a separable tensor (not entangled), whereas if $f(0) \neq f(1)$ the state is entangled.

We have seen that elements in the tensor product space $U \otimes V$ are linear combinations of separable tensors $u \otimes v$, with $u \in U$, $v \in V$. If $\{e_1, e_2, \ldots, e_n\}$ is a basis for U, and $\{f_1, f_2, \ldots, f_m\}$ is a basis for V then we know

$$u = \sum_{i=1}^{n} \alpha_i e_i, \qquad v = \sum_{j=1}^{m} \beta_j f_j,$$

for some α_i, β_j in the underlying field. Thus

$$u \otimes v = \sum_{i=1}^{n} \alpha_i e_i \otimes \sum_{j=1}^{m} \beta_j f_j.$$

This expression can be rewritten using the distributive rules to give

$$u \otimes v = \sum_{i=1}^{n} \sum_{j=1}^{m} \alpha_i \beta_j \ e_i \otimes f_j.$$

This means that any separable tensor can be written as a linear combination of the tensor products, $e_i \otimes f_j$, of the basis vectors, e_i and f_j in the original vector spaces. The number of such products is nm. We see that the 'building blocks' of the separable tensors are the nm products $e_i \otimes f_j$ and because any tensor is a linear combination of separable tensors it follows that the set $\{e_i \otimes f_j\}$ over all i, j, is a basis for the tensor product space. Thus the dimension of the tensor product space is nm and any element can be written

$$\sum_{i=1}^{n}\sum_{j=1}^{m}\beta_{i,j}\, e_i \otimes f_j \qquad \text{where } \beta_{i,j} \in \mathbb{F}.$$

Here we focus initially on the tensor product space $\mathbb{C}^2 \otimes \mathbb{C}^2$ because in quantum computation it is the finite dimensional tensor product spaces $\mathbb{C}^2 \otimes \mathbb{C}^2 \otimes \ldots \mathbb{C}^2$, ($n$ copies), written $\otimes^n \mathbb{C}^2$, that are of primary interest. Consider the following example when $n = 2$.

Example 10.3.5 A basis for $\mathbb{C}^2 \otimes \mathbb{C}^2$.

Suppose we choose the standard basis of $\mathbb{C}^2$

$$e_1 = \begin{pmatrix} 1 \\ 0 \end{pmatrix}, \quad e_2 = \begin{pmatrix} 0 \\ 1 \end{pmatrix}.$$

For any $u, v \in \mathbb{C}^2$ we can then write $u = \begin{pmatrix} u_1 \\ u_2 \end{pmatrix} = u_1 e_1 + u_2 e_2$, $v = \begin{pmatrix} v_1 \\ v_2 \end{pmatrix} = v_1 e_1 + v_2 e_2$.
Then

$$u \otimes v = (u_1 e_1 + u_2 e_2) \otimes (v_1 e_1 + v_2 e_2)$$
$$= u_1 v_1\, e_1 \otimes e_1 + u_1 v_2\, e_1 \otimes e_2 + u_2 v_1\, e_2 \otimes e_1 + u_2 v_2\, e_2 \otimes e_2.$$

Observe that the dimension of $\mathbb{C}^2 \otimes \mathbb{C}^2$ is four – there must be four tensors in the basis:

$$\{e_1 \otimes e_1, \; e_1 \otimes e_2, \; e_2 \otimes e_1, \; e_2 \otimes e_2\}.$$

To facilitate computation, we introduce a matrix notation for this basis, writing the four basis elements, $e_i \otimes e_j$, as:

$$e_1 \otimes e_1 = \begin{pmatrix} 1 \\ 0 \\ 0 \\ 0 \end{pmatrix}, \quad e_1 \otimes e_2 = \begin{pmatrix} 0 \\ 1 \\ 0 \\ 0 \end{pmatrix}, \quad e_2 \otimes e_1 = \begin{pmatrix} 0 \\ 0 \\ 1 \\ 0 \end{pmatrix}, \quad e_2 \otimes e_2 = \begin{pmatrix} 0 \\ 0 \\ 0 \\ 1 \end{pmatrix}.$$

It then follows that we can write

$$\begin{pmatrix} u_1 \\ u_2 \end{pmatrix} \otimes \begin{pmatrix} v_1 \\ v_2 \end{pmatrix} = u_1 v_1 \begin{pmatrix} 1 \\ 0 \\ 0 \\ 0 \end{pmatrix} + u_1 v_2 \begin{pmatrix} 0 \\ 1 \\ 0 \\ 0 \end{pmatrix} + u_2 v_1 \begin{pmatrix} 0 \\ 0 \\ 1 \\ 0 \end{pmatrix} + u_2 v_2 \begin{pmatrix} 0 \\ 0 \\ 0 \\ 1 \end{pmatrix}$$

$$= \begin{pmatrix} u_1 v_1 \\ u_1 v_2 \\ u_2 v_1 \\ u_2 v_2 \end{pmatrix}.$$

In ket notation we write

$$u = u_1 |0\rangle + u_2 |1\rangle, \qquad v = v_1 |0\rangle + v_2 |1\rangle,$$

so that

$$u \otimes v = (u_1|0\rangle + u_2|1\rangle) \otimes (v_1|0\rangle + v_2|1\rangle)$$
$$= u_1v_1|0\rangle \otimes |0\rangle + u_1v_2|0\rangle \otimes |1\rangle + u_2v_1|1\rangle \otimes |0\rangle + u_2v_2|1\rangle \otimes |1\rangle$$
$$= u_1v_1|00\rangle + u_1v_2|01\rangle + u_2v_1|10\rangle + u_2v_2|11\rangle.$$

The four basis elements thus obtained are referred to as the computational basis of $\mathbb{C}^2 \otimes \mathbb{C}^2$.

Definition 10.2 The computational basis of $\mathbb{C}^2 \otimes \mathbb{C}^2$

In matrix notation, the four computational basis elements are

$$e_1 \otimes e_1 = \begin{pmatrix} 1 \\ 0 \\ 0 \\ 0 \end{pmatrix}, \quad e_1 \otimes e_2 = \begin{pmatrix} 0 \\ 1 \\ 0 \\ 0 \end{pmatrix}, \quad e_2 \otimes e_1 = \begin{pmatrix} 0 \\ 0 \\ 1 \\ 0 \end{pmatrix}, \quad e_2 \otimes e_2 = \begin{pmatrix} 0 \\ 0 \\ 0 \\ 1 \end{pmatrix}$$

and in ket notation

$$|00\rangle, \quad |01\rangle, \quad |10\rangle, \quad |11\rangle.$$

We will frequently write this set of basis vectors as $B_{\mathbb{C}^2 \otimes \mathbb{C}^2}$ or $B_{\otimes^2 \mathbb{C}^2}$.

More generally, the tensor product space $\otimes^n \mathbb{C}^2$ has dimension 2^n which becomes huge as n increases. The basis will be written in a similar fashion, as $B_{\otimes^n \mathbb{C}^2} = \{|00\cdots0\rangle, \ldots, |11\cdots1\rangle\}$.

Example 10.3.6

a) Evaluate $|00\rangle\langle00|$.

   b) Evaluate $|01\rangle\langle01|$.

   c) Show that $\displaystyle\sum_{pq=00}^{11} |pq\rangle\langle pq|$ is the 4×4 identity matrix.

Solution

a) Recall that given the ket $|00\rangle = \begin{pmatrix} 1 \\ 0 \\ 0 \\ 0 \end{pmatrix}$ the corresponding bra is $\begin{pmatrix} 1 & 0 & 0 & 0 \end{pmatrix}$.

Then

$$|00\rangle\langle00| = \begin{pmatrix} 1 \\ 0 \\ 0 \\ 0 \end{pmatrix} \begin{pmatrix} 1 & 0 & 0 & 0 \end{pmatrix} = \begin{pmatrix} 1 & 0 & 0 & 0 \\ 0 & 0 & 0 & 0 \\ 0 & 0 & 0 & 0 \\ 0 & 0 & 0 & 0 \end{pmatrix}.$$

b)

$$|01\rangle\langle01| = \begin{pmatrix} 0 \\ 1 \\ 0 \\ 0 \end{pmatrix} \begin{pmatrix} 0 & 1 & 0 & 0 \end{pmatrix} = \begin{pmatrix} 0 & 0 & 0 & 0 \\ 0 & 1 & 0 & 0 \\ 0 & 0 & 0 & 0 \\ 0 & 0 & 0 & 0 \end{pmatrix}.$$

c) By evaluating $|10\rangle\langle10|$ and $|11\rangle\langle11|$ as in parts a) and b) it follows immediately that

$$\sum_{pq=00}^{11} |pq\rangle\langle pq| = \begin{pmatrix} 1 & 0 & 0 & 0 \\ 0 & 1 & 0 & 0 \\ 0 & 0 & 1 & 0 \\ 0 & 0 & 0 & 1 \end{pmatrix}$$

as required.

Exercises

10.5 Suppose $|+\rangle = \frac{1}{\sqrt{2}}(|0\rangle + |1\rangle)$ and $|-\rangle = \frac{1}{\sqrt{2}}(|0\rangle - |1\rangle)$.

    a) Write down an expression for $|+\rangle \otimes |+\rangle = |++\rangle$.

    b) Write down an expression for $|-\rangle \otimes |-\rangle = |--\rangle$.

    c) Deduce that $|++\rangle + |--\rangle = |00\rangle + |11\rangle$.

10.6 Write down the computational basis $B_{\otimes^3 \mathbb{C}^2}$ in both ket notation and matrix form.

10.7 Show that

$$\sum_{z=0}^{3} z|z\rangle\langle z| = \begin{pmatrix} 0 & 0 & 0 & 0 \\ 0 & 1 & 0 & 0 \\ 0 & 0 & 2 & 0 \\ 0 & 0 & 0 & 3 \end{pmatrix}.$$

    (Hint: express the numbers in the kets and bras as their binary equivalents, e.g., for $z = 2$, $z|z\rangle\langle z| = 2|10\rangle\langle 10|$.)

10.4   Inner products and norms on the tensor product space $\mathbb{C}^2 \otimes \mathbb{C}^2$

We have already seen that the inner product on $\mathbb{C}^2$ can be defined (Definition 4.12) as follows:

if $u = \begin{pmatrix} u_1 \\ u_2 \end{pmatrix} \in \mathbb{C}^2$, $v = \begin{pmatrix} v_1 \\ v_2 \end{pmatrix} \in \mathbb{C}^2$, then

$$\langle u, v \rangle = u_1^* v_1 + u_2^* v_2.$$

In general, the result is a complex number (i.e., a scalar). The norm of u is obtained from

$$\|u\|^2 = \langle u, u \rangle = u_1^* u_1 + u_2^* u_2.$$

We now consider how the inner product can be defined on a pair of tensors in $\mathbb{C}^2 \otimes \mathbb{C}^2$.

Definition 10.3 The inner product in the tensor product space
Suppose we have two separable tensors in $\mathbb{C}^2 \otimes \mathbb{C}^2$, say $u \otimes v$ and $w \otimes x$ where $u, v, w, x \in \mathbb{C}^2$. The inner product in the tensor product space $\mathbb{C}^2 \otimes \mathbb{C}^2$ is defined to be

$$\langle u \otimes v, w \otimes x \rangle = \langle u, w \rangle \cdot \langle v, x \rangle.$$

Because each of the individual inner products on the right hand side is a complex number, then the overall quantity on the right is the product of two complex numbers, i.e., another complex number, again a scalar. The $\cdot$ here indicates the multiplication of these scalars. It is possible to show from this definition that all the required properties of an inner product are satisfied. The definition of the norm follows from the inner product.

Definition 10.4 The norm of a tensor product

$$\|u \otimes v\|^2 = \langle u \otimes v, u \otimes v \rangle = \langle u, u \rangle \cdot \langle v, v \rangle = \|u\|^2 \|v\|^2.$$

Example 10.4.1

Find the inner product in $\mathbb{C}^2 \otimes \mathbb{C}^2$ given by

$$\left\langle \begin{pmatrix} 1+\mathrm{i} \\ 2-\mathrm{i} \end{pmatrix} \otimes \begin{pmatrix} 1-\mathrm{i} \\ \mathrm{i} \end{pmatrix}, \begin{pmatrix} \mathrm{i} \\ 2\mathrm{i} \end{pmatrix} \otimes \begin{pmatrix} 3 \\ -2\mathrm{i} \end{pmatrix} \right\rangle.$$

Solution

By definition the inner product equals

$$\left\langle \begin{pmatrix} 1+\mathrm{i} \\ 2-\mathrm{i} \end{pmatrix}, \begin{pmatrix} \mathrm{i} \\ 2\mathrm{i} \end{pmatrix} \right\rangle \cdot \left\langle \begin{pmatrix} 1-\mathrm{i} \\ \mathrm{i} \end{pmatrix}, \begin{pmatrix} 3 \\ -2\mathrm{i} \end{pmatrix} \right\rangle$$

which equals

$$\begin{pmatrix} 1-\mathrm{i} & 2+\mathrm{i} \end{pmatrix} \begin{pmatrix} \mathrm{i} \\ 2\mathrm{i} \end{pmatrix} \cdot \begin{pmatrix} 1+\mathrm{i} & -\mathrm{i} \end{pmatrix} \begin{pmatrix} 3 \\ -2\mathrm{i} \end{pmatrix}.$$

Calculating the individual inner products gives:

$$\Big((1-\mathrm{i})(\mathrm{i}) + (2+\mathrm{i})(2\mathrm{i})\Big) \cdot \Big((1+\mathrm{i})(3) + (-\mathrm{i})(-2\mathrm{i})\Big) = (-1+5\mathrm{i}) \cdot (1+3\mathrm{i}) = -16 + 2\mathrm{i}.$$

Whilst the inner product on the tensor product space is defined on separable tensors in $\mathbb{C}^2 \otimes \mathbb{C}^2$, we can extend the definition in a linear fashion. For example,

$$\langle u \otimes v + u' \otimes v', w \otimes x \rangle = \langle u \otimes v, w \otimes x \rangle + \langle u' \otimes v', w \otimes x \rangle.$$

10.5   Formal construction of the tensor product space

In this section we show how the tensor product space is formally constructed. Given two vector spaces U and V over the same field, $\mathbb{F}$, we begin by introducing the concept of a free vector space generated by the Cartesian product $U \times V$. As we shall see this space is huge, but the introduction of an equivalence relation and equivalence classes enable us to group together elements with certain well-defined properties, and in turn, these equivalence classes become the tensor products.

10.5.1   The free vector space generated by $U \times V$

To construct the required tensor product space we begin with two finite dimensional vector spaces U and V over the same field $\mathbb{F}$. Let U and V have dimensions n and m, respectively. Recall that the Cartesian product $U \times V$ is the set of ordered pairs (u, v) with $u \in U, v \in V$. In general, there will be an infinite number of such ordered pairs even when the underlying vector spaces each have finite dimension. Suppose we make any selection of such ordered pairs and form a finite linear combination of them. Such a finite linear combination would take the form

$$\sum_{i,j} \alpha_{i,j}(u_i, v_j) \qquad (u_i \in U, v_j \in V, \alpha_{i,j} \in \mathbb{F}).$$

The set containing all such linear combinations is called LinSpan $(U \times V)$.

Definition 10.5 The set LinSpan $(U \times V)$

LinSpan $(U \times V)$ is a set containing all finite linear combinations of ordered pairs (u_i, v_j) with $u_i \in U$, $v_j \in V$:

$$LinSpan(U \times V) = \left\{ \sum_{i,j} \alpha_{i,j}(u_i, v_j) \right\} \qquad (\alpha_{i,j} \in \mathbb{F}).$$

Example 10.5.1

Suppose we select just the three ordered pairs

$$(u_1, v_1), (u_1, v_5) \text{ and } (u_3, v_2).$$

A linear combination of these ordered pairs is then

$$a_{1,1}(u_1, v_1) + a_{1,5}(u_1, v_5) + a_{3,2}(u_3, v_2) \quad \text{where} \quad a_{1,1}, a_{1,5}, a_{3,2} \in \mathbb{F}.$$

This linear combination is an element of LinSpan $(U \times V)$.

It is crucial to realise in this discussion that the ordered pair (u_i, v_j) is regarded as a single, 'enclosed' entity. That is, we can't perform any further manipulation on it by, for example, taking the constants inside the parentheses, or by adding ordered pairs as we might do ordinarily. Expressions such as

$$a_{1,1}(u_1, v_1) + a_{1,5}(u_1, v_5) + a_{3,2}(u_3, v_2)$$

are sometimes referred to as 'formal sums', which means we write down what looks like a sum but we cannot add or scalar multiply the components in parentheses. It might be helpful to think of (u_i, v_j) as an impenetrable and unbreakable 'atom'.

The set LinSpan$(U \times V)$ can be endowed with the properties of addition and scalar multiplication in an obvious way which turn it into a vector space called a **free vector space** for which the ordered pairs (u_i, v_j) are the basis vectors.

Example 10.5.2

Suppose $U = V = \mathbb{R}$. Then

$$U \times V = \mathbb{R} \times \mathbb{R} = \{(u, v) : u, v \in \mathbb{R}\}.$$

LinSpan $(\mathbb{R} \times \mathbb{R})$ is a set containing all finite linear combinations of ordered pairs of real numbers. Consider two vectors in LinSpan$(\mathbb{R} \times \mathbb{R})$, say:

$$u_1 = 5(2,3) + 7(6,11) - 3(1,-1),$$

$$u_2 = 2(6,11) + 3(1,-1) + 8(9,8).$$

Note that $(2,3), (6,11), (1,-1)$ and $(9,8)$ are just some of the infinite number of basis elements of LinSpan $(\mathbb{R} \times \mathbb{R})$. Then, to illustrate scalar multiplication and vector addition, observe that

$$4u_1 = 20(2,3) + 28(6,11) - 12(1,-1),$$
$$u_1 + u_2 = 5(2,3) + 9(6,11) + 8(9,8).$$

No further simplification is possible.

Example 10.5.3

Working in the free vector space LinSpan($\mathbb{R} \times \mathbb{R}$), simplify, if possible,

(a) $(5,2) - (3,2) - (2,2)$,

(b) $5(2,4) - (10,20)$,

(c) $12(2,3) - 8(2,3)$.

Solution

(a), (b) In neither case can the expression be simplified. Remember that in the free vector space, each ordered pair (,) is treated as a distinct entity. In particular, $5(2,4) \neq (10,20)$.

(c) $12(2,3) - 8(2,3) = 4(2,3)$.

Example 10.5.4

Consider the following vectors in the free vector space LinSpan($\mathbb{C}^2 \times \mathbb{C}^2$). Each is a linear combination of ordered pairs in $\mathbb{C}^2 \times \mathbb{C}^2$.

$$u_1 = 7 \left(\begin{pmatrix} 1+i \\ 2-i \end{pmatrix}, \begin{pmatrix} 3i \\ 5-2i \end{pmatrix} \right) + 2 \left(\begin{pmatrix} 3-i \\ -8i \end{pmatrix}, \begin{pmatrix} 1+2i \\ 9-2i \end{pmatrix} \right).$$

$$u_2 = 4 \left(\begin{pmatrix} 1+i \\ 2-i \end{pmatrix}, \begin{pmatrix} 3i \\ 5-2i \end{pmatrix} \right) - 2 \left(\begin{pmatrix} 3-i \\ -8i \end{pmatrix}, \begin{pmatrix} 1+2i \\ 9-2i \end{pmatrix} \right).$$

They can be added to give:

$$u_1 + u_2 = 11 \left(\begin{pmatrix} 1+i \\ 2-i \end{pmatrix}, \begin{pmatrix} 3i \\ 5-2i \end{pmatrix} \right)$$

but no further simplification is possible.

The vector space LinSpan($U \times V$) thus defined has every element in $U \times V$ as a basis element. So this space has infinite dimension. In the following sections we shall see how to narrow it down to something more manageable and useful by grouping together elements with certain properties.

10.5.2   An equivalence relation on LinSpan($U \times V$).

We have seen that LinSpan($U \times V$) is a vector space. Suppose that S is a subspace of LinSpan($U \times V$) generated by all linear combinations of the form

1. $(u_1 + u_2, v) - (u_1, v) - (u_2, v)$,

2. $(u, v_1 + v_2) - (u, v_1) - (u, v_2)$,

3. $k(u, v) - (ku, v)$,

4. $k(u, v) - (u, kv)$.

Example 10.5.5

Suppose $V = W = \mathbb{R}$. Consider the vector x in LinSpan ($\mathbb{R} \times \mathbb{R}$):

$$x = (5, 11) - (5, 7) - (5, 4).$$

This vector cannot be simplified. But observe that it is of the form given above (see 2.) and hence $x \in S$.

Example 10.5.6

Suppose $V = W = \mathbb{R}$. Consider the vector y in LinSpan $(\mathbb{R} \times \mathbb{R})$:

$$y = 6(5, 11) - (5, 66).$$

This vector cannot be simplified. But observe that it is of the form given above (see 4.) and hence $x \in S$.

Suppose that x and y are both elements of $\mathrm{LinSpan}(U \times V)$. We shall say that x is equivalent to y, written $x \sim y$ if $x - y \in S$, the aforementioned subspace. The relation $\sim$ is an equivalence relation which induces a partition so that every vector in the set $\mathrm{LinSpan}(U \times V)$, (u, v) say, is in one, and only one, equivalence class or coset written $[(u, v)]$. The set of equivalence classes themselves form another vector space written $\mathrm{LinSpan}(U \times V)/\sim$. Thus this is the quotient space formed by factoring $\mathrm{LinSpan}(U \times V)$ by S.

10.5.3   Definition of the tensor product space

Definition 10.6 *The **tensor product space**, written $U \otimes V$, is defined to be the space of equivalence classes:*

$$U \otimes V = LinSpan(U \times V)/\sim$$

with the equivalence class of (u, v) given by

$$[(u, v)] = \{x \in LinSpan(U \times V) : x \sim (u, v)\}$$

where the equivalence relation $\sim$ is defined by

$$x \sim y \ if \ x - y \in S$$

and S is the subspace of $LinSpan(U \times V)$ spanned by vectors of the form

    1. $(u_1 + u_2, v) - (u_1, v) - (u_2, v)$,   ·

    2. $(u, v_1 + v_2) - (u, v_1) - (u, v_2)$,

    3. $k(u, v) - (ku, v)$,

    4. $k(u, v) - (u, kv)$.

The above equivalence class can be written equivalently as

$$[(u, v)] = \{x \in \mathrm{LinSpan}(U \times V) : x - (u, v) \in S\}$$

or as

$$[(u, v)] = \{x \in \mathrm{LinSpan}(U \times V) : x - (u, v) = s \text{ for some } s \in S\}.$$

The elements of the tensor product space $U \otimes V$ are then written as

$$[(u, v)] \text{ or more usually } u \otimes v.$$

(Note that it is usual to use the symbol $\otimes$ both for the tensor product of individual tensors and in the tensor product space definition.)

We now have every vector in $\text{LinSpan}(U \times V)$ assigned to its equivalence class.

$$u \otimes v = [(u, v)] = \{x \in \text{LinSpan}(U \times V) : x = (u, v) + s \text{ for some } s \in S\}$$
$$= (u, v) + S.$$

Thus the tensor product is written as the sum of a vector and a subspace (see Section 6.13). Because of the way the tensor product space was constructed, it follows that the tensor product $u \otimes v$ obeys the properties given in Table 10.1. We now detail why this is the case.

Property (i) of Table 10.1: $(u_1 + u_2) \otimes v = u_1 \otimes v + u_2 \otimes v$.

By definition

$$(u_1 + u_2) \otimes v = [(u_1 + u_2, v)]$$
$$= (u_1 + u_2, v) + S$$
$$= (u_1 + u_2, v) - s + S \text{ for } \underline{\text{any}} \ s \in S.$$

Suppose now we choose s to be

$$s = (u_1 + u_2, v) - (u_1, v) - (u_2, v)$$

which is clearly in S (Property 1 above). Then

$$(u_1 + u_2) \otimes v = (u_1 + u_2, v) - ((u_1 + u_2, v) - (u_1, v) - (u_2, v)) + S$$
$$= (u_1, v) + (u_2, v) + S$$
$$= ((u_1, v) + S) + ((u_2, v) + S)$$
$$= [(u_1, v)] + [(u_2, v)]$$
$$= u_1 \otimes v + u_2 \otimes v$$

as required.

Property (iii) of Table 10.1: $\lambda(u \otimes v) = (\lambda u) \otimes v$

By definition

$$\lambda(u \otimes v) = \lambda[(u, v)]$$
$$= \lambda((u, v) + S)$$
$$= \lambda(u, v) + S$$
$$= \lambda(u, v) - s + S \text{ for any } s \in S.$$

Choose $s = \lambda(u, v) - (\lambda u, v)$ which is clearly in S (Property 3 above). Then

$$\lambda(u \otimes v) = \lambda(u, v) - (\lambda(u, v) - (\lambda u, v)) + S$$
$$= (\lambda u, v) + S$$
$$= [(\lambda u, v)]$$
$$= (\lambda u) \otimes v$$

as required. The remaining two properties can be verified in a similar way.

10.6 End-of-chapter exercises

1. State the dimension of each of the following tensor product spaces and give their natural bases:

   (a) $\mathbb{R}^2 \otimes \mathbb{R}^2$,

   (b) $\mathbb{R}^3 \otimes \mathbb{R}^2$,

   (c) $\mathbb{R}^2 \otimes \mathbb{R}^4$.

2. Determine whether

$$|00\rangle - |01\rangle + |10\rangle - |11\rangle$$

   is an entangled tensor.

3. Calculate the inner product

$$\left\langle \begin{pmatrix} 4 \\ 5 \end{pmatrix} \otimes \begin{pmatrix} 2 \\ -3 \end{pmatrix}, \begin{pmatrix} 1 \\ -1 \end{pmatrix} \otimes \begin{pmatrix} -2 \\ 2 \end{pmatrix} \right\rangle.$$

4. Calculate the inner product

$$\left\langle \begin{pmatrix} 1-i \\ 3+2i \end{pmatrix} \otimes \begin{pmatrix} 3i \\ 1+i \end{pmatrix}, \begin{pmatrix} 1+i \\ -3i \end{pmatrix} \otimes \begin{pmatrix} 2i \\ 5+i \end{pmatrix} \right\rangle.$$

5. Let

$$B_1 = \frac{1}{\sqrt{2}}(|0\rangle \otimes |0\rangle + |1\rangle \otimes |1\rangle), \quad B_2 = \frac{1}{\sqrt{2}}(|0\rangle \otimes |1\rangle + |1\rangle \otimes |0\rangle),$$

$$B_3 = \frac{1}{\sqrt{2}}(|0\rangle \otimes |0\rangle - |1\rangle \otimes |1\rangle), \quad B_4 = \frac{1}{\sqrt{2}}(|0\rangle \otimes |1\rangle - |1\rangle \otimes |0\rangle).$$

   Evaluate:

   (a) $B_1 + B_3$,

   (b) $B_2 + B_4$,

   (c) $B_1 - B_3$,

   (d) $B_2 - B_4$.

6. Use the distributive properties of the tensor product to show that

$$(|0\rangle + |1\rangle) \otimes (|0\rangle \otimes |0\rangle + |1\rangle \otimes |1\rangle)$$

   can be expressed as

$$|000\rangle + |100\rangle + |011\rangle + |111\rangle.$$

7. If $\{e_1, e_2, \ldots, e_n\}$ is a basis for U, and $\{f_1, f_2, \ldots, f_m\}$ is a basis for V then if $u \in U$, $v \in V$,

$$u = \sum_{i=1}^{n} \alpha_i e_i, \qquad v = \sum_{j=1}^{m} \beta_j f_j,$$

for some α_i, β_j in the underlying field. Use the distributive properties of the tensor product (Table 10.1) to show that any separable tensor can be written

$$u \otimes v = \sum_{i=1}^{n} \sum_{j=1}^{m} \alpha_i \beta_j \ e_i \otimes f_j.$$

8. Express $\displaystyle\sum_{pq=00}^{11} |pq\rangle \langle p\bar{q}|$ as a 4×4 matrix. (Note the sum is over all elements of the set $\{00, 01, 10, 11\}$.)

11

Linear operators and their matrix representations

11.1   Objectives

Linear operators play a prominent role in the study of quantum computation. The principal objective of this chapter is to define a linear operator and demonstrate how one can be represented by a matrix. Two particularly important types of linear operator are unitary operators and self-adjoint operators. Unitary operators determine the time evolution, or dynamics of a quantum system by acting on the state vectors residing in a vector space V. We shall be interested in the way that an operator, $\mathcal{A}$ say, acts on a quantum state vector $|\psi\rangle$ to produce another state vector $|\phi\rangle$, i.e.,

$$\mathcal{A}|\psi\rangle = |\phi\rangle, \qquad |\psi\rangle, |\phi\rangle \in V.$$

That is, unitary operators bring about state-to-state transitions. Furthermore, self-adjoint operators are used to model physical observables. Following the measurement of an observable on a quantum state the post-measurement state is related to the eigenvectors of the operator and the possible measured values are its eigenvalues. We demonstrate how to calculate these eigenvalues and eigenvectors. The relevance of self-adjoint and unitary operators to quantum computation is explored in detail when we study the axioms of quantum computation in Chapter 13. Finally, given two linear operators, $\mathcal{A}$ and $\mathcal{B}$ say, we explain how their tensor product $\mathcal{A} \otimes \mathcal{B}$ is calculated.

11.2   Linear operators

In Chapter 9 we saw that a **linear transformation**, f, is a function or mapping between vector spaces, V and W say, over the same field $\mathbb{F}$, which preserves the linear structure of the vector space V. That is, for any $u, v \in V$ and $k \in \mathbb{F}$,

$$f : V \to W$$
$$f(u + v) = f(u) + f(v)$$
$$f(ku) = kf(u)$$

Now consider the special case when both domain and co-domain are the same vector space V over the field $\mathbb{F}$ which we will henceforth take to be $\mathbb{C}$.

Definition 11.1 Linear operator
*A **linear operator**, $\mathcal{A}$, is a linear transformation from a vector space V, over the field $\mathbb{C}$, to itself:*

$$\mathcal{A} : V \to V$$

DOI: 10.1201/9781003264569-11

with the property that

$$A(u + v) = A(u) + A(v)$$
$$A(ku) = kA(u)$$

for any $u, v \in V$ and $k \in \mathbb{C}$.

Given a basis of V we shall see that the operator A can be represented by a matrix. A calligraphic font has been used to denote the linear operator, but occasionally we shall abuse this notation and simply denote A by its matrix representation, A. (Some authors write the operator as $\hat{A}$ to distinguish it from the matrix A.)

Example 11.2.1

Consider the operator $A : \mathbb{C}^2 \to \mathbb{C}^2$ defined by

$$A \begin{pmatrix} z_1 \\ z_2 \end{pmatrix} = \begin{pmatrix} 2z_1 - 3z_2 \\ 4z_1 + z_2 \end{pmatrix}$$

where $z_1, z_2 \in \mathbb{C}$.

a) Find $A \begin{pmatrix} 1 \\ 2 \end{pmatrix}$.

b) Find $A \begin{pmatrix} 1 + i \\ 2 - i \end{pmatrix}$.

c) Show that A is a linear operator.

Solution

a) $A \begin{pmatrix} 1 \\ 2 \end{pmatrix} = \begin{pmatrix} 2(1) - 3(2) \\ 4(1) + 2 \end{pmatrix} = \begin{pmatrix} -4 \\ 6 \end{pmatrix}$.

b) $A \begin{pmatrix} 1 + i \\ 2 - i \end{pmatrix} = \begin{pmatrix} 2(1 + i) - 3(2 - i) \\ 4(1 + i) + (2 - i) \end{pmatrix} = \begin{pmatrix} -4 + 5i \\ 6 + 3i \end{pmatrix}$.

Clearly the operator A maps vectors in $\mathbb{C}^2$ to other vectors in $\mathbb{C}^2$.

c) Consider two vectors in $\mathbb{C}^2$, $u = \begin{pmatrix} u_1 \\ u_2 \end{pmatrix}$, $v = \begin{pmatrix} v_1 \\ v_2 \end{pmatrix}$, where $u_1, u_2, v_1, v_2 \in \mathbb{C}$. First note that

$$A(u) = A \begin{pmatrix} u_1 \\ u_2 \end{pmatrix} = \begin{pmatrix} 2u_1 - 3u_2 \\ 4u_1 + u_2 \end{pmatrix}.$$

$$A(v) = A \begin{pmatrix} v_1 \\ v_2 \end{pmatrix} = \begin{pmatrix} 2v_1 - 3v_2 \\ 4v_1 + v_2 \end{pmatrix}.$$

Adding u and v:

$$u + v = \begin{pmatrix} u_1 \\ u_2 \end{pmatrix} + \begin{pmatrix} v_1 \\ v_2 \end{pmatrix} = \begin{pmatrix} u_1 + v_1 \\ u_2 + v_2 \end{pmatrix}.$$

Applying the operator $\mathcal{A}$ to $u + v$ we obtain

$$\mathcal{A}(u + v) = \mathcal{A}\left(\begin{array}{c} u_1 + v_1 \\ u_2 + v_2 \end{array}\right) = \left(\begin{array}{c} 2(u_1 + v_1) - 3(u_2 + v_2) \\ 4(u_1 + v_1) + (u_2 + v_2) \end{array}\right)$$

$$= \left(\begin{array}{c} (2u_1 - 3u_2) + (2v_1 - 3v_2) \\ (4u_1 + u_2) + (4v_1 + v_2) \end{array}\right)$$

$$= \left(\begin{array}{c} 2u_1 - 3u_2 \\ 4u_1 + u_2 \end{array}\right) + \left(\begin{array}{c} 2v_1 - 3v_2 \\ 4v_1 + v_2 \end{array}\right)$$

$$= \mathcal{A}(u) + \mathcal{A}(v).$$

The first of the requirements for $\mathcal{A}$ to be a linear operator is therefore satisfied. Now, if $k \in \mathbb{C}$, $ku = k\left(\begin{array}{c} u_1 \\ u_2 \end{array}\right) = \left(\begin{array}{c} ku_1 \\ ku_2 \end{array}\right)$. Then

$$\mathcal{A}(ku) = \mathcal{A}\left(\begin{array}{c} ku_1 \\ ku_2 \end{array}\right)$$

$$= \left(\begin{array}{c} 2ku_1 - 3ku_2 \\ 4ku_1 + ku_2 \end{array}\right)$$

$$= k\left(\begin{array}{c} 2u_1 - 3u_2 \\ 4u_1 + u_2 \end{array}\right)$$

$$= k\mathcal{A}(u).$$

Thus the second of the requirements for $\mathcal{A}$ to be a linear operator is satisfied. We have confirmed that $\mathcal{A}$ is a linear operator.

Exercises

11.1 Consider $\mathcal{A} : \mathbb{R}^2 \to \mathbb{R}^2$ defined by $\mathcal{A}\left(\begin{array}{c} x_1 \\ x_2 \end{array}\right) = \left(\begin{array}{c} x_1 + x_2 + 1 \\ x_1 - x_2 \end{array}\right)$ where $x_1, x_2 \in \mathbb{R}$. Show that $\mathcal{A}$ is not a linear operator.

11.2 Show that the identity operator $\mathcal{I} : \mathbb{C}^2 \to \mathbb{C}^2$ defined by $\mathcal{I}\left(\begin{array}{c} z_1 \\ z_2 \end{array}\right) = \left(\begin{array}{c} z_1 \\ z_2 \end{array}\right)$ where $z_1, z_2 \in \mathbb{C}$ is a linear operator.

11.3 Show that the operator $\mathcal{A} : \mathbb{C}^2 \to \mathbb{C}^2$ defined by $\mathcal{A}\left(\begin{array}{c} z_1 \\ z_2 \end{array}\right) = \left(\begin{array}{c} z_1 \\ 0 \end{array}\right)$ is a linear operator.

11.4 Show that the operator $\mathcal{A} : \mathbb{C}^4 \to \mathbb{C}^4$ defined by $\mathcal{A}\left(\begin{array}{c} z_1 \\ z_2 \\ z_3 \\ z_4 \end{array}\right) = \left(\begin{array}{c} 0 \\ z_2 \\ 0 \\ z_4 \end{array}\right)$ is a linear operator.

11.5 Show that the kernel of the linear operator $\mathcal{A} : \mathbb{R}^2 \to \mathbb{R}^2$ defined by $\mathcal{A}\left(\begin{array}{c} x_1 \\ x_2 \end{array}\right) = \left(\begin{array}{c} x_1 + 2x_2 \\ 3x_1 + 4x_2 \end{array}\right)$ contains only the zero vector $\left(\begin{array}{c} 0 \\ 0 \end{array}\right)$. What is the dimension of $\ker(A)$?

11.6 Show that the kernel of the linear operator $\mathcal{A} : \mathbb{R}^2 \to \mathbb{R}^2$ defined by $\mathcal{A} \begin{pmatrix} x_1 \\ x_2 \end{pmatrix} = \begin{pmatrix} x_1 + 2x_2 \\ 2x_1 + 4x_2 \end{pmatrix}$ is spanned by the vector $\begin{pmatrix} -2 \\ 1 \end{pmatrix}$.

11.7 Consider the linear operators $\mathcal{A} : \mathbb{C}^2 \to \mathbb{C}^2$, $\mathcal{B} : \mathbb{C}^2 \to \mathbb{C}^2$. We define their sum by

$$(\mathcal{A} + \mathcal{B})u = \mathcal{A}u + \mathcal{B}u$$

for $u \in \mathbb{C}^2$, and multiplication by a scalar $k \in \mathbb{C}$ as

$$(k\mathcal{A})u = k(\mathcal{A}u).$$

Let $z_1, z_2 \in \mathbb{C}$. If $\mathcal{A}(z_1, z_2) = (z_1 - z_2, z_1 + z_2)$, $\mathcal{B}(z_1, z_2) = (z_1 + z_2, z_1 - z_2)$, find an expression for the operators $\mathcal{A} + \mathcal{B}$ and $k\mathcal{A}$.

11.8 If $\mathcal{A} : \mathbb{C}^2 \to \mathbb{C}^2$, $\mathcal{B} : \mathbb{C}^2 \to \mathbb{C}^2$, their product $\mathcal{AB}$ is defined by the composition

$$(\mathcal{AB})u = (\mathcal{A} \circ \mathcal{B})u = \mathcal{A}(\mathcal{B}u).$$

With $\mathcal{A}$ and $\mathcal{B}$ defined in the previous question:

   a) find an expression for $\mathcal{AB}$,

   b) find an expression for $\mathcal{BA}$.

Deduce that, in general, $\mathcal{AB} \neq \mathcal{BA}$.

11.3   The matrix representation of a linear operator

Consider a linear operator $\mathcal{A} : V \to V$. Suppose that $\mathcal{E} = \{e_1, e_2, \ldots, e_n\}$ is a basis for V. We can apply $\mathcal{A}$ to each basis vector and the result will be another vector in V. Hence,

$$\mathcal{A}(e_1) = a_{11}e_1 + a_{21}e_2 + \ldots + a_{n1}e_n$$
$$\mathcal{A}(e_2) = a_{12}e_1 + a_{22}e_2 + \ldots + a_{n2}e_n$$

$$\vdots \quad = \quad \vdots$$

$$\mathcal{A}(e_n) = a_{1n}e_1 + a_{2n}e_2 + \ldots + a_{nn}e_n$$

for some $a_{ij} \in \mathbb{C}, i, j = 1, 2, \ldots, n$. The transpose of the matrix of coefficients in the above expressions is the matrix representation of $\mathcal{A}$ in the given basis. Thus

$$A = \begin{pmatrix} a_{11} & a_{12} & \cdots & a_{1n} \\ a_{21} & a_{22} & \cdots & a_{2n} \\ \vdots & \vdots & \ddots & \vdots \\ a_{n1} & a_{n2} & \cdots & a_{nn} \end{pmatrix}.$$

Observe that each column of A is the coordinate vector of the image under $\mathcal{A}$ of a basis vector. It is important to note that changing the basis will change the representation.

Example 11.3.1 Finding a matrix representation of a linear operator

Suppose $z = (z_1, z_2) \in \mathbb{C}^2$. Consider the linear operator defined by

$$\mathcal{A} : \mathbb{C}^2 \to \mathbb{C}^2, \qquad \mathcal{A}(z) = \mathcal{A}(z_1, z_2) = (0, z_2).$$

Thus $\mathcal{A}$ acts on a row vector in $\mathbb{C}^2$ to produce another row vector in the same vector space. Suppose we choose $\mathcal{E} = \{e_1, e_2\}$ where $e_1 = (1, 0)$, $e_2 = (0, 1)$ as the basis of $\mathbb{C}^2$, i.e., the standard basis. Then, $z = (z_1, z_2)$ can be expressed in terms of this basis as

$$z = (z_1, z_2) = z_1(1, 0) + z_2(0, 1) = z_1 e_1 + z_2 e_2.$$

Thus the coordinate vector (see Chapter 6) corresponding to z in the stated basis is then

$$[z]_{\mathcal{E}} = \begin{pmatrix} z_1 \\ z_2 \end{pmatrix}.$$

Applying the operator $\mathcal{A}$ to each basis vector,

$$\mathcal{A}(e_1) = \mathcal{A}(1, 0) = (0, 0) = 0e_1 + 0e_2,$$

$$\mathcal{A}(e_2) = \mathcal{A}(0, 1) = (0, 1) = 0e_1 + 1e_2,$$

and the transpose of the matrix of coefficients gives the required matrix representation:

$$A = \begin{pmatrix} 0 & 0 \\ 0 & 1 \end{pmatrix}.$$

We can now use this matrix representation to carry out the transformation. Any vector z is transformed by multiplying its coordinate vector $[z]_{\mathcal{E}}$ by the matrix A. Thus

$$A[z]_{\mathcal{E}} = A \begin{pmatrix} z_1 \\ z_2 \end{pmatrix} = \begin{pmatrix} 0 & 0 \\ 0 & 1 \end{pmatrix} \begin{pmatrix} z_1 \\ z_2 \end{pmatrix} = \begin{pmatrix} 0 \\ z_2 \end{pmatrix}.$$

The result, $\begin{pmatrix} 0 \\ z_2 \end{pmatrix}$, is the coordinate vector following the transformation. We emphasise again that if we change the underlying basis, the matrix representation will change. Consider the following example.

Example 11.3.2 Finding a matrix representation of a linear operator

Suppose we have the same operator as in Example 11.3.1 but this time we choose the basis $\bar{\mathcal{E}} = \{\bar{e}_1, \bar{e}_2\}$ where $\bar{e}_1 = (\frac{1}{\sqrt{2}}, \frac{1}{\sqrt{2}})$, $\bar{e}_2 = (\frac{1}{\sqrt{2}}, -\frac{1}{\sqrt{2}})$. We first express $z = (z_1, z_2)$ in terms of this basis: let

$$(z_1, z_2) = k_1 \bar{e}_1 + k_2 \bar{e}_2 = k_1(\frac{1}{\sqrt{2}}, \frac{1}{\sqrt{2}}) + k_2(\frac{1}{\sqrt{2}}, -\frac{1}{\sqrt{2}})$$

Equating components

$$z_1 = \frac{1}{\sqrt{2}} k_1 + \frac{1}{\sqrt{2}} k_2, \qquad z_2 = \frac{1}{\sqrt{2}} k_1 - \frac{1}{\sqrt{2}} k_2$$

and solving for k_1 and k_2 we obtain $k_1 = \frac{\sqrt{2}(z_1 + z_2)}{2}$, $k_2 = \frac{\sqrt{2}(z_1 - z_2)}{2}$. Then

$$z = (z_1, z_2) = \frac{\sqrt{2}(z_1 + z_2)}{2} \bar{e}_1 + \frac{\sqrt{2}(z_1 - z_2)}{2} \bar{e}_2.$$

Thus the coordinate vector corresponding to z in the basis $\overline{\mathcal{E}}$ is

$$[z]_{\overline{\mathcal{E}}} = \begin{pmatrix} \frac{\sqrt{2}(z_1+z_2)}{2} \\ \frac{\sqrt{2}(z_1-z_2)}{2} \end{pmatrix}.$$

Applying the operator $\mathcal{A}$ to each basis vector

$$\mathcal{A}(\overline{e}_1) = \mathcal{A}(\frac{1}{\sqrt{2}}, \frac{1}{\sqrt{2}}) = (0, \frac{1}{\sqrt{2}})$$

which we express in terms of the basis $\overline{\mathcal{E}} = \{\overline{e}_1, \overline{e}_2\}$ where $\overline{e}_1 = (\frac{1}{\sqrt{2}}, \frac{1}{\sqrt{2}})$, $\overline{e}_2 = (\frac{1}{\sqrt{2}}, -\frac{1}{\sqrt{2}})$:

$$(0, \frac{1}{\sqrt{2}}) = \frac{1}{2}\overline{e}_1 - \frac{1}{2}\overline{e}_2.$$

Likewise,

$$\mathcal{A}(\overline{e}_2) = \mathcal{A}(\frac{1}{\sqrt{2}}, -\frac{1}{\sqrt{2}}) = (0, -\frac{1}{\sqrt{2}}) = -\frac{1}{2}\overline{e}_1 + \frac{1}{2}\overline{e}_2$$

and the transpose of the matrix of coefficients gives the required matrix representation:

$$A = \begin{pmatrix} \frac{1}{2} & -\frac{1}{2} \\ -\frac{1}{2} & \frac{1}{2} \end{pmatrix}.$$

As above, the matrix representation acts on column vectors:

$$Az = A[z]_{\overline{\mathcal{E}}} = A \begin{pmatrix} \frac{\sqrt{2}(z_1+z_2)}{2} \\ \frac{\sqrt{2}(z_1-z_2)}{2} \end{pmatrix} = \begin{pmatrix} \frac{1}{2} & -\frac{1}{2} \\ -\frac{1}{2} & \frac{1}{2} \end{pmatrix} \begin{pmatrix} \frac{\sqrt{2}(z_1+z_2)}{2} \\ \frac{\sqrt{2}(z_1-z_2)}{2} \end{pmatrix} = \begin{pmatrix} \frac{\sqrt{2}}{2}z_2 \\ -\frac{\sqrt{2}}{2}z_2 \end{pmatrix}.$$

The result is the coordinate vector with respect to the basis $\overline{\mathcal{E}}$. We emphasise again that changing the basis changes the matrix representation.

Example 11.3.3 The identity operator, $\mathcal{I}$, on $\mathbb{C}^2$.

Consider the linear operator $\mathcal{I} : \mathbb{C}^2 \to \mathbb{C}^2$ by $\mathcal{I}(z_1, z_2) = (z_1, z_2)$, where $z_1, z_2 \in \mathbb{C}$. The underlying field is $\mathbb{C}$. Note that the output vector is the same as the input, hence the name **identity** operator. Suppose we choose as a basis for $\mathbb{C}^2$ the vectors $e_1 = (1,0), e_2 = (0,1)$. Then it follows immediately that the matrix representation is the identity matrix
$I = \begin{pmatrix} 1 & 0 \\ 0 & 1 \end{pmatrix}.$

Exercises

11.9 Suppose $z = (z_1, z_2) \in \mathbb{C}^2$ and consider the linear operator $\mathcal{A} : \mathbb{C}^2 \to \mathbb{C}^2$, $\mathcal{A}(z) = \mathcal{A}(z_1, z_2) = (z_2, z_1)$. Find the matrix representation of $\mathcal{A}$ in the standard basis and in the basis $\overline{\mathcal{E}} = \{\overline{e}_1, \overline{e}_2\}$ where $\overline{e}_1 = (\frac{1}{\sqrt{2}}, \frac{1}{\sqrt{2}})$ and $\overline{e}_2 = (\frac{1}{\sqrt{2}}, -\frac{1}{\sqrt{2}})$.

11.4 The matrix representation of a linear operator when the underlying basis is orthonormal

Consider again a linear operator $\mathcal{A} : V \to V$. Suppose this time that we choose a basis for V, $\mathcal{E} = \{e_1, e_2, \ldots, e_n\}$, which is orthonormal so that $\langle e_i | e_j \rangle = \delta_{ij}$ with respect to the inner product $\langle u, v \rangle = \sum_{i=1}^{n} u_i^* v_i$. As before, we can apply $\mathcal{A}$ to each basis vector and the result will be another vector in V. Hence,

$$\mathcal{A}(e_1) = a_{11} e_1 + a_{21} e_2 + \ldots + a_{n1} e_n$$
$$\mathcal{A}(e_2) = a_{12} e_1 + a_{22} e_2 + \ldots + a_{n2} e_n$$
$$\vdots \quad = \quad \vdots$$
$$\mathcal{A}(e_n) = a_{1n} e_1 + a_{2n} e_2 + \ldots + a_{nn} e_n.$$

Now consider taking the inner product of each of the above expressions with each basis vector in turn: for example

$$\langle e_1 | \mathcal{A} e_1 \rangle = a_{11} \langle e_1 | e_1 \rangle + a_{21} \langle e_1 | e_2 \rangle + \ldots + a_{n1} \langle e_1 | e_n \rangle = a_{11}$$
$$\langle e_2 | \mathcal{A} e_1 \rangle = a_{11} \langle e_2 | e_1 \rangle + a_{21} \langle e_2 | e_2 \rangle + \ldots + a_{n1} \langle e_2 | e_n \rangle = a_{21}$$

and so on. Note $\langle e_i | e_j \rangle = \delta_{ij} = \begin{cases} 0 & i \neq j \\ 1 & i = j \end{cases}$.

Likewise,

$$\langle e_1 | \mathcal{A} e_2 \rangle = a_{12} \langle e_1 | e_1 \rangle + a_{22} \langle e_1 | e_2 \rangle + \ldots + a_{n2} \langle e_1 | e_n \rangle = a_{12}$$
$$\langle e_2 | \mathcal{A} e_2 \rangle = a_{12} \langle e_2 | e_1 \rangle + a_{22} \langle e_2 | e_2 \rangle + \ldots + a_{n2} \langle e_2 | e_n \rangle = a_{22}.$$

Observe that $a_{ij} = \langle e_i | \mathcal{A} e_j \rangle$. Hence, the matrix A is given by

$$A = \begin{pmatrix} \langle e_1 | \mathcal{A} e_1 \rangle & \langle e_1 | \mathcal{A} e_2 \rangle & \cdots & \langle e_1 | \mathcal{A} e_n \rangle \\ \langle e_2 | \mathcal{A} e_1 \rangle & \langle e_2 | \mathcal{A} e_2 \rangle & \cdots & \langle e_2 | \mathcal{A} e_n \rangle \\ \vdots & \vdots & \ddots & \vdots \\ \langle e_n | \mathcal{A} e_1 \rangle & \langle e_n | \mathcal{A} e_2 \rangle & \cdots & \langle e_n | \mathcal{A} e_n \rangle \end{pmatrix}.$$

This representation shows the advantage of choosing an orthonormal basis for V.

Exercises

11.10 Consider the linear operator $\mathcal{A} : \mathbb{C}^2 \to \mathbb{C}^2, \mathcal{A}(z) = \mathcal{A}(z_1, z_2) = (0, z_2)$. Let an orthonormal basis of $\mathbb{C}^2$ be $\{\bar{e}_1, \bar{e}_2\}$ where $\bar{e}_1 = (\frac{1}{\sqrt{2}}, \frac{1}{\sqrt{2}})$, $\bar{e}_2 = (\frac{1}{\sqrt{2}}, -\frac{1}{\sqrt{2}})$. With the inner product on $\mathbb{C}^2$ defined as $\langle u, v \rangle = \sum_{i=1}^{2} u_i^* v_i$ calculate:

a) $\mathcal{A}\bar{e}_1$ b) $\mathcal{A}\bar{e}_2$ c) $\langle \bar{e}_1, \mathcal{A}\bar{e}_1 \rangle$ d) $\langle \bar{e}_1, \mathcal{A}\bar{e}_2 \rangle$

e) $\langle \bar{e}_2, \mathcal{A}\bar{e}_1 \rangle$ f) $\langle \bar{e}_2, \mathcal{A}\bar{e}_2 \rangle$

and hence write down the matrix representation of $\mathcal{A}$ in the given orthonormal basis.

11.5   Eigenvalues and eigenvectors of linear operators

Consider a linear operator $\mathcal{A}$ on a vector space V, and particular, non-zero, state vectors $|\psi\rangle \in V$ which satisfy

$$\mathcal{A}|\psi\rangle = \lambda|\psi\rangle \qquad \text{where } \lambda \in \mathbb{C}.$$

Definition 11.2 The eigenvalues and eigenvectors of a linear operator $\mathcal{A}$
Given a vector space V over the field $\mathbb{C}$ and a linear operator $\mathcal{A} : V \to V$, then a non-zero state vector $|\psi\rangle$ is an **eigenvector** *or* **eigenstate** *of $\mathcal{A}$ with* **eigenvalue** $\lambda \in \mathbb{C}$ *if*

$$\mathcal{A}|\psi\rangle = \lambda|\psi\rangle.$$

These properties are analogous to those of eigenvalues and eigenvectors of matrices discussed in Chapter 7. In fact we can use the matrix representation of a linear operator to find its eigenvalues and eigenvectors.

Example 11.5.1

Find the eigenvalues and corresponding eigenvectors of the linear operator $\mathcal{A} : \mathbb{C}^2 \to \mathbb{C}^2$, $\mathcal{A}(z_1, z_2) = (0, z_2)$, where $z_1, z_2 \in \mathbb{C}$, which has matrix representation, with respect to the standard basis, given by

$$A = \begin{pmatrix} 0 & 0 \\ 0 & 1 \end{pmatrix}.$$

Solution

The eigenvalues of this matrix (and of the operator) are found by solving $|A - \lambda I| = 0$, i.e.,

$$\begin{vmatrix} -\lambda & 0 \\ 0 & 1-\lambda \end{vmatrix} = \lambda^2 - \lambda = \lambda(\lambda - 1) = 0$$

so the eigenvalues are $\lambda = 0, 1$. The corresponding (normalised) eigenvectors of A are readily shown to be $(1, 0)^T$ and $(0, 1)^T$, respectively.

Exercises

11.11 Consider the linear operator $\mathcal{A} : \mathbb{C}^2 \to \mathbb{C}^2$, $\mathcal{A}(z_1, z_2) = (z_1, -z_2)$.

   a) Show that the matrix representation of $\mathcal{A}$ with respect to the standard basis is $A = \begin{pmatrix} 1 & 0 \\ 0 & -1 \end{pmatrix}$.

   b) Find the eigenvalues and corresponding eigenvectors of A.

11.12 Consider the linear operator $\mathcal{A} : \mathbb{C}^2 \to \mathbb{C}^2$, $\mathcal{A}(z_1, z_2) = (z_2, -z_1)$. Choose the orthonormal basis of $\mathbb{C}^2$ to be $\{\bar{e}_1, \bar{e}_2\}$ where $\bar{e}_1 = (\frac{1}{\sqrt{2}}, \frac{1}{\sqrt{2}})$, $\bar{e}_2 = (\frac{1}{\sqrt{2}}, -\frac{1}{\sqrt{2}})$. Find the matrix representation of $\mathcal{A}$ in this basis and determine its eigenvalues and eigenvectors.

11.6 The adjoint and self-adjoint linear operators

Consider the vector space $\mathbb{C}^n$ over the field $\mathbb{C}$. Suppose $\mathcal{A}$ is a linear operator on this space. Suppose further that for $u, v \in \mathbb{C}^n$ an inner product is defined by $\langle u, v \rangle = \sum_{i=1}^{n} u_i^* v_i$ for $u_i, v_i \in \mathbb{C}$. For every linear operator $\mathcal{A}$ there exists another, called the **adjoint** and denoted $\mathcal{A}^\dagger$, which is defined as follows:

Definition 11.3 The adjoint operator, $\mathcal{A}^\dagger$
For every linear operator $\mathcal{A}$, the adjoint, $\mathcal{A}^\dagger$, is a linear operator such that

$$\langle u, \mathcal{A}v \rangle = \langle \mathcal{A}^\dagger u, v \rangle$$

for any $u, v \in \mathbb{C}^n$.

Example 11.6.1

Consider the linear operator $\mathcal{S} : \mathbb{C}^2 \to \mathbb{C}^2$ defined by $\mathcal{S}(z_1, z_2) = (az_1 + bz_2, cz_1 + dz_2)$, where $a, b, c, d, z_1, z_2 \in \mathbb{C}$. Determine the adjoint operator $\mathcal{S}^\dagger$.

Solution

Let the adjoint operator, $\mathcal{S}^\dagger$, be defined by

$$\mathcal{S}^\dagger : \mathbb{C}^2 \to \mathbb{C}^2, \qquad \mathcal{S}^\dagger(z_1, z_2) = (a'z_1 + b'z_2, c'z_1 + d'z_2).$$

Then

$$\begin{aligned}
\langle (z_1, z_2), \mathcal{S}(w_1, w_2) \rangle &= \langle (z_1, z_2), (aw_1 + bw_2, cw_1 + dw_2) \rangle \\
&= z_1^*(aw_1 + bw_2) + z_2^*(cw_1 + dw_2).
\end{aligned}$$

Also,

$$\begin{aligned}
\langle (\mathcal{S}^\dagger(z_1, z_2), (w_1, w_2) \rangle &= \langle (a'z_1 + b'z_2, c'z_1 + d'z_2), (w_1, w_2) \rangle \\
&= (a'^* z_1^* + b'^* z_2^*)w_1 + (c'^* z_1^* + d'^* z_2^*)w_2.
\end{aligned}$$

For these to be equal, we require

$$a'^* = a, \ b'^* = c, \ c'^* = b, \ d'^* = d.$$

Thus

$$\mathcal{S}^\dagger(z_1, z_2) = (a^* z_1 + c^* z_2, b^* z_1 + d^* z_2).$$

Example 11.6.2

Find the matrix representations, with respect to the standard basis, of the linear operators $\mathcal{S}$ and $\mathcal{S}^\dagger$ of Example 11.6.1. Show that the matrix $S^\dagger$ is the adjoint of the matrix S as defined in Chapter 5.

Solution

Applying S to each basis vector:

$$S(1,0) = (a,c) = a(1,0) + c(0,1), \quad S(0,1) = (b,d) = b(1,0) + d(0,1).$$

The transpose of the matrix of coefficients gives the required matrix representation:

$$S = \begin{pmatrix} a & b \\ c & d \end{pmatrix}.$$

Likewise, applying $S^\dagger$ to each basis vector:

$$S^\dagger(1,0) = (a^*, b^*) = a^*(1,0) + b^*(0,1), \quad S^\dagger(0,1) = (c^*, d^*) = c^*(1,0) + d^*(0,1).$$

The transpose of the matrix of coefficients gives the required matrix representation:

$$S^\dagger = \begin{pmatrix} a^* & c^* \\ b^* & d^* \end{pmatrix}.$$

Referring to Chapter 5, we observe that $S^\dagger$ is the adjoint (that is, the conjugate transpose) of S.

Exercises

11.13 Given that $S : \mathbb{C}^2 \to \mathbb{C}^2$, $S(z_1, z_2) = (-iz_1 + (1+i)z_2, 4z_1 - 2iz_2)$ verify that the adjoint operator $S^\dagger$ is given by $S^\dagger(z_1, z_2) = (iz_1 + 4z_2, (1-i)z_1 + 2iz_2)$.

Linear operators $\mathcal{A}$ which have the property that they are equal to their adjoint, i.e $\mathcal{A}^\dagger = \mathcal{A}$, are described as **self-adjoint**.

Definition 11.4 Self-adjoint operator
If a linear operator $\mathcal{A}$ is equal to its adjoint, i.e. $\mathcal{A}^\dagger = \mathcal{A}$, then the operator is said to be self-adjoint. In such cases,
$$\langle u, \mathcal{A}v \rangle = \langle \mathcal{A}u, v \rangle.$$

Self-adjoint operators are important in quantum computation because they are related to quantum measurement. This aspect is discussed in detail in Chapter 13. We note, without proof, the following properties of a self-adjoint operator, $\mathcal{A}$, on a finite-dimensional complex inner product space:

Theorem 11.1 *Given a self-adjoint operator $\mathcal{A}$, on a finite-dimensional complex inner product space,*

    1. the eigenvalues are real,

    2. eigenvectors corresponding to distinct eigenvalues are orthogonal,

    3. if the eigenvalues are all distinct then the eigenvectors form a basis of the underlying vector space.

Note that even when some of the eigenvalues are repeated, it is possible to construct an orthonormal basis of the underlying vector space.

Example 11.6.3

Suppose the inner product is defined by $\langle u, v \rangle = \sum_i u_i^* v_i$. Show that the linear operator $\mathcal{A}$ of Example 11.3.1 is self-adjoint, that is $\langle u, \mathcal{A}v \rangle = \langle \mathcal{A}u, v \rangle$.

Solution

We use the previous matrix representation $A = \begin{pmatrix} 0 & 0 \\ 0 & 1 \end{pmatrix}$ with respect to the standard basis $\mathcal{E}$.

Let $[u]_\mathcal{E} = \begin{pmatrix} u_1 \\ u_2 \end{pmatrix}$, $[v]_\mathcal{E} = \begin{pmatrix} v_1 \\ v_2 \end{pmatrix}$. Then

$$A[u]_\mathcal{E} = \begin{pmatrix} 0 & 0 \\ 0 & 1 \end{pmatrix} \begin{pmatrix} u_1 \\ u_2 \end{pmatrix} = \begin{pmatrix} 0 \\ u_2 \end{pmatrix},$$

$$A[v]_\mathcal{E} = \begin{pmatrix} 0 & 0 \\ 0 & 1 \end{pmatrix} \begin{pmatrix} v_1 \\ v_2 \end{pmatrix} = \begin{pmatrix} 0 \\ v_2 \end{pmatrix}.$$

Then

$$\langle u, Av \rangle = \begin{pmatrix} u_1^* & u_2^* \end{pmatrix} \begin{pmatrix} 0 \\ v_2 \end{pmatrix} = u_2^* v_2.$$

Similarly,

$$\langle Au, v \rangle = \begin{pmatrix} 0 & u_2^* \end{pmatrix} \begin{pmatrix} v_1 \\ v_2 \end{pmatrix} = u_2^* v_2.$$

Thus $\langle u, Av \rangle = \langle Au, v \rangle$ and so $\mathcal{A}$ is a self-adjoint operator. Thus the operator of Example 11.3.1 is self-adjoint and is represented in the basis $\{|0\rangle, |1\rangle\}$ by the matrix $A = \begin{pmatrix} 0 & 0 \\ 0 & 1 \end{pmatrix}$. It has eigenvalues 0 and 1 with eigenvectors $(1,0)$ and $(0,1)$, respectively. These facts will be used repeatedly in the quantum computations which follow.

11.7   Unitary operators

Consider a linear operator $\mathcal{U} : V \to V$, and its adjoint $\mathcal{U}^\dagger$. We have seen that the adjoint operator $\mathcal{U}^\dagger$ satisfies

$$\langle u, \mathcal{U}v \rangle = \langle \mathcal{U}^\dagger u, v \rangle.$$

Definition 11.5 Unitary operator
If a linear operator $\mathcal{U}$ is such that

$$\mathcal{U}^\dagger = \mathcal{U}^{-1}$$

*the operator is said to be **unitary**.*

So a unitary operator has the property that its adjoint is equal to its inverse. Note that if an operator $\mathcal{A}$ is both self-adjoint and unitary, then $\mathcal{A} = \mathcal{A}^\dagger = \mathcal{A}^{-1}$ in which case we say that $\mathcal{A}$ is **self-inverse**.

Unitary operators play the role in quantum computation that Boolean functions play in digital computation, that is they operate on quantum states to effect time evolution. In quantum computation a state vector must have norm equal to 1, and thus in the time evolution of the state this norm must be preserved. The following theorem shows that unitary operators have this desired effect.

Theorem 11.2 Preservation of the inner product
If a unitary operator $\mathcal{U}$ acts on vectors u and v in a vector space V the value of the inner product of u and v remains unchanged:

$$\langle \mathcal{U}u, \mathcal{U}v \rangle = \langle u, v \rangle.$$

Hence, the norm of a vector is preserved under the unitary transformation.

Proof

$$
\begin{aligned}
\langle \mathcal{U}u, \mathcal{U}v \rangle &= \langle \mathcal{U}^\dagger \mathcal{U}u, v \rangle && \text{using the property of the adjoint} \\
&= \langle \mathcal{U}^{-1} \mathcal{U}u, v \rangle && \text{because } \mathcal{U} \text{ is unitary} \\
&= \langle \mathcal{I}u, v \rangle && \text{because } \mathcal{U}^{-1}\mathcal{U} \text{ is the identity operator} \\
&= \langle u, v \rangle.
\end{aligned}
$$

We have shown that when the operator is unitary, it preserves the value of the inner product. Recall that the norm-squared of v is given by $\langle v, v \rangle = \|v\|^2$. Hence, from the previous result,

$$\langle \mathcal{U}v, \mathcal{U}v \rangle = \langle v, v \rangle$$
$$\|\mathcal{U}v\|^2 = \|v\|^2.$$

Thus when a unitary operator acts on a vector, the norm of the vector is not changed. ∎

In quantum computation, state vectors with unit norm will transform under a unitary operator to another state vector also with a unit norm.

Example 11.7.1

Consider the linear operator $\mathcal{A} : \mathbb{C}^3 \to \mathbb{C}^3$ with matrix representation

$$A = \begin{pmatrix} 0 & 1 & 0 \\ 0 & 0 & 1 \\ 1 & 0 & 0 \end{pmatrix}.$$

Show that this matrix is unitary but not self-adjoint.

Solution

The adjoint matrix is given by the conjugate transpose $A^\dagger = \begin{pmatrix} 0 & 0 & 1 \\ 1 & 0 & 0 \\ 0 & 1 & 0 \end{pmatrix}$. Observe immediately that $A \neq A^\dagger$ so A is not self-adjoint. It is straightforward to verify that

$$AA^\dagger = \begin{pmatrix} 0 & 1 & 0 \\ 0 & 0 & 1 \\ 1 & 0 & 0 \end{pmatrix} \begin{pmatrix} 0 & 0 & 1 \\ 1 & 0 & 0 \\ 0 & 1 & 0 \end{pmatrix} = \begin{pmatrix} 1 & 0 & 0 \\ 0 & 1 & 0 \\ 0 & 0 & 1 \end{pmatrix} = I.$$

Likewise, $A^\dagger A = I$, and hence, $\mathcal{A}$ is a unitary operator.

Example 11.7.2

The Pauli operator $\mathcal{X} : \mathbb{C}^2 \to \mathbb{C}^2$ is defined such that if $z = (z_1, z_2)$ then

$$\mathcal{X}(z) = \mathcal{X}(z_1, z_2) = (z_2, z_1).$$

Show that this operator is unitary and self-adjoint.

Solution

The matrix representation of $\mathcal{X}$ in the standard basis is $X = \begin{pmatrix} 0 & 1 \\ 1 & 0 \end{pmatrix}$. Taking the conjugate transpose it is clear that the adjoint of X is $X^\dagger = \begin{pmatrix} 0 & 1 \\ 1 & 0 \end{pmatrix} = X$ and hence X is self-adjoint. Moreover, $X^{-1} = \frac{1}{-1} \begin{pmatrix} 0 & -1 \\ -1 & 0 \end{pmatrix} = \begin{pmatrix} 0 & 1 \\ 1 & 0 \end{pmatrix} = X^\dagger$ and hence X is unitary.

Exercises

11.14 Find the matrix representation, with respect to the standard basis, of the linear operator (Pauli operator) $\Sigma_2 : \mathbb{C}^2 \to \mathbb{C}^2$ defined by $\Sigma_2(z_1, z_2) = (-iz_2, iz_1)$. Show that this operator is self-adjoint and unitary.

11.15 Show that the product of two unitary matrices is a unitary matrix.

11.8   Linear operators on tensor product spaces

We have already seen how a linear operator $\mathcal{A}$ say, acts on vectors in a vector space, for example $\mathbb{C}^2$, to produce another vector in the same space. We now start to explore how such operators act on vectors in the tensor product space $\mathbb{C}^2 \otimes \mathbb{C}^2$.

Definition 11.6 Linear operators on tensor product spaces
If $\mathcal{A} : \mathbb{C}^2 \to \mathbb{C}^2$ and $\mathcal{B} : \mathbb{C}^2 \to \mathbb{C}^2$, then $\mathcal{A} \otimes \mathcal{B} : \mathbb{C}^2 \otimes \mathbb{C}^2 \to \mathbb{C}^2 \otimes \mathbb{C}^2$ is given by

$$(\mathcal{A} \otimes \mathcal{B})(|\psi\rangle \otimes |\phi\rangle) = \mathcal{A}|\psi\rangle \otimes \mathcal{B}|\phi\rangle$$

where $|\psi\rangle, |\phi\rangle \in \mathbb{C}^2$.

We shall assume that the operators are self-adjoint because such operators are related to the measurement of observables. We now demonstrate how the tensor product can be calculated in practice. Assume the standard computational basis. Let the matrix representations of $\mathcal{A}$ and $\mathcal{B}$ be

$$A = \begin{pmatrix} a_{11} & a_{12} \\ a_{21} & a_{22} \end{pmatrix}, \quad B = \begin{pmatrix} b_{11} & b_{12} \\ b_{21} & b_{22} \end{pmatrix}$$

and let $|\psi\rangle = \begin{pmatrix} \psi_1 \\ \psi_2 \end{pmatrix}$, $|\phi\rangle = \begin{pmatrix} \phi_1 \\ \phi_2 \end{pmatrix}$. Then

$$A|\psi\rangle = \begin{pmatrix} a_{11} & a_{12} \\ a_{21} & a_{22} \end{pmatrix} \begin{pmatrix} \psi_1 \\ \psi_2 \end{pmatrix} = \begin{pmatrix} a_{11}\psi_1 + a_{12}\psi_2 \\ a_{21}\psi_1 + a_{22}\psi_2 \end{pmatrix},$$

$$B|\phi\rangle = \begin{pmatrix} b_{11} & b_{12} \\ b_{21} & b_{22} \end{pmatrix} \begin{pmatrix} \phi_1 \\ \phi_2 \end{pmatrix} = \begin{pmatrix} b_{11}\phi_1 + b_{12}\phi_2 \\ b_{21}\phi_1 + b_{22}\phi_2 \end{pmatrix}.$$

It follows that (see e.g., Example 10.3.5)

$$A|\psi\rangle \otimes B|\phi\rangle = \begin{pmatrix} a_{11}\psi_1 + a_{12}\psi_2 \\ a_{21}\psi_1 + a_{22}\psi_2 \end{pmatrix} \otimes \begin{pmatrix} b_{11}\phi_1 + b_{12}\phi_2 \\ b_{21}\phi_1 + b_{22}\phi_2 \end{pmatrix}$$

$$= \begin{pmatrix} (a_{11}\psi_1 + a_{12}\psi_2)(b_{11}\phi_1 + b_{12}\phi_2) \\ (a_{11}\psi_1 + a_{12}\psi_2)(b_{21}\phi_1 + b_{22}\phi_2) \\ (a_{21}\psi_1 + a_{22}\psi_2)(b_{11}\phi_1 + b_{12}\phi_2) \\ (a_{21}\psi_1 + a_{22}\psi_2)(b_{21}\phi_1 + b_{22}\phi_2) \end{pmatrix}$$

$$= \begin{pmatrix} a_{11}b_{11}\psi_1\phi_1 + a_{11}b_{12}\psi_1\phi_2 + a_{12}b_{11}\psi_2\phi_1 + a_{12}b_{12}\psi_2\phi_2 \\ a_{11}b_{21}\psi_1\phi_1 + a_{11}b_{22}\psi_1\phi_2 + a_{12}b_{21}\psi_2\phi_1 + a_{12}b_{22}\psi_2\phi_2 \\ a_{21}b_{11}\psi_1\phi_1 + a_{21}b_{12}\psi_1\phi_2 + a_{22}b_{11}\psi_2\phi_1 + a_{22}b_{12}\psi_2\phi_2 \\ a_{21}b_{21}\psi_1\phi_1 + a_{21}b_{22}\psi_1\phi_2 + a_{22}b_{21}\psi_2\phi_1 + a_{22}b_{22}\psi_2\phi_2 \end{pmatrix}$$

$$= \begin{pmatrix} a_{11}b_{11} & a_{11}b_{12} & a_{12}b_{11} & a_{12}b_{12} \\ a_{11}b_{21} & a_{11}b_{22} & a_{12}b_{21} & a_{12}b_{22} \\ a_{21}b_{11} & a_{21}b_{12} & a_{22}b_{11} & a_{22}b_{12} \\ a_{21}b_{21} & a_{21}b_{22} & a_{22}b_{21} & a_{22}b_{22} \end{pmatrix} \begin{pmatrix} \psi_1\phi_1 \\ \psi_1\phi_2 \\ \psi_2\phi_1 \\ \psi_2\phi_2 \end{pmatrix}.$$

We require this expression to equal $(A \otimes B)(|\psi\rangle \otimes |\phi\rangle)$. Now

$$|\psi\rangle \otimes |\phi\rangle = \begin{pmatrix} \psi_1\phi_1 \\ \psi_1\phi_2 \\ \psi_2\phi_1 \\ \psi_2\phi_2 \end{pmatrix}.$$

Thus

$$A \otimes B \quad \text{must equal} \quad \begin{pmatrix} a_{11}b_{11} & a_{11}b_{12} & a_{12}b_{11} & a_{12}b_{12} \\ a_{11}b_{21} & a_{11}b_{22} & a_{12}b_{21} & a_{12}b_{22} \\ a_{21}b_{11} & a_{21}b_{12} & a_{22}b_{11} & a_{22}b_{12} \\ a_{21}b_{21} & a_{21}b_{22} & a_{22}b_{21} & a_{22}b_{22} \end{pmatrix}.$$

And so

$$A \otimes B = \begin{pmatrix} a_{11} & a_{12} \\ a_{21} & a_{22} \end{pmatrix} \otimes \begin{pmatrix} b_{11} & b_{12} \\ b_{21} & b_{22} \end{pmatrix} = \begin{pmatrix} a_{11}b_{11} & a_{11}b_{12} & a_{12}b_{11} & a_{12}b_{12} \\ a_{11}b_{21} & a_{11}b_{22} & a_{12}b_{21} & a_{12}b_{22} \\ a_{21}b_{11} & a_{21}b_{12} & a_{22}b_{11} & a_{22}b_{12} \\ a_{21}b_{21} & a_{21}b_{22} & a_{22}b_{21} & a_{22}b_{22} \end{pmatrix}$$

revealing that $A \otimes B$ can be calculated from

$$A \otimes B = \begin{pmatrix} a_{11} & a_{12} \\ a_{21} & a_{22} \end{pmatrix} \otimes \begin{pmatrix} b_{11} & b_{12} \\ b_{21} & b_{22} \end{pmatrix}$$

$$= \begin{pmatrix} a_{11} \begin{Bmatrix} b_{11} & b_{12} \\ b_{21} & b_{22} \end{Bmatrix} & a_{12} \begin{Bmatrix} b_{11} & b_{12} \\ b_{21} & b_{22} \end{Bmatrix} \\ a_{21} \begin{Bmatrix} b_{11} & b_{12} \\ b_{21} & b_{22} \end{Bmatrix} & a_{22} \begin{Bmatrix} b_{11} & b_{12} \\ b_{21} & b_{22} \end{Bmatrix} \end{pmatrix}.$$

Example 11.8.1

If $X = \begin{pmatrix} 0 & 1 \\ 1 & 0 \end{pmatrix}$ and $Y = \begin{pmatrix} 0 & -i \\ i & 0 \end{pmatrix}$ find $X \otimes Y$ and $Y \otimes X$.

Solution

$$X \otimes Y = \begin{pmatrix} 0 & 1 \\ 1 & 0 \end{pmatrix} \otimes \begin{pmatrix} 0 & -i \\ i & 0 \end{pmatrix}$$

$$= \begin{pmatrix} 0\begin{pmatrix} 0 & -i \\ i & 0 \end{pmatrix} & 1\begin{pmatrix} 0 & -i \\ i & 0 \end{pmatrix} \\ 1\begin{pmatrix} 0 & -i \\ i & 0 \end{pmatrix} & 0\begin{pmatrix} 0 & -i \\ i & 0 \end{pmatrix} \end{pmatrix}$$

$$= \begin{pmatrix} 0 & 0 & 0 & -i \\ 0 & 0 & i & 0 \\ 0 & -i & 0 & 0 \\ i & 0 & 0 & 0 \end{pmatrix}.$$

$$Y \otimes X = \begin{pmatrix} 0 & -i \\ i & 0 \end{pmatrix} \otimes \begin{pmatrix} 0 & 1 \\ 1 & 0 \end{pmatrix}$$

$$= \begin{pmatrix} 0\begin{pmatrix} 0 & 1 \\ 1 & 0 \end{pmatrix} & -i\begin{pmatrix} 0 & 1 \\ 1 & 0 \end{pmatrix} \\ i\begin{pmatrix} 0 & 1 \\ 1 & 0 \end{pmatrix} & 0\begin{pmatrix} 0 & 1 \\ 1 & 0 \end{pmatrix} \end{pmatrix}$$

$$= \begin{pmatrix} 0 & 0 & 0 & -i \\ 0 & 0 & -i & 0 \\ 0 & i & 0 & 0 \\ i & 0 & 0 & 0 \end{pmatrix}.$$

Example 11.8.2

If $H = \frac{1}{\sqrt{2}} \begin{pmatrix} 1 & 1 \\ 1 & -1 \end{pmatrix}$ and $I = \begin{pmatrix} 1 & 0 \\ 0 & 1 \end{pmatrix}$ find $H \otimes I$.

Solution

$$H \otimes I = \frac{1}{\sqrt{2}} \begin{pmatrix} 1 & 1 \\ 1 & -1 \end{pmatrix} \otimes \begin{pmatrix} 1 & 0 \\ 0 & 1 \end{pmatrix}$$

$$= \frac{1}{\sqrt{2}} \begin{pmatrix} 1\begin{pmatrix} 1 & 0 \\ 0 & 1 \end{pmatrix} & 1\begin{pmatrix} 1 & 0 \\ 0 & 1 \end{pmatrix} \\ 1\begin{pmatrix} 1 & 0 \\ 0 & 1 \end{pmatrix} & -1\begin{pmatrix} 1 & 0 \\ 0 & 1 \end{pmatrix} \end{pmatrix}$$

$$= \frac{1}{\sqrt{2}} \begin{pmatrix} 1 & 0 & 1 & 0 \\ 0 & 1 & 0 & 1 \\ 1 & 0 & -1 & 0 \\ 0 & 1 & 0 & -1 \end{pmatrix}.$$

Example 11.8.3

Consider the following two linear operators:

the Pauli $\mathcal{X}$ operator is given by $\mathcal{X} : \mathbb{C}^2 \to \mathbb{C}^2$ which on the standard computational basis states is defined as $\mathcal{X}|0\rangle = |1\rangle$, $\mathcal{X}|1\rangle = |0\rangle$.

the Pauli $\mathcal{Z}$ operator is given by $\mathcal{Z} : \mathbb{C}^2 \to \mathbb{C}^2$ which on the standard computational basis states is defined as $\mathcal{Z}|0\rangle = |0\rangle$, $\mathcal{Z}|1\rangle = -|1\rangle$.

Find $(\mathcal{X} \otimes \mathcal{Z})|\psi\rangle$ when $|\psi\rangle = |1\rangle|1\rangle$.

Solution

$$
\begin{aligned}
(\mathcal{X} \otimes \mathcal{Z})|\psi\rangle &= (\mathcal{X} \otimes \mathcal{Z})(|1\rangle|1\rangle) \\
&= \mathcal{X}|1\rangle \otimes \mathcal{Z}|1\rangle \\
&= |0\rangle \otimes (-|1\rangle) \\
&= -|0\rangle|1\rangle \\
&= -|01\rangle.
\end{aligned}
$$

We illustrate how the same result can be obtained using matrices. The matrix representations (in the standard computational basis of $\mathbb{C}^2$) of $\mathcal{X}$ and $\mathcal{Z}$ are

$$
X = \begin{pmatrix} 0 & 1 \\ 1 & 0 \end{pmatrix}, \qquad Z = \begin{pmatrix} 1 & 0 \\ 0 & -1 \end{pmatrix}.
$$

Then

$$
X \otimes Z = \begin{pmatrix} 0 & 0 & 1 & 0 \\ 0 & 0 & 0 & -1 \\ 1 & 0 & 0 & 0 \\ 0 & -1 & 0 & 0 \end{pmatrix}.
$$

Now

$$
|\psi\rangle = |1\rangle|1\rangle
$$

$$
= \begin{pmatrix} 0 \\ 1 \end{pmatrix} \otimes \begin{pmatrix} 0 \\ 1 \end{pmatrix} = \begin{pmatrix} 0 \\ 0 \\ 0 \\ 1 \end{pmatrix}
$$

and so

$$
(X \otimes Z)|\psi\rangle = \begin{pmatrix} 0 & 0 & 1 & 0 \\ 0 & 0 & 0 & -1 \\ 1 & 0 & 0 & 0 \\ 0 & -1 & 0 & 0 \end{pmatrix} \begin{pmatrix} 0 \\ 0 \\ 0 \\ 1 \end{pmatrix} = \begin{pmatrix} 0 \\ -1 \\ 0 \\ 0 \end{pmatrix} = -\begin{pmatrix} 0 \\ 1 \\ 0 \\ 0 \end{pmatrix} = -|01\rangle.
$$

Example 11.8.4

Given the Hadamard transform $H = \frac{1}{\sqrt{2}} \begin{pmatrix} 1 & 1 \\ 1 & -1 \end{pmatrix}$,

a) calculate the matrix representation of $H^{\otimes 2} = H \otimes H$.

b) evaluate $(H \otimes H)(|0\rangle \otimes |0\rangle)$ by explicitly evaluating the tensor products.

c) evaluate $(H \otimes H)(|0\rangle \otimes |0\rangle)$ using the matrix found in part a).

Solution

a)

$$
\begin{aligned}
H \otimes H &= \frac{1}{\sqrt{2}} \begin{pmatrix} 1 & 1 \\ 1 & -1 \end{pmatrix} \otimes \frac{1}{\sqrt{2}} \begin{pmatrix} 1 & 1 \\ 1 & -1 \end{pmatrix} \\
&= \frac{1}{2} \begin{pmatrix} 1 & 1 \\ 1 & -1 \end{pmatrix} \otimes \begin{pmatrix} 1 & 1 \\ 1 & -1 \end{pmatrix} \\
&= \frac{1}{2} \begin{pmatrix} 1\begin{pmatrix} 1 & 1 \\ 1 & -1 \end{pmatrix} & 1\begin{pmatrix} 1 & 1 \\ 1 & -1 \end{pmatrix} \\ 1\begin{pmatrix} 1 & 1 \\ 1 & -1 \end{pmatrix} & -1\begin{pmatrix} 1 & 1 \\ 1 & -1 \end{pmatrix} \end{pmatrix} \\
&= \frac{1}{2} \begin{pmatrix} 1 & 1 & 1 & 1 \\ 1 & -1 & 1 & -1 \\ 1 & 1 & -1 & -1 \\ 1 & -1 & -1 & 1 \end{pmatrix}.
\end{aligned}
$$

b) From the definition of how the tensor product of operators acts on vectors in tensor product spaces, we have

$$
\begin{aligned}
(H \otimes H)(|0\rangle \otimes |0\rangle) &= H|0\rangle \otimes H|0\rangle \\
&= \frac{1}{\sqrt{2}}(|0\rangle + |1\rangle) \otimes \frac{1}{\sqrt{2}}(|0\rangle + |1\rangle) \\
&= \frac{1}{2}(|00\rangle + |01\rangle + |10\rangle + |11\rangle) \\
&= \frac{1}{2}\left(\begin{pmatrix} 1 \\ 0 \\ 0 \\ 0 \end{pmatrix} + \begin{pmatrix} 0 \\ 1 \\ 0 \\ 0 \end{pmatrix} + \begin{pmatrix} 0 \\ 0 \\ 1 \\ 0 \end{pmatrix} + \begin{pmatrix} 0 \\ 0 \\ 0 \\ 1 \end{pmatrix} \right) \\
&= \frac{1}{2} \begin{pmatrix} 1 \\ 1 \\ 1 \\ 1 \end{pmatrix}.
\end{aligned}
$$

c) Noting that $|0\rangle \otimes |0\rangle = \begin{pmatrix} 1 \\ 0 \\ 0 \\ 0 \end{pmatrix}$ we have

$$
(H \otimes H)(|0\rangle \otimes |0\rangle) = \frac{1}{2} \begin{pmatrix} 1 & 1 & 1 & 1 \\ 1 & -1 & 1 & -1 \\ 1 & 1 & -1 & -1 \\ 1 & -1 & -1 & 1 \end{pmatrix} \begin{pmatrix} 1 \\ 0 \\ 0 \\ 0 \end{pmatrix} = \frac{1}{2} \begin{pmatrix} 1 \\ 1 \\ 1 \\ 1 \end{pmatrix}.
$$

Example 11.8.5

Suppose $A = \begin{pmatrix} 0 & 0 \\ 0 & 1 \end{pmatrix}$ and $I = \begin{pmatrix} 1 & 0 \\ 0 & 1 \end{pmatrix}$. Evaluate $A \otimes I \otimes I$.

Solution

By a natural extension of the foregoing, we have

$$A \otimes I \otimes I = \begin{pmatrix} 0 & 0 \\ 0 & 1 \end{pmatrix} \otimes \begin{pmatrix} 1 & 0 \\ 0 & 1 \end{pmatrix} \otimes \begin{pmatrix} 1 & 0 \\ 0 & 1 \end{pmatrix}$$

$$= \begin{pmatrix} 0 & 0 \\ 0 & 1 \end{pmatrix} \otimes \begin{pmatrix} 1 & 0 & 0 & 0 \\ 0 & 1 & 0 & 0 \\ 0 & 0 & 1 & 0 \\ 0 & 0 & 0 & 1 \end{pmatrix}$$

$$= \begin{pmatrix} 0\begin{pmatrix} 1 & 0 & 0 & 0 \\ 0 & 1 & 0 & 0 \\ 0 & 0 & 1 & 0 \\ 0 & 0 & 0 & 1 \end{pmatrix} & 0\begin{pmatrix} 1 & 0 & 0 & 0 \\ 0 & 1 & 0 & 0 \\ 0 & 0 & 1 & 0 \\ 0 & 0 & 0 & 1 \end{pmatrix} \\ 0\begin{pmatrix} 1 & 0 & 0 & 0 \\ 0 & 1 & 0 & 0 \\ 0 & 0 & 1 & 0 \\ 0 & 0 & 0 & 1 \end{pmatrix} & 1\begin{pmatrix} 1 & 0 & 0 & 0 \\ 0 & 1 & 0 & 0 \\ 0 & 0 & 1 & 0 \\ 0 & 0 & 0 & 1 \end{pmatrix} \end{pmatrix}$$

$$= \begin{pmatrix} 0 & 0 & 0 & 0 & 0 & 0 & 0 & 0 \\ 0 & 0 & 0 & 0 & 0 & 0 & 0 & 0 \\ 0 & 0 & 0 & 0 & 0 & 0 & 0 & 0 \\ 0 & 0 & 0 & 0 & 0 & 0 & 0 & 0 \\ 0 & 0 & 0 & 0 & 1 & 0 & 0 & 0 \\ 0 & 0 & 0 & 0 & 0 & 1 & 0 & 0 \\ 0 & 0 & 0 & 0 & 0 & 0 & 1 & 0 \\ 0 & 0 & 0 & 0 & 0 & 0 & 0 & 1 \end{pmatrix}.$$

Exercises

11.16 Evaluate each of $(H \otimes H)|01\rangle$, $(H \otimes H)|10\rangle$, $(H \otimes H)|11\rangle$, by explicitly evaluating the tensor products and by using the matrix in Example 11.8.4.

11.17 Given $x_1, x_2 \in \{0,1\}$ and the Hadamard transform, H, show that

$$(H \otimes H)(|x_1\rangle \otimes |x_2\rangle)$$

can be expressed as

$$\frac{1}{2}\left\{ \sum_{y=0}^{1}(-1)^{y \cdot x_1}|y\rangle \otimes \sum_{y=0}^{1}(-1)^{y \cdot x_2}|y\rangle \right\}$$

and then as

$$\frac{1}{2}\bigotimes_{i=1}^{2}\sum_{y=0}^{1}(-1)^{y \cdot x_i}|y\rangle.$$

(Note $\bigotimes_{i=1}^{n}$ represents the n-fold tensor product.) Using this result, confirm the answers to Exercise 11.16.

11.18 Consider again Exercise 11.17. Now let $|y\rangle, y = 0, 1, 2, 3$, be the decimal representation of $|00\rangle, |01\rangle, |10\rangle, |11\rangle$. Let $x = (x_1, x_2)$. Show that the expressions for the tensor product $(H \otimes H)(|x_1\rangle \otimes |x_2\rangle)$ given therein are equivalent to

$$\frac{1}{2} \sum_{y=0}^{3} (-1)^{x \cdot y} |y\rangle$$

where $x \cdot y$ is the scalar product $\sum x_i y_i$ and multiplication and addition are performed modulo 2. (Note that when evaluating the scalar product $x \cdot y$, use the binary form for y, that is $y \in \{(0,0), (0,1), (1,0), (1,1)\}$.)

11.19 Let $Z = \begin{pmatrix} 1 & 0 \\ 0 & -1 \end{pmatrix}$ and $H = \frac{1}{\sqrt{2}} \begin{pmatrix} 1 & 1 \\ 1 & -1 \end{pmatrix}$,

    a) find $(Z \otimes H)(|0\rangle \otimes |0\rangle)$ b) find $(Z \otimes H)(|0\rangle \otimes |1\rangle)$

    c) find $(Z \otimes H)(|1\rangle \otimes |0\rangle)$ d) find $(Z \otimes H)(|1\rangle \otimes |1\rangle)$.

Knowing the eigenbasis of self-adjoint operators of the form $\mathcal{A} \otimes \mathcal{B}$ will be essential for the study of quantum measurement. Consider the following example.

Example 11.8.6 The eigenbasis of the operators $\mathcal{A} \otimes \mathcal{I}$ and $\mathcal{I} \otimes \mathcal{A}$.

Suppose $\mathcal{A}$ and $\mathcal{I}$ have matrix representations

$$A = \begin{pmatrix} 0 & 0 \\ 0 & 1 \end{pmatrix}, \qquad I = \begin{pmatrix} 1 & 0 \\ 0 & 1 \end{pmatrix}.$$

Then

$$A \otimes I = \begin{pmatrix} 0 & 0 & 0 & 0 \\ 0 & 0 & 0 & 0 \\ 0 & 0 & 1 & 0 \\ 0 & 0 & 0 & 1 \end{pmatrix}, \qquad I \otimes A = \begin{pmatrix} 0 & 0 & 0 & 0 \\ 0 & 1 & 0 & 0 \\ 0 & 0 & 0 & 0 \\ 0 & 0 & 0 & 1 \end{pmatrix}.$$

Now the eigenvalues of $A \otimes I$ are 0 and 1. The eigenvectors corresponding to 0 are $|0\rangle \otimes |0\rangle = \begin{pmatrix} 1 \\ 0 \\ 0 \\ 0 \end{pmatrix}$ and $|0\rangle \otimes |1\rangle = \begin{pmatrix} 0 \\ 1 \\ 0 \\ 0 \end{pmatrix}$. Those corresponding to 1 are $|1\rangle \otimes |0\rangle = \begin{pmatrix} 0 \\ 0 \\ 1 \\ 0 \end{pmatrix}$ and

$|1\rangle \otimes |1\rangle = \begin{pmatrix} 0 \\ 0 \\ 0 \\ 1 \end{pmatrix}$.

The eigenvalues of $I \otimes A$ are also 0 and 1. The eigenvectors corresponding to 0 are $|0\rangle \otimes |0\rangle = \begin{pmatrix} 1 \\ 0 \\ 0 \\ 0 \end{pmatrix}$ and $|1\rangle \otimes |0\rangle = \begin{pmatrix} 0 \\ 0 \\ 1 \\ 0 \end{pmatrix}$. Those corresponding to 1 are $|0\rangle \otimes |1\rangle = \begin{pmatrix} 0 \\ 1 \\ 0 \\ 0 \end{pmatrix}$

and $|1\rangle \otimes |1\rangle = \begin{pmatrix} 0 \\ 0 \\ 0 \\ 1 \end{pmatrix}$.

Thus the eigenbasis of both operators is $|00\rangle, |01\rangle, |10\rangle$ and $|11\rangle$.

11.9   End-of-chapter exercises

1. Show that $A : \mathbb{R}^3 \to \mathbb{R}^3$ defined by $A(x_1, x_2, x_3) = (2x_1 - x_2 + x_3, -x_2 - x_3, x_1)$ where $x_1, x_2, x_3 \in \mathbb{R}$ is a linear operator.

2. Show that $A : \mathbb{R}^2 \to \mathbb{R}^2$ defined by $A(x_1, x_2) = (x_1^2, x_2^2)$ where $x_1, x_2 \in \mathbb{R}$ is not a linear operator.

3. Show that $A : \mathbb{R} \to \mathbb{R}$ defined by $A(x) = 2x - 1$ where $x \in \mathbb{R}$ is not a linear operator.

4. Show that $A : \mathbb{C}^2 \to \mathbb{C}^2$ defined by $A(z_1, z_2) = (2z_1, z_2)$ where $z_1, z_2 \in \mathbb{C}$ is a linear operator.

5. Show that $|0\rangle\langle 0| + |1\rangle\langle 1| = \mathcal{I}$, the identity operator.

6. Suppose $|+\rangle = \frac{1}{\sqrt{2}}(|0\rangle + |1\rangle)$ and $|-\rangle = \frac{1}{\sqrt{2}}(|0\rangle - |1\rangle)$. Show that

$$|+\rangle\langle +| + |-\rangle\langle -| = \mathcal{I}.$$

7. Given $A = \begin{pmatrix} 0 & 0 \\ 0 & 1 \end{pmatrix}$, evaluate $(A \otimes I)|11\rangle$, that is $\begin{pmatrix} 0 & 0 & 0 & 0 \\ 0 & 0 & 0 & 0 \\ 0 & 0 & 1 & 0 \\ 0 & 0 & 0 & 1 \end{pmatrix} \begin{pmatrix} 0 \\ 0 \\ 0 \\ 1 \end{pmatrix}$.

8. Given $A = \begin{pmatrix} 0 & 0 \\ 0 & 1 \end{pmatrix}$, show that

$$(I \otimes A)|11\rangle = \begin{pmatrix} 0 & 0 & 0 & 0 \\ 0 & 1 & 0 & 0 \\ 0 & 0 & 0 & 0 \\ 0 & 0 & 0 & 1 \end{pmatrix} \begin{pmatrix} 0 \\ 0 \\ 0 \\ 1 \end{pmatrix} = \begin{pmatrix} 0 \\ 0 \\ 0 \\ 1 \end{pmatrix} = |11\rangle.$$

Deduce that $|11\rangle$ is an eigenvector of $I \otimes A$. State the corresponding eigenvalue.

9. Consider the linear operator $A : \mathbb{C}^2 \to \mathbb{C}^2, A(z) = A(z_1, z_2) = (z_1, 0)$. Let an orthonormal basis of $\mathbb{C}^2$ be $\{\bar{e}_1, \bar{e}_2\}$ where $\bar{e}_1 = (\frac{1}{\sqrt{2}}, \frac{1}{\sqrt{2}})$, $\bar{e}_2 = (\frac{1}{\sqrt{2}}, -\frac{1}{\sqrt{2}})$. With the inner product on $\mathbb{C}^2$ defined as $\langle u, v \rangle = \sum_{i=1}^{2} u_i^* v_i$ calculate:

   a) $A\bar{e}_1$ b) $A\bar{e}_2$ c) $\langle \bar{e}_1, A\bar{e}_1 \rangle$ d) $\langle \bar{e}_1, A\bar{e}_2 \rangle$

   e) $\langle \bar{e}_2, A\bar{e}_1 \rangle$ f) $\langle \bar{e}_2, A\bar{e}_2 \rangle$

   and hence write down the matrix representation of A in the given orthonormal basis.

10. Consider the linear operator $A : \mathbb{C}^2 \to \mathbb{C}^2, A(z) = A(z_1, z_2) = (z_2, z_1)$. Let an orthonormal basis of $\mathbb{C}^2$ be $\{\bar{e}_1, \bar{e}_2\}$ where $\bar{e}_1 = (\frac{1}{\sqrt{2}}, \frac{1}{\sqrt{2}})$, $\bar{e}_2 = (\frac{1}{\sqrt{2}}, -\frac{1}{\sqrt{2}})$. With the inner product on $\mathbb{C}^2$ defined as $\langle u, v \rangle = \sum_{i=1}^{2} u_i^* v_i$ calculate:

   a) $A\bar{e}_1$ b) $A\bar{e}_2$ c) $\langle \bar{e}_1, A\bar{e}_1 \rangle$ d) $\langle \bar{e}_1, A\bar{e}_2 \rangle$

   e) $\langle \bar{e}_2, A\bar{e}_1 \rangle$ f) $\langle \bar{e}_2, A\bar{e}_2 \rangle$

   and hence write down the matrix representation of A in the given orthonormal basis.

11. Let A and B be Hermitian operators.

    a) Show that the commutator $[A, B] = AB - BA$ is not Hermitian.

    b) Show that $i[A, B]$ is Hermitian.

12. In Example 11.8.5 we calculated the matrix representation of $A \otimes I \otimes I$ where $A = \begin{pmatrix} 0 & 0 \\ 0 & 1 \end{pmatrix}$ and $I = \begin{pmatrix} 1 & 0 \\ 0 & 1 \end{pmatrix}$. Use this representation to confirm that $|111\rangle$ is an eigenvector of $A \otimes I \otimes I$ with eigenvalue 1.

13. Explore whether eigenvalues and eigenvectors are basis independent.

14. Let $B = \begin{pmatrix} 1 & 0 \\ 0 & 0 \end{pmatrix}$.

    a) Find $B \otimes B$.

    b) Show that $|00\rangle = \begin{pmatrix} 1 \\ 0 \\ 0 \\ 0 \end{pmatrix}$ is an eigenvector of $B \otimes B$ with eigenvalue 1.

Part II

Foundations of quantum-gate computation

12

Introduction to Part II

12.1 Objectives

This chapter overviews some physical aspects of computation in preparation, and as a foundation, for the presentation of the formal axioms of quantum computation that follow. In particular:

1. the concepts of state, dynamics, observables and measurement are introduced in both the classical and quantum domains,

2. the Stern-Gerlach quantum experiment is discussed to provide a basis for the introduction of the qubit (the quantum equivalent of the digital bit).

12.2 Computation and physics

Computation may be mathematically defined, but it is a physical process requiring 'machines' for its implementation. The machines may be mechanical and simple, for example the abacus, or of a more complex mechanical nature (for example, the mechanical Pascaline and Babbage machines). Modern digital computers are electro-mechanical in nature, and based on the von Neumann architecture; they are founded on the laws of (i) classical electro-magnetism (Maxwell's equations) and (ii) the non-relativistic classical-mechanics of Newton, Lagrange and Hamilton. Digital computers have proved to be immensely powerful, they continue to dominate the world of general computation, and will occupy this position for years to come.

12.3 Physical systems

Dynamic physical systems, including both the classical and quantum cases, incorporate:

1. state spaces,

2. time-evolution functions, and

3. 'measurement' processes.

In the next section, we view digital computation from this perspective which we will, later, compare directly with the quantum case.

DOI: 10.1201/9781003264569-12

12.4   An overview of digital computation

12.4.1   Digital computer states

The **state space** of a digital computer comprises its set of registers; and its **state**, at time t, is defined to be the data in its registers at time t.

12.4.2   Digital computer dynamics

Dynamics are defined to be the changes of state in time. A program (derived from an algorithm) for a digital computer moves the computer through a series of states for each valid input. When the process terminates the output is available. The changes of state, determined by the program, are realised by classical Boolean gates.

A digital machine has only a finite number of possible states; these can therefore be listed as

$$s_1, \ldots, s_k.$$

It follows that the dynamic path determined by a specific coding of an algorithm, with a given input, may be represented as a sequence of, time-ordered, machine states:

$$s_1^*, s_2^*, \cdots, s_p^*$$

where $s_i^* \in \{s_1, \ldots, s_k\}$ for each $i \in \{1, \ldots, p\}$. Hence there are Boolean functions $f_0, \ldots, f_{p-1}$ such that

$$x \xrightarrow{f_0} s_1^* \xrightarrow{f_1} s_2^* \xrightarrow{f_2} \cdots \xrightarrow{f_{p-2}} s_{p-1}^* \xrightarrow{f_{p-1}} s_p^*$$

where x is a Boolean string that incorporates the input data, and f_0 is a Boolean function that gives rise to the initial machine state, s_1^*, from the program and the data provided as input.

Equivalently, we can use the notation of function composition and write:

$$(f_{p-1} \circ \cdots \circ f_2 \circ f_1 \circ f_0)(x).$$

12.4.3   Digital computer 'measurement'

In digital computation extracting the output of a computation is a trivial process. For comparison (later) with the quantum case, we extend the functional representation of a digital computation to include an 'output', or 'measurement', process.

We have

$$x \xrightarrow{f_0} s_1^* \xrightarrow{f_1} s_2^* \xrightarrow{f_2} \cdots \xrightarrow{f_{p-2}} s_{p-1}^* \xrightarrow{f_{p-1}} s_p^* \xrightarrow{\mu_D} y$$

where μ_D is the Boolean function that 'projects' the contents of the registers, of the final state s_p^*, comprising the desired output, y, of the computation.

That is, the complete process may be written as:

$$y = \mu_D \circ (f_{p-1} \circ \cdots \circ f_2 \circ f_1 \circ f_0)(x) \tag{12.1}$$

The function μ_D is not a computation, in the sense that it does not change data in the registers of the final state s_p^*; rather, it projects the contents of the registers that comprise the output data of the computation.

The function μ_D is trivial, and is included here only for the purposes of comparison with the quantum case. The equivalent quantum 'operator', which we denote by μ_Q, plays the same role as that of μ_D in the digital case, i.e., it determines, by projection from the final (quantum) state, the output of the computation. The quantum operator μ_Q is said to 'measure' the final quantum state; the quantum measurement process is, unlike the digital case, non-trivial, as we shall see.

12.5   The emergence of quantum computation

In the 1980s, the physicist Richard Feynman expressed the view that, in order to model the quantum world adequately, vastly more powerful computers than those likely to be implemented using digital technology would be needed – and he suggested that computers based on quantum physics (rather than the classical physics of the digital world) might provide solutions.

In recent years, a number of quantum approaches to computation have evolved and their, potential, power has been demonstrated by the development of quantum algorithms of massively greater efficiency than known digital algorithms for the same problem. Shor's quantum algorithm for the factorisation of integers is regarded as the most significant to date; it demonstrates that integers can be factorised 'efficiently' on a quantum-mechanical computer. Given that many current encryption algorithms depend on the inability to factorise large integers efficiently on a digital computer, Shor's algorithm threatens the current approach to encryption that we all depend upon. However, the threat is not immediate as, at this time, no 'stable' quantum computer of the 'size' required has been physically realised (i.e., engineered).

12.6   Observables and measurement

12.6.1   Observables

Observables, in both classical and quantum mechanics, correspond to 'measurable' properties such as mass, position, momentum, energy etc. Measurements result from experiments designed to produce values for observables. In both the classical and quantum cases, the values measured are real numbers.

12.6.2   Measurement

In quantum mechanics the 'sharp end' of the measurement device is, inevitably, of the same 'size' or 'scale' as the system under measurement. It should not therefore be too surprising that quantum measurement may disturb the state. The bull-in-the-china-shop analogy is helpful – the state of the china is unlikely to be the same, post-inspection by the bull! Although such large-scale analogies impart some understanding, they do not explain the complexities and/or subtleties of the quantum measurement process. In comparing classical and quantum measurement we make the following observations:

1. In classical physics, measurement processes are usually far less intrusive and may normally be performed with minimal, or no, change of state. Where changes of state do occur, they are predictable and not stochastic. This is unlike the quantum case where distinct post-measurement states, with known a priori probabilities, can occur.

2. The measurement processes, in both the classical and quantum cases, produce real numbers as output. The classical measurement process is deterministic – in the sense that measuring systems prepared in the same state will produce identical measured values for all observables (i.e., mass, momentum etc.). This is not true in quantum mechanics; the measurement of identically prepared systems can produce distinct outcomes – again with known a priori probabilities.

The Stern-Gerlach experiment, discussed below, illustrates the issues raised in both (1) and (2) above. Later we also use the Stern-Gerlach experiment to introduce the concept of a 'qubit', this being the quantum equivalent of the bit in classical digital computing.

12.7   The Stern-Gerlach quantum experiment

The full generalities of quantum mechanics are not required for quantum computation. In particular, vector spaces of infinite dimension and the associated spectral complexities of operators on such spaces can be avoided. Below we identify a minimal quantum physics, specifically the Stern-Gerlach experiment, required to underpin the axioms of quantum computation and to illustrate the processes of quantum measurement and dynamics. The Stern-Gerlach experiment illustrates some of the essential differences between classical and quantum mechanics. The experiment is illustrated in Figure 12.1, which shows a stream of quantum particles, all prepared in the same way, passing through a magnetic 'measuring' device. As is shown, the input particle stream splits into two distinct output streams – each comprising 50% of the particles. This is very distinct from what would be expected if the particles were classical in nature – specifically classical particles would emerge from the magnetic field all displaying similar behaviour – dependent upon the nature of the field. The quantum 'observable' under measurement in the Stern-Gerlach experiment is intrinsic 'quantum spin', and the interpretation of the experiment is that the spin state of each of the particles in the input stream is a linear 'superposition' of the two states that occur post-measurement. If we denote the output spin states by $|\uparrow\rangle$ and $|\downarrow\rangle$ then the spin state of the particles in the input stream is represented quantum mechanically as the superimposed state $\frac{1}{\sqrt{2}}|\uparrow\rangle + \frac{1}{\sqrt{2}}|\downarrow\rangle$.

This experiment also illustrates a fundamental aspect of quantum measurement – specifically that the measurement process may change the state of the particles. We have:

1. all particles in the input stream are in the 'pure' quantum state $\frac{1}{\sqrt{2}}|\uparrow\rangle + \frac{1}{\sqrt{2}}|\downarrow\rangle$,

2. following measurement 50% of the particles are in the 'pure' state $|\uparrow\rangle$ and 50% in the 'pure' state $|\downarrow\rangle$.

We denote the measured values distinguishing the experimental outcomes as $+1$ and -1 for spin-up and spin-down respectively with corresponding post-measurement states $|\uparrow\rangle$ and $|\downarrow\rangle$.

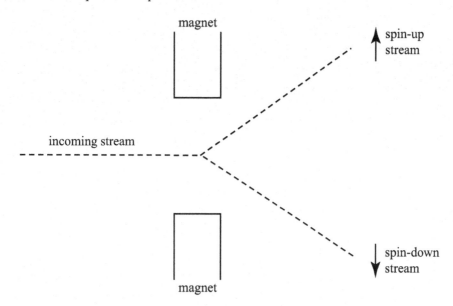

FIGURE 12.1
A Stern-Gerlach configuration.

Hence the Stern-Gerlach measurement changes the state of each of the particles in the input stream. This is not always the case in quantum measurement processes and in quantum computation, it is necessary to understand the conditions under which a change-of-state occurs under measurement – and the conditions under which it doesn't. In addition, it is often necessary to understand precisely what the post-measurement state is. These issues are addressed later in the statements of the axioms for quantum computation.

It should be stressed that the states above are representative of the spin states of the incoming and outgoing streams, and should not be confused with a complete 'wave function' representation, which would model additional physical properties that are not required in an introductory treatment of quantum computation.

The Stern-Gerlach experiment (see, for example, the reference [1], p252) dates back to the early 1920s and was one of the first experiments to demonstrate the quantum properties of physical systems.

13

Axioms for quantum computation

13.1 Objectives

The purpose of Sections 13.2, 13.3 and 13.4 is to impart some physical insight into the formal axioms of quantum computation presented in Section 13.6. Axioms are, by definition, unprovable; their purpose being to provide a foundation for a particular theory. To avoid the introduction of further physics, we use the Stern-Gerlach experiment to provide support for the current approach to information representation in quantum computation. It should be stated that other approaches to providing support for the same axioms exist. Readers familiar with the axioms of quantum mechanics may skip much of Sections 13.2, 13.3 and 13.4 and move straight to Section 13.6. Section 13.6 also includes two important special cases of the general measurement axiom, these being sufficient for many of the quantum measurements performed in later chapters.

13.2 Quantum state spaces for computation

Some understanding of the state spaces, time-evolution functions and measurement processes of quantum mechanics is required in quantum computation. In the case of digital computing, we have seen that the state space comprises the digital registers, changes of state are implemented by Boolean functions and measurement is a simple projection function.

In general quantum mechanics, the state spaces are vector spaces of infinite dimension. However, a complete understanding of the generalities of quantum mechanics is not required for quantum computation. We begin by considering the state space sub-domain required for computation. The considerations of the time-evolution and measurement processes will follow in Sections 13.3 and 13.4.

For quantum computation:

1. the underlying state spaces are finite dimensional (this is a significant simplification),

2. many of the foundations (in particular, the coverage of the current text) may be fully understood without the introduction of so-called 'mixed' quantum states, it being sufficient to consider only the pure states of quantum systems. A mixed quantum state may, for example, relate to a gas of quantum particles where the precise state of each individual particle cannot be known and statistical estimates are necessary. This should not be confused with pure states that are superpositions in a given basis. For example, the input state of the particles in the Stern-Gerlach experiment is not a mixed state; it is a superposition of pure states which is itself

DOI: 10.1201/9781003264569-13

pure. The interpretation is that the particle is in both states at the same time and there is no quantum mechanical uncertainty relating to the nature of the state.

However, it is necessary to understand the structure of the state spaces of multi-particle quantum systems; these being the tensor product of the component systems. Hence in some ways the quantum mechanics of the computational domain are relatively simple, but in other ways, e.g., the requirement for multi-particle state spaces, it is not the simplest quantum sub-domain. It is within the context of the above that the axioms for quantum computation, presented in Section 13.6, should be viewed.

13.2.1   The 2-dimensional vector space representation of 'spin' and quantum bits

The interpretation of the Stern-Gerlach experiment is that the incoming stream of particles have two internal quantum 'spin' states, which we denote using Dirac ket notation as $|\uparrow\rangle$ and $|\downarrow\rangle$. Given that a fifty-fifty post-measurement outcome is obtained, the incoming spin state is

$$\frac{1}{\sqrt{2}}|\uparrow\rangle + \frac{1}{\sqrt{2}}|\downarrow\rangle$$

or equivalently,

$$\left(\frac{1}{2}\right)^{\frac{1}{2}}|\uparrow\rangle + \left(\frac{1}{2}\right)^{\frac{1}{2}}|\downarrow\rangle.$$

The amplitudes (i.e., the multipliers) of $|\uparrow\rangle$ and $|\downarrow\rangle$ are equal in this case because the post-measurement states $|\uparrow\rangle$ and $|\downarrow\rangle$ are equally likely for all incoming particles. Generally, in quantum mechanics, amplitudes occurring in states such as the above need not be equal. For example, if a sixty-forty distribution of a pair of states $|a\rangle$ and $|b\rangle$ is observed post-measurement, then the initial state of the system under measurement would be expressed as:

$$\left(\frac{3}{5}\right)^{\frac{1}{2}}|a\rangle + \left(\frac{2}{5}\right)^{\frac{1}{2}}|b\rangle.$$

Hence, in general, amplitudes α_a and α_b in pre-measurement states of the form

$$\alpha_a|a\rangle + \alpha_b|b\rangle, \quad \text{where } \alpha_a{}^2 + \alpha_b{}^2 = 1 \tag{13.1}$$

are interpreted as follows:

- $\alpha_a{}^2$ is the probability of post-measurement state $|a\rangle$ occurring, and

- $\alpha_b{}^2$ is the probability of post-measurement state $|b\rangle$ occurring.

We have seen expressions like Equation 13.1 before. If we think of $|a\rangle$ and $|b\rangle$ as vectors in a vector space then Equation 13.1 is a linear combination of $|a\rangle$ and $|b\rangle$, and if $|a\rangle$ and $|b\rangle$ are orthogonal unit vectors then any linear combination $\alpha_a|a\rangle + \alpha_b|b\rangle$, with $\alpha_a{}^2 + \alpha_b{}^2 = 1$, is a unit vector.

Returning to the specific case of the Stern-Gerlach experiment, it is clear that we are dealing with particles that have two distinct internal spin states $|\uparrow\rangle$ and $|\downarrow\rangle$, and that this has some similarity with the classical bit (binary digit) of digital computation. The bit is the basis of information representation in classical computation. Hence, if we could entrap (i.e., locally constrain) 2-state quantum systems, their states – e.g., $|\uparrow\rangle, |\downarrow\rangle$ – could form a basis for the quantum-mechanical representation of information, performing the same role as the binary digits $\{0, 1\}$ of digital computation.

Of particular interest is that, in the quantum case, linear combinations $\alpha_\uparrow|\uparrow\rangle + \alpha_\downarrow|\downarrow\rangle$, with $\alpha_\uparrow^2 + \alpha_\downarrow^2 = 1$, are valid pure states distinct from $|\uparrow\rangle$ and $|\downarrow\rangle$. This generalises the classical case where $\{0,1\}$ is closed under the algebraic operations $\oplus$ and $\wedge$ (under which $\{0,1\}$ determines a vector space over the field $\mathbb{F}_2$ – see Chapter 1) and is partially responsible for the increased power of quantum computation.

In summary, the above suggests the existence of a 2-dimensional vector space of quantum states that could, possibly, be used as the basis of an approach to computation – replacing the digital pair $\{0,1\}$ of classical computation with the unit vectors (or 'rays' – see Sections 13.2.2, 13.2.4 and 13.2.5) of a, suitably defined, 2-dimensional vector space. We call such quantum equivalents of the classical bits $\{0,1\}$ quantum bits, or qubits.

13.2.2 The case for quantum bit (qubit) representation in $\mathbb{C}^2$

From the discussion of the Stern-Gerlach experiment, it is clear that

1. real-number multipliers, relating to post-measurement outcome probabilities (i.e., the $\alpha_\uparrow$ and $\alpha_\downarrow$ in $\alpha_\uparrow|\uparrow\rangle + \alpha_\downarrow|\downarrow\rangle$) can occur in the representation of pure quantum states,

2. the suggested 2-dimensional vector space representation must have an inner product, enabling the concepts of unit and orthogonal vectors to be defined.

That real numbers can occur implies, by a theorem of Frobenius, that the fields over which the 2-dimensional space is defined are restricted to be either $\mathbb{R}$ or $\mathbb{C}$. If non-commutativity of multiplication is permitted then Frobenius' theorem admits a further alternative – i.e., Hamilton's 'quaternions'. Jauch [1], p131, discusses these alternatives (i.e., $\mathbb{R}$ and the quaternions) to the generally accepted choice of the complex number field $\mathbb{C}$ and provides references to the literature on this issue. Here, from this point, we focus on the development of quantum theory in vector spaces over $\mathbb{C}$ and interpret superpositions such as

$$\alpha_\uparrow|\uparrow\rangle + \alpha_\downarrow|\downarrow\rangle$$

to be unit vectors in $\mathbb{C}^2$. It follows that $\alpha_\uparrow \in \mathbb{C}$ and $\alpha_\downarrow \in \mathbb{C}$ are such that $|\alpha_\uparrow|^2 + |\alpha_\downarrow|^2 = 1$ and that the superpositions are unit vectors with respect to the inner product

$$(z_1, z_2) \cdot (w_1, w_2) = z_1\overline{w_1} + z_2\overline{w_2}$$

on $\mathbb{C}^2$. (You should be aware, as we noted in Chapter 4, an alternative definition of the inner product, in which conjugation is performed on the first of the vectors in the products on the right-hand side, is favoured by many physicists and computer scientists). The row vectors $(1,0), (0,1)$ determine an orthonormal basis of $\mathbb{C}^2$ and can be realised as $|\uparrow\rangle = (1,0)$ and $|\downarrow\rangle = (0,1)$. Alternatively, using Dirac ket notation and column vectors, the standard basis of $\mathbb{C}^2$ is often written in the form $\begin{pmatrix} 1 \\ 0 \end{pmatrix} = |0\rangle$, and $\begin{pmatrix} 0 \\ 1 \end{pmatrix} = |1\rangle$.

13.2.3 Quantum bits – or qubits

In quantum computation the unit vectors or, more precisely, the rays of $\mathbb{C}^2$ (see below) replace the bits $\{0,1\}$ of digital computation and are called **qubits**.

13.2.4 Global phase

We note, again from the Stern-Gerlach experiment, that distinct unit vectors in $\mathbb{C}^2$ can represent the same quantum spin state. Specifically, we have a representation of the incoming particle states as

$$\frac{1}{\sqrt{2}}|\uparrow\rangle + \frac{1}{\sqrt{2}}|\downarrow\rangle$$

with the post-measurement probabilities of $(\frac{1}{\sqrt{2}})^2 = \frac{1}{2}$ for both possible outcomes. Accepting that the amplitudes $\alpha_\uparrow, \alpha_\downarrow$ are, generally, complex numbers, we note that

$$\frac{i}{\sqrt{2}}|\uparrow\rangle + \frac{i}{\sqrt{2}}|\downarrow\rangle$$

(with $i^2 = -1$) is, therefore, also a valid representation of the state $\frac{1}{\sqrt{2}}|\uparrow\rangle + \frac{1}{\sqrt{2}}|\downarrow\rangle$ because $|\frac{i}{\sqrt{2}}|^2 = \frac{1}{2}$. More generally we have that

$$\frac{e^{i\theta}}{\sqrt{2}}|\uparrow\rangle + \frac{e^{i\theta}}{\sqrt{2}}|\downarrow\rangle$$

for all $\theta \in \mathbb{R}$ are valid unit vector representations of the same state. That is, if $(z_1, z_2) \in \mathbb{C}^2$ is a unit vector then so is $e^{i\theta}(z_1, z_2)$ for any $\theta \in \mathbb{R}$, and the vectors $e^{i\theta}(z_1, z_2)$ represent the same quantum state for all $\theta \in \mathbb{R}$.

Generally, in quantum mechanics, the vectors $|\psi\rangle$ and $z|\psi\rangle$, for any non-zero $z \in \mathbb{C}$, represent the same pure state (Mackey [2], p76). That is, pure quantum states are 'rays' in the underlying vector space and global multipliers of quantum states, (i.e., the z in $z|\psi\rangle$) may be 'ignored'; equivalently all such z may be replaced by 1 without changing the physical properties of the state.

We note that $\frac{e^{i\theta_1}}{\sqrt{2}}|\uparrow\rangle + \frac{e^{i\theta_2}}{\sqrt{2}}|\downarrow\rangle$, for $\theta_1 \neq \theta_2$, is not on the same ray in $\mathbb{C}^2$ as $\frac{1}{\sqrt{2}}|\uparrow\rangle + \frac{1}{\sqrt{2}}|\downarrow\rangle$, i.e., local phase factors cannot be ignored.

Rays in vector spaces determine projective spaces and, to assist their understanding, a short introduction is given below. Readers may, at least initially, omit this material and simply regard qubits as unit vectors in $\mathbb{C}^2$ – this being sufficient for much of what follows in this text. We will, primarily, represent states as unit vectors in $\mathbb{C}^2$, but occasionally, in the sections on quantum algorithms, global phase multipliers will arise – which, for the reasons given above, may be ignored.

13.2.5   Projective spaces and the Bloch sphere

First we consider real projective spaces, before moving on to complex projective spaces, which are more relevant to quantum computation.

1. Real projective spaces

The projective space of $\mathbb{R}^{n+1}$ is the n-dimensional space of lines passing through the origin of $\mathbb{R}^{n+1}$. The lines are the 1-dimensional subspaces of $\mathbb{R}^{n+1}$ and are also referred to as rays.

Consider the case $n = 1$ for which we obtain a 1-dimensional space of 1-dimensional subspaces (i.e., the lines in $\mathbb{R}^2$ through $(0,0)$). We note that if (x, y) lies on a ray in $\mathbb{R}^2$, then (x, y) identifies the ray and $(-x, -y)$ lies on the same ray – see Figure 13.1.

The rays do not determine equivalence classes of points in $\mathbb{R}^2$ because their intersections are not empty – as each ray passes through the origin $(0,0) \in \mathbb{R}^2$. If instead we remove the origin from $\mathbb{R}^2$ (Figure 13.2) and define, on $\mathbb{R}^2 \setminus \{(0,0)\}$, the equivalence relation

$$(x, y) \sim (x^*, y^*) \quad \text{iff} \quad (x^*, y^*) = \lambda(x, y) \qquad \text{for some } \lambda \in \mathbb{R} \setminus \{0\}$$

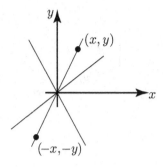

FIGURE 13.1

The projective space of 1-dimensional subspaces (or rays) in $\mathbb{R}^2$. Points (x, y) and $(-x, -y)$ lie on the same ray.

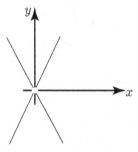

FIGURE 13.2

Rays in $\mathbb{R}^2 \setminus \{0, 0\}$.

then the rays, i.e., the elements, $[(x, y)]$, of

$$(\mathbb{R}^2 \setminus \{(0, 0)\})/ \sim$$

are such that

$$[(x, y)] \cap [(x^*, y^*)] = \emptyset \text{ if } (x, y) \nsim (x^*, y^*)$$

and

$$[(x, y)] = [(x^*, y^*)] \text{ if } (x, y) \sim (x^*, y^*)$$

and determine a partition of $\mathbb{R}^2 \setminus \{(0, 0)\}$ – see Figures 13.3 and 13.4.

Rays may be identified in $\mathbb{R}^2 \setminus \{(0, 0)\}$ by

1. any vector (x, y) on the ray – see Figure 13.3 (this is not so for rays in $\mathbb{R}^2$ as then each ray passes through the origin $(0, 0)$). There are, clearly, infinitely many representative points for a given ray – imposing the condition that the representative be a unit vector reduces the options.

2. a unit vector that lies on the ray – this leads to just two choices, specifically $\dfrac{(x, y)}{\sqrt{x^2 + y^2}}$ and $\dfrac{(-x, -y)}{\sqrt{x^2 + y^2}}$ where (x, y) is any point on the ray.

Hence, rays in $\mathbb{R}^2 \setminus \{(0, 0)\}$ may be identified by either of two unit vectors.

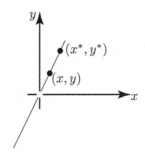

FIGURE 13.3
$(x, y) \sim (x^*, y^*)$.

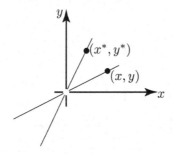

FIGURE 13.4
$(x, y) \not\sim (x^*, y^*)$.

The relevance of this to quantum computing is that, as we have seen in Section 13.2.4, qubits are more precisely defined as elements of the ray space

$$(\mathbb{C}^2 \setminus \{(0, 0)\})/ \sim$$

rather than unit vectors in $\mathbb{C}^2$.

In the above, $\sim$ is an equivalence relation similar to that defined earlier on $\mathbb{R}^2 \setminus \{(0, 0)\}$. It is therefore of interest to investigate the space $(\mathbb{C}^2 \setminus \{(0, 0)\})/ \sim$ (which is known as the Bloch sphere). We do this below.

2. A complex projective space – the Bloch sphere

The vector space $\mathbb{C}^2$ is of particular significance in quantum computation. Generally, in quantum mechanics, the wave functions $|\psi\rangle$ and $z|\psi\rangle$ (for $z \in \mathbb{C}, z \neq 0$) define the same ray in the underlying vector space – and hence the same quantum state. The Bloch sphere (introduced below) is, essentially, a 'parametrisation' of the quantum states $(\mathbb{C}^2 \setminus \{(0, 0)\})/ \sim$; i.e., the pure qubit states.

Up to global phase the pure states of $\mathbb{C}^2$, in an orthonormal basis $\{|0\rangle, |1\rangle\}$, take the form

$$|\psi\rangle = \alpha|0\rangle + \beta|1\rangle$$

where $\alpha, \beta \in \mathbb{C}$ with $|\alpha|^2 + |\beta|^2 = 1$. Hence for some $\eta_0, \eta_1 \in [0, 2\pi)$ and $r_0, r_1 \in \mathbb{R}$, with $r_0 > 0$ and $r_1 > 0$, we have

$$|\psi\rangle = r_0 e^{i\eta_0}|0\rangle + r_1 e^{i\eta_1}|1\rangle$$

where $r_0^2 + r_1^2 = 1$. (See Section 3.6 on the exponential form of a complex number.) We write $r_0 = \cos(\theta/2)$ and $r_1 = \sin(\theta/2)$; and the equivalence of $|\psi\rangle$ and $z|\psi\rangle$ then gives

$$
\begin{aligned}
|\psi\rangle &= \cos(\theta/2)e^{i\eta_0}|0\rangle + \sin(\theta/2)e^{i\eta_1}|1\rangle \\
&\sim \cos(\theta/2)|0\rangle + \sin(\theta/2)e^{i(\eta_1-\eta_0)}|1\rangle.
\end{aligned}
$$

Writing $\varphi = \eta_1 - \eta_0$, we have

$$
\begin{aligned}
|\psi\rangle &= \cos(\theta/2)\,|0\rangle + \sin(\theta/2)e^{i\varphi}\,|1\rangle \\
&= \cos(\theta/2)\,|0\rangle + (\sin(\theta/2)\cos\varphi + i\sin(\theta/2)\sin\varphi)\,|1\rangle \\
&= \cos(\theta/2)\,|0\rangle + \sin(\theta/2)\cos\varphi\,|1\rangle + i\sin(\theta/2)\sin\varphi\,|1\rangle.
\end{aligned}
$$

The vector

$$(\sin(\theta/2)\cos\varphi,\ \sin(\theta/2)\sin\varphi,\ \cos(\theta/2))$$

lies on the unit sphere, $x^2 + y^2 + z^2 - 1 = 0$, of $\mathbb{R}^3$. That is, each pure qubit state corresponds to a point on the surface of the unit sphere centred at the origin of $\mathbb{R}^3$. This 'parametrisation' of the pure states of $\mathbb{C}^2$ is known as the Bloch sphere depicted in Figure 13.5.

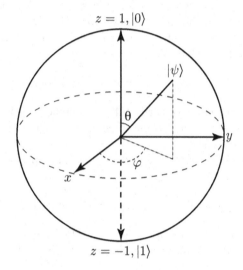

FIGURE 13.5
The Bloch sphere, $x = \sin\theta\cos\varphi$, $y = \sin\theta\sin\varphi$, $z = \cos\theta$.

To assist the interpretation of the Bloch sphere representation, we evaluate the positions of a number of pure states on the surface of the sphere. For

$$|\psi\rangle = \cos(\theta/2)\,|0\rangle + \sin(\theta/2)\cos\varphi\,|1\rangle + i\sin(\theta/2)\sin\varphi\,|1\rangle$$

we have

- for $\theta = 0$, the north-pole of the sphere, $(0, 0, 1)$, is representative of the vector $|\psi\rangle = |0\rangle$,

- for $\theta = \pi$ and $\varphi = 0$, the south-pole, $(0, 0, -1)$, of the sphere, is representative of the vector $|\psi\rangle = |1\rangle$,

- for $\theta = \pi/2$ and $\varphi = 0$, the positive x-radial vector, $(1, 0, 0)$, of the sphere, is representative of the vector $|\psi\rangle = \frac{1}{\sqrt{2}}(|0\rangle + |1\rangle)$,

- for $\theta = \pi/2$ and $\varphi = \pi$, the negative x-radial vector, $(-1,0,0)$, of the sphere, is representative of the vector $|\psi\rangle = \frac{1}{\sqrt{2}}(|0\rangle - |1\rangle)$,

- for $\theta = \pi/2$ and $\varphi = \pi/2$, the positive y-radial vector, $(0,1,0)$, of the sphere, is representative of the vector $|\psi\rangle = \frac{1}{\sqrt{2}}(|0\rangle + i|1\rangle)$,

- for $\theta = \pi/2$ and $\varphi = 3\pi/2$, the negative y-radial vector, $(0,-1,0)$, of the sphere, is representative of the vector $|\psi\rangle = \frac{1}{\sqrt{2}}(|0\rangle - i|1\rangle)$.

We note, for example, that although the states $|0\rangle$ and $|1\rangle$ are orthogonal in the vector space $\mathbb{C}^2$, they are not orthogonal when represented in $\mathbb{R}^3$ on the Bloch sphere. The representation is therefore far from being conformal or geometric. It should be regarded as a 'parametrisation' of the pure quantum states of $\mathbb{C}^2$.

13.2.6   Multi-qubit state spaces

Accepting that the unit vectors of $\mathbb{C}^2$ form the basis of information representation in quantum computation we need, following the digital case, to define n-qubit quantum registers. Considerations of the requirements of quantum measurement and the ability to have sufficient structure to model qubit-qubit interactions appropriately, lead naturally to the conclusion that the underlying vector space for n-qubit registers, or systems, is the tensor product space of the constituent qubits, that is,

$$\otimes^n \mathbb{C}^2 = \mathbb{C}^2 \otimes \mathbb{C}^2 \otimes \cdots \mathbb{C}^2, \quad n \text{ copies.}$$

We should note that it is not easy to justify fully this choice a priori; but it will be seen as suitable when specific concepts, such as entanglement, are introduced in the presentation of quantum computation to follow.

It follows from the discussion of tensor product spaces in Chapter 10 that an orthonormal basis of $\mathbb{C}^2 \otimes \mathbb{C}^2$ is given by

$$\{|0\rangle \otimes |0\rangle, |0\rangle \otimes |1\rangle, |1\rangle \otimes |0\rangle, |1\rangle \otimes |1\rangle\}.$$

For brevity we often write $|x\rangle \otimes |y\rangle$, for $x, y \in \{0,1\}$, as $|x\rangle|y\rangle$ or even as $|xy\rangle$. So, for example, the state $|0\rangle \otimes |0\rangle$ will be written as simply $|00\rangle$ and the orthonormal basis of $\mathbb{C}^2 \otimes \mathbb{C}^2$ is then

$$\{|00\rangle, |01\rangle, |10\rangle, |11\rangle\}.$$

On occasions, especially when n is large, the basis may be written using decimal equivalents

$$\{|0\rangle, |1\rangle, |2\rangle, |3\rangle\}.$$

Clearly, care must be taken to read the specific context under consideration.

Similarly, orthonormal bases of $\otimes^n \mathbb{C}^2$ may be constructed. It follows that the dimension of the vector space $\otimes^n \mathbb{C}^2$ is 2^n with basis

$$\{|00\cdots 0\rangle, \ldots, |11\cdots 1\rangle\}.$$

13.3 Quantum observables and measurement for computation

Appealing again to the Stern-Gerlach experiment, we illustrate the quantum model of the measurement of observables on quantum states. We note that the self-adjoint operator $A : \mathbb{C}^2 \to \mathbb{C}^2$ defined by

$$A = \begin{pmatrix} 1 & 0 \\ 0 & -1 \end{pmatrix}$$

has the two distinct eigenvalues of 1 and -1 with eigenvectors, respectively

$$(1,0)^T = |\uparrow\rangle = |0\rangle$$

and

$$(0,1)^T = |\downarrow\rangle = |1\rangle.$$

Hence, the spectral properties of the self-adjoint operator A are representative of the two possible outcomes of Stern-Gerlach experiment, i.e.,

1. the real value 1 measured with post-measurement state $|\uparrow\rangle$,

2. the real value -1 measured with post-measurement state $|\downarrow\rangle$.

Specifically, the operator A models the potential outcomes of the measurement of the 'intrinsic spin' observable on the input state:

$$|\phi\rangle = \frac{1}{\sqrt{2}}|\uparrow\rangle + \frac{1}{\sqrt{2}}|\downarrow\rangle.$$

We recall that the pure state $|\phi\rangle$ is a linear superposition of the two possible outcomes, incorporating (in the amplitudes) the probabilities of each outcome. The interpretation being that the particle is simultaneously in both states prior to measurement.

We stress that the self-adjoint operator A is not applied to the input state $|\phi\rangle$, it simply models the possible outputs of

1. the measured value (1 or -1),

2. the corresponding post-measurement state ($|0\rangle$ or $|1\rangle$ respectively)

following a measurement process.

This generalises to the following: to each physically measurable observable on a quantum system there corresponds a self-adjoint operator, on the underlying vector space, the eigenvalue-eigenvector pairs of which relate to the measurable values of the observable and the resulting post-measurement state. The precise nature of the post-measurement state depends on the degeneracy of the related eigenvalue - the full details of this are given in Axiom 3 of Section 13.6 below.

13.4 Quantum dynamics for computation

The following observations may be used to support the axiom of unitary evolution of quantum states:

1. given that all quantum states are represented by unit vectors, state-to-state transitions must be unitary (because linear unit vector to unit vector transformations are unitary – see Section 11.7),

2. for readers, more familiar with general quantum mechanics, we note that the dynamics of non-relativistic quantum systems are determined by Schrödinger's famous equation:

$$i\hbar \frac{\partial |\psi_t\rangle}{\partial t} = \mathcal{H}|\psi_t\rangle$$

where $\mathcal{H}$ is a self-adjoint operator on the underlying vector space of states V, $|\psi_t\rangle \in V$, $\hbar$ is Planck's constant, $i = (-1)^{\frac{1}{2}}$ and t is the time variable. The operator $\mathcal{H}$ is representative of the energy observable. Schrödinger's equation may be integrated, to the form:

$$|\psi_t\rangle = U_t|\psi_0\rangle,$$

where U_t is a unitary operator on V and $|\psi_0\rangle$ is the system state at $t = 0$. The unitary operator may be expressed, as a function of the quantum Hamiltonian $\mathcal{H}$, as $U_t = e^{-\frac{it}{\hbar}\mathcal{H}}$.

From (1) and (2) it follows that unitary operators play the role, in quantum computation, that Boolean functions play in digital computation; i.e., they implement the steps in the computation from input to output.

13.5   Orthogonal projection in a complex inner product space

Before formally summarising the axioms of quantum computation, we take a moment to discuss the concept of orthogonal projection in a complex inner product space. We begin with, for the purposes of ease of visualisation, the real vector space $\mathbb{R}^3$ with the usual inner product of two vectors (x, y, z) and (x^*, y^*, z^*) given by

$$\langle (x, y, z), (x^*, y^*, z^*)\rangle = xx^* + yy^* + zz^*$$

and standard orthonormal basis

$$i = (1, 0, 0), \ j = (0, 1, 0), \ k = (0, 0, 1).$$

It is clear from Figure 13.6 that the orthogonal projection of the point $(x, y, z) \in \mathbb{R}^3$ onto the xy-plane is $(x, y, 0)$; i.e., the dashed 'projection' line from the point (x, y, z) to the point $(x, y, 0)$ is parallel to the z-axis and orthogonal to the xy-plane. Denoting the projected vector by $P_{xy}(x, y, z)$, mathematically we have

$$\begin{aligned} P_{xy}(x, y, z) &= \langle i, (x, y, z)\rangle i + \langle j, (x, y, z)\rangle j. \\ &= (x, y, 0). \end{aligned}$$

We note that P_{xy} is a linear operator on $\mathbb{R}^3$ with the property $P_{xy}^2 = P_{xy}$.

More generally (and in standard rather than Dirac notation): let $\phi_0, \ldots, \phi_n$ be an orthonormal basis of an $(n+1)$-dimensional complex inner product space $\mathcal{H}$ with inner product $\langle, \rangle$. Any subset $\{\chi_1, \ldots, \chi_k\}$ of k elements of $\{\phi_0, \ldots, \phi_n\}$ clearly defines a k-dimensional subspace of $\mathcal{H}$, which we denote by $\mathcal{H}_\chi$.

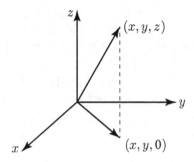

FIGURE 13.6
Orthogonal projection in $\mathbb{R}^3$.

For any $\psi \in \mathcal{H}$ the orthogonal projection, $P_\chi \psi$, of ψ onto the subspace $\mathcal{H}_\chi$ is, generalising the $\mathbb{R}^3$ case, given by

$$P_\chi \psi = \langle \chi_1, \psi \rangle \chi_1 + \cdots + \langle \chi_k, \psi \rangle \chi_k.$$

13.6 A summary of the axioms for quantum computation

From the above we state the following **axioms**, or **postulates**, for quantum computation.

Axiom 1 *The underlying vector space of an n-qubit quantum computer is the n-fold tensor-product space*

$$\otimes^n \mathbb{C}^2$$

and the pure states are the rays of $\otimes^n \mathbb{C}^2$, i.e., elements of the space $\otimes^n \mathbb{C}^2 / \sim$.

Axiom 2 *The observables of a quantum system with underlying state space $\otimes^n \mathbb{C}^2$ correspond to the self-adjoint operators on $\otimes^n \mathbb{C}^2$.*

Axiom 3 *Following the measurement of an observable $\mathcal{O}$ on the state $|\psi\rangle \in \otimes^n \mathbb{C}^2$, the post-measurement state is the normalisation of the vector $P|\psi\rangle$, where P is the orthogonal projection of $|\psi\rangle$ onto the subspace of $\otimes^n \mathbb{C}^2$ generated by the eigenstates of $\mathcal{O}$ that*

    1. occur in the linear-superposition of $|\psi\rangle$, and

    2. are compatible with the outcome of the measurement of $\mathcal{O}$.

The probability of the outcome obtained is $\|P|\psi\rangle\|^2$.

Axiom 4 *Unitary operators implement the computational steps in quantum computation.*

It should be noted that

1. Axiom 3 is due to von Neumann and Lüders, and relates to the measurement of general pure quantum states. For readers with no prior knowledge of quantum mechanics, it is the most difficult of the axioms to understand. It is therefore helpful to consider two special cases:

(i) Where the eigenvalues of the associated self-adjoint operator are non-degenerate – i.e., to each eigenvalue of the operator there is only one eigendirection. In this case, the measurement axiom simplifies to Axiom 3.1, see Section 13.7 below.

(ii) Where the state is an eigenstate of the observable to be measured, in which case the measurement axiom simplifies to Axiom 3.2, see Section 13.7 below.

2. A measurement is not a computational step in a quantum computation – it is a physical process modelled by a self-adjoint operator on the underlying quantum state space.

3. It follows from Axiom 4 that quantum computation is reversible – this is not true of the current implementations of digital computation.

4. Quantum measurement processes are not reversible.

13.7 Special cases of the measurement axiom

Case 1: In the case where an observable $\mathcal{O}$ is non-degenerate, Axiom 3 may be stated as

Axiom 3.1 Following the measurement of $\mathcal{O}$ on the quantum state

$$|\chi\rangle = \alpha_0|\phi_0\rangle + \alpha_1|\phi_1\rangle + \cdots + \alpha_n|\phi_n\rangle; \quad \text{where} \quad \sum_{i=0}^{n}|\alpha_i|^2 = 1$$

in an $(n+1)$-dimensional complex inner product space (expressed in an orthonormal basis $\{|\phi_0\rangle, \ldots |\phi_n\rangle\}$ determined by $\mathcal{O}$) the post-measurement state is one of the eigenvectors

$$\{|\phi_0\rangle, \ldots, |\phi_n\rangle\}$$

of $\mathcal{O}$, and the possible measured values of $\mathcal{O}$ are its eigenvalues

$$\{\lambda_0, \ldots, \lambda_n\}.$$

When λ_i is measured for $\mathcal{O}$, the post-measurement state is $|\phi_i\rangle$ and this occurs with probability $|\alpha_i|^2$.

Case 2: In the case where the state, $|\psi\rangle$, to be measured is an eigenstate of the observable $\mathcal{O}$ (which may be degenerate) we have

Axiom 3.2 Following the measurement of $\mathcal{O}$ on the eigenstate $|\psi\rangle$, the measured value is the eigenvalue corresponding to $|\psi\rangle$ and the post-measurement state is $|\psi\rangle$.

13.8 A formal comparison with digital computation

We noted in Section 12.4.3 that digital computations may be expressed functionally as

$$y = \mu_D \circ (f_{p-1} \circ \cdots \circ f_2 \circ f_1 \circ f_0)(x) \qquad \text{(D)}$$

and, from the axioms above for quantum computing, we may now express quantum computations in the form

$$|y_s\rangle = (U_{q-1} \cdots U_2\, U_1\, U_0)|x\rangle$$

where $|x\rangle$ and $|y_s\rangle$ are the initial and final quantum states of the computation and $U_0, \ldots, U_{q-1}$ are unitary operators. The inclusion of a quantum measurement process, μ_Q, leads to

$$y = \mu_Q|y_s\rangle = \mu_Q(U_{q-1} \cdots U_2\, U_1\, U_0)|x\rangle$$

or

$$y = \mu_Q(U_{q-1} \cdots U_2\, U_1\, U_0)|x\rangle \tag{Q}$$

where y is the output, in real numbers, of the quantum computation.

Relation (Q) is structurally similar to relation (D) for the digital case. We should however note that the quantum measurement process, μ_Q, is not an operator on the quantum state vector; it is a physical measurement process, modelled by an appropriate self-adjoint operator on the underlying quantum state space (refer to quantum Axioms 3, 3.1 and 3.2 above).

13.9 Gates

In both digital and quantum computation, fundamental computational steps are often referred to as 'gates'. In the digital case, gates are implemented by (non-invertible) Boolean functions and, by Axiom 4, it follows that in the quantum case gates are implemented by unitary operators – which are invertible. Hence digital computation is not invertible, i.e., the input cannot be determined from the output, but in the quantum case this is not so. We will meet invertible digital gates later – when we consider the quantum emulation of digital computations. However, current digital computers are not based on invertible digital technology.

Exercises

13.1 Discuss the identity operator, which is self-adjoint, as an observable on the quantum states of $\otimes^n \mathbb{C}^2$.

13.2 Let A and B be self-adjoint operators on $\otimes^n \mathbb{C}^2$ with identical non-degenerate eigenvectors. Show that the measurement of A followed by the measurement of B, on any state, is equivalent, in the sense that the measured values obtained are identical, to the measurement of B followed by the measurement of A.

14

Quantum measurement 1

14.1 Objectives

In this chapter, the simplest quantum measurement situation, i.e., where the state is an eigenvector of the observable to be measured, is considered. The states are therefore non-superimposed and measurement Axiom 3.2 of Section 13.7 in the previous chapter is applicable. We consider states in $\mathbb{C}^2, \mathbb{C}^2 \otimes \mathbb{C}^2$ and $\mathbb{C}^2 \otimes \mathbb{C}^2 \otimes \mathbb{C}^2$. The generalisation to $\otimes^n \mathbb{C}^2$ is then clear.

14.2 Measurement using Axiom 3.2

The applications of Axiom 3.2, discussed below, are applied in Chapters 15 and 17 to the quantum emulation of familiar digital gates (Chapter 15) and, more generally, to arbitrary Boolean functions (Chapter 17).

14.2.1 Measurement of non-superimposed states in $\mathbb{C}^2$

Recall that a superimposed state in $\mathbb{C}^2$ takes the form $|\psi\rangle = \alpha_0|0\rangle + \alpha_1|1\rangle$ where $\alpha_0, \alpha_1 \in \mathbb{C}$. But for non-superimposed states in $\mathbb{C}^2$ there are just two states to measure, i.e., $|\psi\rangle = |0\rangle$ and $|\psi\rangle = |1\rangle$. Consider the observable A, that is, the self-adjoint operator, with matrix representation

$$A = \begin{pmatrix} 0 & 0 \\ 0 & 1 \end{pmatrix}.$$

The eigenvalues of A are 0 and 1 with eigenvectors $|0\rangle = (1,0)^T$ and $|1\rangle = (0,1)^T$ respectively. Hence, measurement Axiom 3.2 is applicable and it follows that if A is measured on the state

$$|\psi\rangle = |0\rangle$$

then, with probability 1, the real value of 0 will be obtained for A and the post-measurement state $|0\rangle$ will result. Similarly measuring A on the state

$$|\psi\rangle = |1\rangle$$

results, with probability 1, in the value of 1 for A and the post-measurement state $|1\rangle$ will result. Hence in this simple case, quantum measurement is deterministic.

We note that because $\begin{pmatrix} 0 & 0 \\ 0 & 1 \end{pmatrix} = \begin{pmatrix} 0 \\ 1 \end{pmatrix} \begin{pmatrix} 0 & 1 \end{pmatrix}$ the operator A can be expressed, in Dirac notation, as

$$A = |1\rangle\langle 1|$$

DOI: 10.1201/9781003264569-14

and we have

$$A|0\rangle = |1\rangle\langle 1|0\rangle$$
$$= 0|0\rangle$$

and

$$A|1\rangle = |1\rangle\langle 1|1\rangle$$
$$= |1\rangle$$
$$= 1|1\rangle.$$

We do not concern ourselves with the physical realisation of measurement; the underlying assumption of quantum mechanics is that if A is self-adjoint, then there is some physical means of implementing the associated observable.

14.2.2   Measurement of non-superimposed states in $\mathbb{C}^2 \otimes \mathbb{C}^2$

Here we measure the states $|x\rangle|y\rangle \in \mathbb{C}^2 \otimes \mathbb{C}^2$, i.e., states without superposition; specifically the four states $|0\rangle|0\rangle, |0\rangle|1\rangle, |1\rangle|0\rangle$ and $|1\rangle|1\rangle$ of $\mathbb{C}^2 \otimes \mathbb{C}^2$.

Again we apply Axiom 3.2 and, later, illustrate its application to the quantum *cnot* function and the Peres half-adder. To understand the quantum *cnot* and half-adder gates it is necessary to measure simple 2-qubit quantum states. The measurement of 'simple' n-qubit states follows by generalisation of the 2-qubit case.

If A and B are self-adjoint operators on $\mathbb{C}^2$ the operator $A \otimes B : \mathbb{C}^2 \otimes \mathbb{C}^2 \to \mathbb{C}^2 \otimes \mathbb{C}^2$ defined by $(A \otimes B)(|x\rangle \otimes |y\rangle) = A|x\rangle \otimes B|y\rangle$ is self-adjoint.

Two qubit states are elements of the tensor product space $\mathbb{C}^2 \otimes \mathbb{C}^2$. We measure an observable A on the first qubit using the self-adjoint operator $A \otimes I$ on $\mathbb{C}^2 \otimes \mathbb{C}^2$ and, similarly, $I \otimes A$, is used to measure the second qubit. Here, I denotes the identity operator on $\mathbb{C}^2$ – which is self-adjoint.

We will measure the quantum *cnot* output using the operators $A \otimes I$ and $I \otimes A$ where A is as defined earlier. That is,

$$A = \begin{pmatrix} 0 & 0 \\ 0 & 1 \end{pmatrix}.$$

It follows that

$$A \otimes I = \begin{pmatrix} 0 & 0 & 0 & 0 \\ 0 & 0 & 0 & 0 \\ 0 & 0 & 1 & 0 \\ 0 & 0 & 0 & 1 \end{pmatrix}$$

and

$$I \otimes A = \begin{pmatrix} 0 & 0 & 0 & 0 \\ 0 & 1 & 0 & 0 \\ 0 & 0 & 0 & 0 \\ 0 & 0 & 0 & 1 \end{pmatrix}.$$

We also have

$$|0\rangle \otimes |0\rangle = \begin{pmatrix} 1 \\ 0 \\ 0 \\ 0 \end{pmatrix}, |0\rangle \otimes |1\rangle = \begin{pmatrix} 0 \\ 1 \\ 0 \\ 0 \end{pmatrix}, |1\rangle \otimes |0\rangle = \begin{pmatrix} 0 \\ 0 \\ 1 \\ 0 \end{pmatrix}, |1\rangle \otimes |1\rangle = \begin{pmatrix} 0 \\ 0 \\ 0 \\ 1 \end{pmatrix}.$$

Hence,

Proposition 14.1

   (i) The eigenvalues of the operator $A \otimes I$ are 0 and 1. The eigenvectors with respect to 0 are $|0\rangle \otimes |0\rangle$ and $|0\rangle \otimes |1\rangle$ and the eigenvectors with respect to 1 are $|1\rangle \otimes |0\rangle$ and $|1\rangle \otimes |1\rangle$.

   (ii) The eigenvalues of the operator $I \otimes A$ are 0 and 1. The eigenvectors with respect to 0 are $|0\rangle \otimes |0\rangle$ and $|1\rangle \otimes |0\rangle$ and those with respect to 1 are $|0\rangle \otimes |1\rangle$ and $|1\rangle \otimes |1\rangle$.

Measuring the first qubit of the states $|0\rangle|0\rangle, \ldots, |1\rangle|1\rangle$:

We measure A on the first qubit of the simple (i.e., non-superimposed) state

$$|\psi\rangle = |x\rangle|y\rangle \quad \text{where} \quad x, y \in \mathbb{B}$$

using the operator $A \otimes I$. From Proposition 14.1 and the quantum measurement postulate, we have

1. If $|\psi\rangle = |0\rangle|0\rangle$ then, because $A \otimes I|0\rangle|0\rangle = 0|0\rangle|0\rangle$, 0 will be measured for A on $|\psi\rangle$ with probability 1 and the post-measurement state will be $|0\rangle|0\rangle$.

2. If $|\psi\rangle = |0\rangle|1\rangle$ then, because $A \otimes I|0\rangle|1\rangle = 0 \, |0\rangle|1\rangle$, 0 will be measured for A on $|\psi\rangle$ with probability 1 and the post-measurement state will be $|0\rangle|1\rangle$.

3. If $|\psi\rangle = |1\rangle|0\rangle$ then, because $A \otimes I|1\rangle|0\rangle = 1 \, |1\rangle|0\rangle$, 1 will be measured for A on $|\psi\rangle$ with probability 1 and the post-measurement state will be $|1\rangle|0\rangle$.

4. If $|\psi\rangle = |1\rangle|1\rangle$ then, because $A \otimes I|1\rangle|1\rangle = 1 \, |1\rangle|1\rangle$, 1 will be measured for A on $|\psi\rangle$ with probability 1 and the post-measurement state will be $|1\rangle|1\rangle$.

Measuring the second qubit of the states $|0\rangle|0\rangle, \ldots, |1\rangle|1\rangle$:

Measuring A on the second qubit of $|\psi\rangle = |x\rangle|y\rangle$, for $x, y \in \mathbb{B}$, using the operator $I \otimes A$ we have

1. If $|\psi\rangle = |0\rangle|0\rangle$ then, because $I \otimes A|0\rangle|0\rangle = 0 \, |0\rangle|0\rangle$, 0 will be measured for A on $|\psi\rangle$ with probability 1 and the post-measurement state will be $|0\rangle|0\rangle$.

2. If $|\psi\rangle = |0\rangle|1\rangle$ then, because $I \otimes A|0\rangle|1\rangle = 1 \, |0\rangle|1\rangle$, 1 will be measured for A on $|\psi\rangle$ with probability 1 and the post-measurement state will be $|0\rangle|1\rangle$.

3. If $|\psi\rangle = |1\rangle|0\rangle$ then, because $I \otimes A|1\rangle|0\rangle = 0 \, |1\rangle|0\rangle$, 0 will be measured for A on $|\psi\rangle$ with probability 1 and the post-measurement state will be $|1\rangle|0\rangle$.

4. If $|\psi\rangle = |1\rangle|1\rangle$ then because $I \otimes A|1\rangle|1\rangle = 1 \, |1\rangle|1\rangle$, 1 will be measured for A on $|\psi\rangle$ with probability 1 and the post-measurement state will be $|1\rangle|1\rangle$.

As in Section 14.2.1 we note that the above measurement outcomes on non-superimposed quantum states are deterministic.

14.2.3   Measurement of non-superimposed states in $\mathbb{C}^2 \otimes \mathbb{C}^2 \otimes \mathbb{C}^2$

We first note that Proposition 14.1 is easily generalised to the observables $A \otimes I \otimes I$, $I \otimes A \otimes I$ and $I \otimes I \otimes A$ for measuring 3-qubit states of the form $|x\rangle|y\rangle|z\rangle$. We leave the generalisation of Proposition 14.1 and the completion of the below as exercises. From the quantum measurement postulate, we have

Measuring the first qubit of the states $|0\rangle|0\rangle|0\rangle, \ldots, |1\rangle|1\rangle|1\rangle$:

Measuring A on the first qubit of the simple (i.e., non-superimposed) state

$$|\psi\rangle = |x\rangle|y\rangle|z\rangle \quad \text{where} \quad x, y, z \in \mathbb{B}$$

using the operator $A \otimes I \otimes I$, we have

- If $|\psi\rangle = |0\rangle|0\rangle|0\rangle$ then, because $A \otimes I \otimes I|0\rangle|0\rangle|0\rangle = 0|0\rangle|0\rangle|0\rangle$, 0 will be measured for A with probability 1 and the post-measurement state will be $|0\rangle|0\rangle|0\rangle$.

- If $|\psi\rangle = |0\rangle|0\rangle|1\rangle$ then, because $A \otimes I \otimes I|0\rangle|0\rangle|1\rangle = 0|0\rangle|0\rangle|1\rangle$, 0 will be measured for A with probability 1 and the post-measurement state will be $|0\rangle|0\rangle|1\rangle$.

$\vdots$

- If $|\psi\rangle = |1\rangle|1\rangle|1\rangle$ then, because $A \otimes I \otimes I|1\rangle|1\rangle|1\rangle = 1|1\rangle|1\rangle|1\rangle$, 1 will be measured for A with probability 1 and the post-measurement state will be $|1\rangle|1\rangle|1\rangle$.

Measurement of A on the 2nd and 3rd qubits of states of the form $|x\rangle|y\rangle|z\rangle$ is left as an exercise. The measurement of simple, i.e., non-superimposed, n-qubit states is the obvious generalisation of the cases $n = 1, 2$ and 3 discussed above.

Exercises

14.1 For $|\psi\rangle = |x\rangle|y\rangle|z\rangle$ where $x, y, z \in \mathbb{B}$ use the operator $A \otimes I \otimes I$ to perform measurement of A on the first qubit. In each case, give the measured value, its probability and the post-measurement state.

14.2 Use the operators $I \otimes A \otimes I$ and $I \otimes I \otimes A$ to measure the 2nd and 3rd qubits of states of the form $|\psi\rangle = |x\rangle|y\rangle|z\rangle$ where $x, y, z \in \mathbb{B}$. In each case, give the measured value, its probability and the post-measurement state.

15

Quantum information processing 1: the quantum emulation of familiar invertible digital gates

15.1 Objectives

Using Section 14.2 we are now able to define quantum equivalents of a number of familiar digital gates. It will be seen that the quantum gates, with appropriately restricted state vector inputs, produce exactly the same numerical outputs as their digital counterparts.

It follows from the axioms of Section 13.6 that quantum computation is invertible (or reversible). Hence, although current digital technology is not based on invertible Boolean functions, in order to emulate digital computation quantum mechanically, it is necessary to first construct invertible representations of Boolean functions. In this chapter we consider only some of the more familiar examples – up to and including the Peres half-adder. In Chapter 17, we show how to construct invertible representations of arbitrary Boolean functions and discuss their quantum emulations.

15.2 On the graphical representation of quantum gates

As in the digital case, quantum-gate computation is based on fundamental gates. Having discussed quantum states and their measurement, we can now consider the transformation of quantum states, by 'gates', for computational purposes. Some quantum gates have parallels in the digital domain – and it is these that we discuss initially. Quantum gates without digital equivalents are discussed later.

In the graphical representation of quantum gates and circuits, the horizontal lines (or 'wires') indicate the time-ordered application of the unitary gates to the input qubits. There are no physical wires. In the digital case the wires are physical – in addition to indicating the temporal-ordering of application. Wired representations of gates are suitable and unambiguous in the digital case. However, in the quantum case so-called 'entangled states', which we discuss later, can occur, and additional features are required in the diagrams to represent them.

15.3   A 1-bit/qubit gate

15.3.1   The digital *not* gate

As was described in detail in Section 2.4, the digital *not* gate is a function $not_D : \mathbb{B} \to \mathbb{B}$ which may be specified by

Digital not_D:
Input   : $x \in \mathbb{B}$
Output   : $1 \oplus x$

i.e., $not_D(x) = 1 \oplus x = \bar{x}$. Clearly, on input 0 the gate produces 1, and on input 1 the output is 0. As noted in Chapter 2 this function is bijective and hence is invertible.

15.3.2   The quantum *not* gate, not_Q, on $B_{\mathbb{C}^2} = \{|x\rangle : x \in \mathbb{B}\}$

Because the digital gate not_D is invertible, a quantum equivalent, denoted not_Q, is easy to construct. Denote the basis states $|0\rangle$ and $|1\rangle$ of $\mathbb{C}^2$ by $B_{\mathbb{C}^2} = \{|x\rangle : x \in \mathbb{B}\}$. The not_Q gate may be specified, initially, on the pair of basis vectors by

Quantum *not*, not_Q:
Input   : $|x\rangle, \ x \in \mathbb{B}$
Output (quantum)   : $|1 \oplus x\rangle$
Output (post-μ_Q)   : $1 \oplus x$

The quantum *not* gate, not_Q, is defined, on the basis states of $\mathbb{C}^2$, by:

$$not_Q|x\rangle = |1 \oplus x\rangle, \quad \text{where } x \in \{0, 1\}.$$

We have $not_Q|0\rangle = |1\rangle$ and $not_Q|1\rangle = |0\rangle$ and the gate may be represented as the 'wired' circuit shown in Figure 15.1.

$$|x\rangle \ \text{———} \ not_Q \ \text{———} \ |1 \oplus x\rangle$$

FIGURE 15.1
The 1-qubit quantum *not* gate, not_Q, on $\{|0\rangle, |1\rangle\}$, i.e., $x \in \mathbb{B}$.

Recall that the eigenvalue-eigenvector pairs of the observable $A = |1\rangle\langle 1|$ are $0, |0\rangle$ and $1, |1\rangle$. Hence, measuring not_Q with the observable $A = |1\rangle\langle 1|$, we see that:

$$A|1 \oplus x\rangle = |1\rangle\langle 1|1 \oplus x\rangle$$
$$= \begin{cases} |1\rangle\langle 1|1\rangle = 1|1\rangle & \text{when } x = 0 \\ |1\rangle\langle 1|0\rangle = 0|0\rangle & \text{when } x = 1 \end{cases}$$

i.e., following Section 14.2.1, where the measurement of the simple states $|0\rangle$ and $|1\rangle$ is discussed, we have:

- when $x = 0$: 1 will be measured and the post-measurement state will be $|1\rangle$,

- when $x = 1$: 0 will be measured and the post-measurement state will be $|0\rangle$.

Hence,

- on the states $|0\rangle$ and $|1\rangle$ the output is deterministic and the quantum gate performs in exactly the same way as its digital counterpart.

However, in the quantum case, the more general (superimposed) input states:

$$\alpha_0|0\rangle + \alpha_1|1\rangle$$

are valid. We discuss this more general case in Section 16.3.

15.4   2-bit/qubit gates

We consider two digital *cnot* or *xor* gates labelled Digital 1 and Digital 2. The first is not invertible whereas the second is.

15.4.1   The non-invertible digital *cnot*, or *xor*, gate

As described in detail in Section 2.5, the digital *cnot* gate may be specified in the following way:

Digital 1 (classical, non-invertible):
Input: $(x, y) \in \mathbb{B}^2$
Output: $x \oplus y \in \mathbb{B}$

and is usually represented diagrammatically as shown in Figure 15.2.

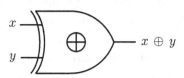

FIGURE 15.2
The classical Boolean *cnot* gate.

We have

Input (x,y)	Output $x \oplus y$
$(0,0)$	0
$(0,1)$	1
$(1,0)$	1
$(1,1)$	0

The gate is clearly the Boolean function $f_6(x, y) = x \oplus y$ (see Table 2.13); and given that it is not invertible we seek an invertible representation from which a quantum equivalent gate may be defined. In Chapter 17 we consider a general technique for constructing invertible representations of non-invertible Boolean functions. Below we introduce Feynman's invertible form of f_6.

15.4.2   The invertible digital *cnot*, or Feynman F_D, gate

The Feynman gate, $F_D : \mathbb{B}^2 \to \mathbb{B}^2$, first introduced in Section 2.5, is an invertible representation of the digital *cnot* gate and is specified by:

Digital 2 (Feynman, invertible):
Input: $(x, y) \in \mathbb{B}^2$
Output: $(x, y \oplus x) \in \mathbb{B}^2$

Functionally we have

$$F_D(x, y) = (x, x \oplus y)$$

and

Input (x, y)	Output $(x, x \oplus y)$
$(0, 0)$	$(0, 0)$
$(0, 1)$	$(0, 1)$
$(1, 0)$	$(1, 1)$
$(1, 1)$	$(1, 0)$

Observe that F_D is bijective and hence is invertible. It may be represented graphically in either of the two ways shown in Figure 15.3. It can be shown that the Feynman gate F_D is self-inverse (see Exercises).

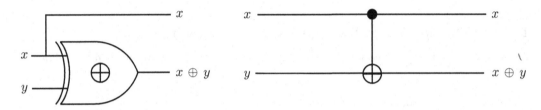

FIGURE 15.3
Two graphical representations of the Feynman invertible *cnot* gate.

The Feynman gate above is, essentially, the only invertible digital gate with 2-inputs and 2-outputs; it is this form of the *cnot* gate that may be used to define a quantum equivalent.

15.4.3   The quantum *cnot*, or Feynman F_Q, gate on
$$B_{\otimes^2 \mathbb{C}^2} = \{|x\rangle |y\rangle : x, y \in \mathbb{B}\}$$

The invertible digital gate, F_D, enables a quantum equivalent to be defined on the basis states

$$B_{\otimes^2 \mathbb{C}^2} = \{|0\rangle |0\rangle, |0\rangle |1\rangle, |1\rangle |0\rangle, |1\rangle |1\rangle\}$$

of $\mathbb{C}^2 \otimes \mathbb{C}^2$ by:

Quantum 1:
Input: $|x\rangle \otimes |y\rangle \in B_{\otimes^2 \mathbb{C}^2}$
Output (quantum): $|x\rangle |y \oplus x\rangle \in B_{\otimes^2 \mathbb{C}^2}$
Output (post-measurement of 2nd qubit): $y \oplus x \in \mathbb{B}$

or

Quantum 2:
Input: $|x\rangle \otimes |y\rangle \in B_{\otimes^2 \mathbb{C}^2}$
Output (quantum): $|x\rangle |y \oplus x\rangle \in B_{\otimes^2 \mathbb{C}^2}$
Output (post-measurement of 1st and 2nd qubits): $(x, y \oplus x) \in \mathbb{B}^2$.

These specifications may be seen as differing, in both the digital and quantum cases, only in the measurement processes. For Quantum 1 only the second output qubit is measured, whereas for Quantum 2 both the first and second qubits are measured. This gate is the function $F_Q : B_{\otimes^2 \mathbb{C}^2} \to B_{\otimes^2 \mathbb{C}^2}$ defined by

$$F_Q(|x\rangle|y\rangle) = |x\rangle|y \oplus x\rangle$$

and represented diagrammatically as shown in Figure 15.4.

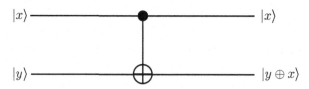

FIGURE 15.4
The quantum *cnot* gate: input $|x\rangle \otimes |y\rangle$, output $|x\rangle \otimes |y \oplus x\rangle$.

We have

Input $\|x\rangle \otimes \|y\rangle$	Output $\|x\rangle \otimes \|y \oplus x\rangle$	μ_Q
$\|0\rangle \otimes \|0\rangle$	$\|0\rangle \otimes \|0\rangle$	$(0,0)$
$\|0\rangle \otimes \|1\rangle$	$\|0\rangle \otimes \|1\rangle$	$(0,1)$
$\|1\rangle \otimes \|0\rangle$	$\|1\rangle \otimes \|1\rangle$	$(1,1)$
$\|1\rangle \otimes \|1\rangle$	$\|1\rangle \otimes \|0\rangle$	$(1,0)$

In the above, μ_Q denotes the measurement of $A = |1\rangle\langle 1|$ on qubits 1 and 2, using the operators $A \otimes I$ and $I \otimes A$ respectively. The measurement outputs, above, follow from the full discussion of the measurement of such output states in Section 14.2.2. Clearly the outputs of F_Q, on the domain $B_{\otimes^2 \mathbb{C}^2}$, are precisely those of the equivalent digital gate F_D.

For Quantum 2 we can write

$$(x, y \oplus x) = \mu_Q \circ F_Q(|x\rangle|y\rangle)$$

where μ_Q denotes a quantum measurement process.

15.5 3-bit/qubit gates

15.5.1 The digital Toffoli (or *ccnot*) gate

In the sense that the Feynman digital gate, F_D, is an invertible form of the digital function $f_6(x, y) = y \oplus x$, the Toffoli gate $T_D : \mathbb{B}^3 \to \mathbb{B}^3$, first introduced in Section 2.5.3 and defined by

$$T_D(x, y, z) = (x, \ y, \ z \oplus (x \wedge y)),$$

provides an invertible representation of the digital gate $f_1(x, y) = x \wedge y$ (see Table 2.13); i.e., we have

$$T_D(x, y, 0) = (x, \ y, \ x \wedge y) = (x, \ y, \ f_1(x, y)).$$

T_D is often represented diagrammatically as shown in Figure 15.5. It can be shown that the digital Toffoli gate is self-inverse (see Exercises).

With $z = 0$ the Toffoli gate gives the $\wedge$ function, and with $x \wedge y = 1$ it gives the *not* function, i.e., we have $T_D(1, 1, z) = (1, 1, \overline{z})$. We have, as a consequence, Lemma 15.1:

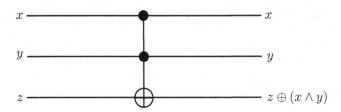

FIGURE 15.5
The digital Toffoli 3-bit, invertible gate.

Lemma 15.1 *The Toffoli function defines a complete (or universal) set for the generation of invertible (or reversible) versions of the Boolean functions $f : \mathbb{B}^2 \to \mathbb{B}$.*

Hence, the Toffoli function can perform any computation on a digital computer in a reversible manner. We also have

$$T_D(x, y, 1) = (x, y, \overline{x \wedge y}) = (x, y, \overline{f_1}(x, y))$$

and

$$T_D(x, 1, 0) = (x, 1, x) \qquad \text{or} \qquad T_D(1, x, 0) = (1, x, x)$$

i.e., the Toffoli function can copy. Further, we note $T_D(x, x, 0) = (x, x, x)$.

15.5.2 The quantum Toffoli (or *ccnot*) gate on $B_{\otimes^3 \mathbb{C}^2} = \{|x\rangle|y\rangle|z\rangle : x, y, z \in \mathbb{B}\}$

The invertible digital gate T_D enables a quantum equivalent, denoted T_Q, to be constructed; we define $T_Q : B_{\otimes^3 \mathbb{C}^2} \to B_{\otimes^3 \mathbb{C}^2}$ to be

$$T_Q(|x\rangle|y\rangle|z\rangle) = |x\rangle|y\rangle|z \oplus (x \wedge y)\rangle$$

and represent it diagrammatically as shown in Figure 15.6.

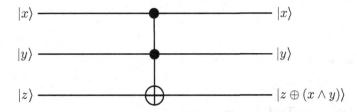

FIGURE 15.6
The quantum Toffoli 3-bit, invertible gate on $B_{\otimes^3 \mathbb{C}^2}$.

We have

Input $|x\rangle|y\rangle|z\rangle$	Output $|x\rangle|y\rangle|z \oplus (x \wedge y)\rangle$	μ_Q
$|0\rangle|0\rangle|0\rangle$	$|0\rangle|0\rangle|0\rangle$	$(0,0,0)$
$|0\rangle|0\rangle|1\rangle$	$|0\rangle|0\rangle|1\rangle$	$(0,0,1)$
$|0\rangle|1\rangle|0\rangle$	$|0\rangle|1\rangle|0\rangle$	$(0,1,0)$
$|0\rangle|1\rangle|1\rangle$	$|0\rangle|1\rangle|1\rangle$	$(0,1,1)$
$|1\rangle|0\rangle|0\rangle$	$|1\rangle|0\rangle|0\rangle$	$(1,0,0)$
$|1\rangle|0\rangle|1\rangle$	$|1\rangle|0\rangle|1\rangle$	$(1,0,1)$
$|1\rangle|1\rangle|0\rangle$	$|1\rangle|1\rangle|1\rangle$	$(1,1,1)$
$|1\rangle|1\rangle|1\rangle$	$|1\rangle|1\rangle|0\rangle$	$(1,1,0)$

where the measurement outputs of the final column follow from the discussion of the measurement of such output states in Section 14.2.3. Clearly the outputs of T_Q, on the domain $B_{\otimes^3 \mathbb{C}^2}$ are precisely those of the equivalent digital gate T_D. Having constructed a quantum equivalent of the digital Toffoli gate T_D we can conclude, from Lemma 15.1, that:

- all digital computations may be performed on a quantum computer.

15.5.3 The digital Peres (invertible half-adder) gate

The digital Peres gate, $P_D : \mathbb{B}^3 \to \mathbb{B}^3$, defined by

$$P_D(x, y, z) = (x, \ y \oplus x, \ z \oplus (x \wedge y))$$

is an example of a 3-input invertible Boolean gate and is an invertible representation of the non-invertible digital half-adder, $(s, c) : \mathbb{B}^2 \to \mathbb{B}^2$, defined by:

$$(s, c)(x, y) = (x \oplus y, \ x \wedge y).$$

(Here, the s and c refer to the sum and carry digits of the half-adder). Note that P_D may be represented as shown in Figure 15.7.

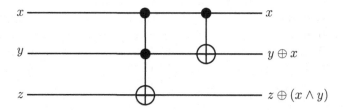

FIGURE 15.7
Circuit for the 3-bit (invertible) Peres half-adder.

The Peres gate P_D is invertible but, unlike F_D and T_D, it is not self-inverse. The Peres gate may be viewed as a product of the digital Toffoli and Feynman gates. We have:

$$\begin{aligned}
P_D(x, y, z) &= F_D \times i \circ T_D(x, y, z) \\
&= F_D \times i(x, y, z \oplus (x \wedge y)) \\
&= (F_D(x, y), z \oplus (x \wedge y)) \\
&= (x, x \oplus y, z \oplus (x \wedge y)).
\end{aligned}$$

15.5.4 The quantum Peres gate on $B_{\otimes^3 \mathbb{C}^2} = \{|x\rangle |y\rangle |z\rangle : x, y, z \in \mathbb{B}\}$

The invertible digital half-adder, P_D, enables a quantum half-adder to be constructed. We define, on $B_{\otimes^3 \mathbb{C}^2}$, the function

$$P_Q(|x\rangle |y\rangle |z\rangle) = |x\rangle |y \oplus x\rangle |z \oplus (x \wedge y)\rangle$$

which is represented diagrammatically as in Figure 15.8.

The quantum Peres gate is the product of two operators on $B_{\otimes^3 \mathbb{C}^2}$: the quantum Toffoli $U_1 = T_Q$ and the operator $U_2 = F_Q \otimes I$, that is,

$$\begin{aligned}
P_Q(|x\rangle |y\rangle |z\rangle) &= |x\rangle |y \oplus x\rangle |z \oplus x \wedge y\rangle \\
&= F_Q \otimes I \ T_Q |x\rangle |y\rangle |z\rangle \\
&= U_2 \, U_1 |x\rangle |y\rangle |z\rangle
\end{aligned}$$

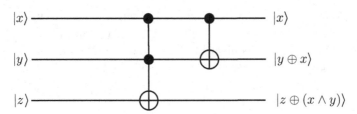

FIGURE 15.8
Circuit for the quantum Peres half-adder.

and

$$(x, y \oplus x, z \oplus x \wedge y) \quad = \quad \mu_Q \circ F_Q \otimes i \, T_Q |x\rangle |y\rangle |z\rangle$$
$$= \quad \mu_Q \circ U_2 \, U_1 |x\rangle |y\rangle |z\rangle.$$

which should be compared with the digital equivalent of the previous sub-section. We have:

Input $|x\rangle|y\rangle|z\rangle$	Output $|x\rangle|y \oplus x\rangle|z \oplus (x \wedge y)\rangle$	μ_Q
$|0\rangle|0\rangle|0\rangle$	$|0\rangle|0\rangle|0\rangle$	$(0,0,0)$
$|0\rangle|0\rangle|1\rangle$	$|0\rangle|0\rangle|1\rangle$	$(0,0,1)$
$|0\rangle|1\rangle|0\rangle$	$|0\rangle|1\rangle|0\rangle$	$(0,1,0)$
$|0\rangle|1\rangle|1\rangle$	$|0\rangle|1\rangle|1\rangle$	$(0,1,1)$
$|1\rangle|0\rangle|0\rangle$	$|1\rangle|1\rangle|0\rangle$	$(1,1,0)$
$|1\rangle|0\rangle|1\rangle$	$|1\rangle|1\rangle|1\rangle$	$(1,1,1)$
$|1\rangle|1\rangle|0\rangle$	$|1\rangle|0\rangle|1\rangle$	$(1,0,1)$
$|1\rangle|1\rangle|1\rangle$	$|1\rangle|0\rangle|0\rangle$	$(1,0,0)$

The real output, in the quantum case, follows by measuring the output qubits in the way described in Section 14.2.3; clearly the outputs of P_Q, on the domain $B_{\otimes^3 \mathbb{C}^2}$, are precisely those of the equivalent digital gate P_D.

Exercises

15.1 Show that the digital Feynman gate F_D is self-inverse i.e., that $F_D^{-1} = F_D$.

15.2 Show that the digital Toffoli gate T_D is self-inverse; i.e., we have $T_D^{-1} = T_D$.

15.3 Show that the inverse of the Peres gate is:

$$P_D^{-1}(x, y, z) = (x, \, x \oplus y, \, z \oplus (x \wedge \overline{y})).$$

16

Unitary extensions of the gates not_Q, F_Q, T_Q and P_Q: more general quantum inputs

16.1 Objectives

In Chapter 15, quantum emulations of some familiar invertible digital gates were discussed; specifically the *not*, *cnot*, Toffoli and Peres gates, each of which is significant in quantum computation, were considered. The emulations were defined on basis elements only – i.e., on $B_{\mathbb{C}^2}$ or $B_{\mathbb{C}^2 \otimes \mathbb{C}^2}$ or $B_{\mathbb{C}^2 \otimes \mathbb{C}^2 \otimes \mathbb{C}^2}$ etc. On these restricted inputs the quantum emulations were seen to produce outputs identical to their digital equivalents. Unlike the digital gates the quantum emulations may be extended beyond the basis sets to the entire tensor product spaces so defined – i.e., to $\mathbb{C}^2, \mathbb{C}^2 \otimes \mathbb{C}^2, \mathbb{C}^2 \otimes \mathbb{C}^2 \otimes \mathbb{C}^2$ etc. Such generalised inputs to the quantum gates lead to more general quantum outputs to which the simple deterministic quantum measurement processes of Chapter 15 are not applicable. The quantum gates defined on the extended domains are shown to be unitary, and are denoted in the same way as their restrictions to the basis vectors. We first consider a sufficient condition for an operator to be unitary; it has applications in our discussions below and in Chapter 17.

16.2 A lemma on unitary operators

Let V be an n-dimensional vector space over $\mathbb{C}$ with inner product $\langle , \rangle$ and let $b_1, \ldots, b_n$ be an orthonormal basis of V. We have

Lemma 16.1 *If* $T : V \to V$ *is a linear operator on* V *and* T *permutes the orthonormal basis* $b_1, \ldots, b_n$ *of* V, *then* T *is unitary on* V; *i.e.,* $\|Tv\| = \|v\|$ *for all* $v \in V$, *equivalently* $\langle Tv, Tv \rangle = \langle v, v \rangle$ *for all* $v \in V$.

Proof

For $v = \alpha_1 b_1 + \ldots + \alpha_n b_n \in V$ we have

$$
\begin{aligned}
Tv &= \alpha_1 Tb_1 + \ldots + \alpha_n Tb_n \\
&= \alpha_1 b_{\pi(1)} + \ldots + \alpha_n b_{\pi(n)} \text{ for some } \pi \in S_n.
\end{aligned}
$$

Here S_n is the permutation group of n elements (see Section 8.3). Further

$$
\begin{aligned}
\|v\|^2 = \|\alpha_1 b_1 + \ldots + \alpha_n b_n\|^2 &= \langle \alpha_1 b_1 + \ldots + \alpha_n b_n, \alpha_1 b_1 + \ldots + \alpha_n b_n \rangle \\
&= |\alpha_1|^2 + \ldots + |\alpha_n|^2
\end{aligned}
$$

DOI: 10.1201/9781003264569-16

and

$$\|Tv\|^2 = \|\alpha_1 b_{\pi(1)} + \ldots + \alpha_n b_{\pi(n)}\|^2 \;\; = \;\; \langle \alpha_1 b_{\pi(1)} + \ldots + \alpha_n b_{\pi(n)}, \alpha_1 b_{\pi(1)} + \ldots + \alpha_n b_{\pi(n)} \rangle$$
$$= \;\; |\alpha_1|^2 + \ldots + |\alpha_n|^2$$

hence $\|Tv\|^2 = \|v\|^2$, as required, and T is unitary on V.

■

16.3   The not_Q gate on $\mathbb{C}^2$

On the general 1-qubit inputs $\alpha_0|0\rangle + \alpha_1|1\rangle$ we have, by linear extension of not_Q to the whole of $\mathbb{C}^2$, that:

$$not_Q(\alpha_0|0\rangle + \alpha_1|1\rangle) \;\; = \;\; \alpha_0 not_Q(|0\rangle) + \alpha_1 not_Q(|1\rangle)$$
$$= \;\; \alpha_0|1\rangle + \alpha_1|0\rangle,$$

equivalently represented as shown in Figure 16.1.

$$\alpha_0|0\rangle + \alpha_1|1\rangle \;\; \rule{2cm}{0.4pt} \;\; not_Q \;\; \rule{2cm}{0.4pt} \;\; \alpha_0|1\rangle + \alpha_1|0\rangle$$

FIGURE 16.1
The 1-qubit quantum negation gate on $\mathbb{C}^2$, where $|\alpha_0|^2 + |\alpha_1|^2 = 1$.

Recall from Section 14.2 that the eigenvalue-eigenvector pairs of the observable $A = |1\rangle\langle 1|$ are $0, |0\rangle$ and $1, |1\rangle$. Measuring the quantum output of the negation gate, using the operator $|1\rangle\langle 1|$, with this general input produces:

- 0 with probability $|\alpha_1|^2$ and

- 1 with probability $|\alpha_0|^2$.

That is, the output of the gate not_Q, following measurement, is generally non-deterministic.

The quantum negation gate on the whole of $\mathbb{C}^2$ may be specified as:

Input : $\alpha_0|0\rangle + \alpha_1|1\rangle$ where $|\alpha_0|^2 + |\alpha_1|^2 = 1$

Output(quantum) : $\alpha_0|1\rangle + \alpha_1|0\rangle$

Output(post$-\mu_Q$) :   1 and post-measurement state $|1\rangle$, with probability $|\alpha_0|^2$, or

0 and post-measurement state $|0\rangle$, with probability $|\alpha_1|^2$.

As not_Q permutes the basis $\{|0\rangle, |1\rangle\}$ it follows from Lemma 16.1 that not_Q is unitary on $\mathbb{C}^2$, and its matrix, in this basis is the Pauli matrix:

$$\sigma_1 = \begin{pmatrix} 0 & 1 \\ 1 & 0 \end{pmatrix}$$

i.e.,

$$\begin{pmatrix} 0 & 1 \\ 1 & 0 \end{pmatrix} \begin{pmatrix} \alpha_0 \\ \alpha_1 \end{pmatrix} = \begin{pmatrix} \alpha_1 \\ \alpha_0 \end{pmatrix}.$$

and, in particular,

$$\begin{pmatrix} 0 & 1 \\ 1 & 0 \end{pmatrix} \begin{pmatrix} 1 \\ 0 \end{pmatrix} = \begin{pmatrix} 0 \\ 1 \end{pmatrix}, \;\; \text{and} \;\; \begin{pmatrix} 0 & 1 \\ 1 & 0 \end{pmatrix} \begin{pmatrix} 0 \\ 1 \end{pmatrix} = \begin{pmatrix} 1 \\ 0 \end{pmatrix},$$

equivalently $\sigma_1|0\rangle = |1\rangle$ and $\sigma_1|1\rangle = |0\rangle$.

16.4 The Feynman F_Q gate on $\otimes^2\mathbb{C}^2$

In Chapter 15, F_Q is only defined on the basis states $B_{\otimes^2\mathbb{C}^2} = \{|x\rangle|y\rangle : x,y \in \mathbb{B}\}$ of the space $\mathbb{C}^2 \otimes \mathbb{C}^2$. We extend F_Q to a linear operator on the whole of $\mathbb{C}^2 \otimes \mathbb{C}^2$ by linear extension. That is, we define F_Q on the general state

$$|\psi\rangle = \alpha_{00}|0\rangle|0\rangle + \alpha_{01}|0\rangle|1\rangle + \alpha_{10}|1\rangle|0\rangle + \alpha_{11}|1\rangle|1\rangle$$

of $\mathbb{C}^2 \otimes \mathbb{C}^2$ by

$$F_Q|\psi\rangle \equiv \alpha_{00}F_Q|0\rangle|0\rangle + \alpha_{01}F_Q|0\rangle|1\rangle + \alpha_{10}F_Q|1\rangle|0\rangle + \alpha_{11}F_Q|1\rangle|1\rangle.$$

Note that F_Q on the basis states has been defined in Section 15.4.3. It follows from Lemma 16.1 that F_Q is unitary on $\otimes^2\mathbb{C}^2$. The representation of F_Q in the basis $B_{\otimes^2\mathbb{C}^2}$ is the unitary matrix

$$\begin{pmatrix} 1 & 0 & 0 & 0 \\ 0 & 1 & 0 & 0 \\ 0 & 0 & 0 & 1 \\ 0 & 0 & 1 & 0 \end{pmatrix}.$$

16.5 The quantum Toffoli gate on $\otimes^3\mathbb{C}^2$

The extension of the quantum Toffoli gate to $\otimes^3\mathbb{C}^2$ is defined similarly to that of F_Q. Given the general state of $\mathbb{C}^2 \otimes \mathbb{C}^2 \otimes \mathbb{C}^2$:

$$|\psi\rangle = \alpha_{000}|0\rangle|0\rangle|0\rangle + \cdots + \alpha_{111}|1\rangle|1\rangle|1\rangle$$

the quantum Toffoli gate is defined to be

$$T_Q|\psi\rangle = \alpha_{000}T_Q|0\rangle|0\rangle|0\rangle + \cdots + \alpha_{111}T_Q|1\rangle|1\rangle|1\rangle,$$

with T_Q on the basis states having been defined in Section 15.5.2. The representation of F_Q in the basis $B_{\otimes^3\mathbb{C}^2}$ is the unitary matrix:

$$\begin{pmatrix} 1 & 0 & 0 & 0 & 0 & 0 & 0 & 0 \\ 0 & 1 & 0 & 0 & 0 & 0 & 0 & 0 \\ 0 & 0 & 1 & 0 & 0 & 0 & 0 & 0 \\ 0 & 0 & 0 & 1 & 0 & 0 & 0 & 0 \\ 0 & 0 & 0 & 0 & 1 & 0 & 0 & 0 \\ 0 & 0 & 0 & 0 & 0 & 1 & 0 & 0 \\ 0 & 0 & 0 & 0 & 0 & 0 & 0 & 1 \\ 0 & 0 & 0 & 0 & 0 & 0 & 1 & 0 \end{pmatrix}.$$

16.6 The quantum Peres gate on $\otimes^3\mathbb{C}^2$

This gate is extended to $\otimes^3\mathbb{C}^2$ in the same way as the Toffoli gate above and its representation in the basis $B_{\otimes^3\mathbb{C}^2}$ is the unitary matrix:

$$
\begin{pmatrix}
1 & 0 & 0 & 0 & 0 & 0 & 0 & 0 \\
0 & 1 & 0 & 0 & 0 & 0 & 0 & 0 \\
0 & 0 & 1 & 0 & 0 & 0 & 0 & 0 \\
0 & 0 & 0 & 1 & 0 & 0 & 0 & 0 \\
0 & 0 & 0 & 0 & 0 & 0 & 0 & 1 \\
0 & 0 & 0 & 0 & 0 & 0 & 1 & 0 \\
0 & 0 & 0 & 0 & 1 & 0 & 0 & 0 \\
0 & 0 & 0 & 0 & 0 & 1 & 0 & 0
\end{pmatrix}.
$$

16.7 Summation expressions for the unitary extensions

16.7.1 On $\mathbb{C}^2$

The unitary extensions of the identity function i_Q and the not_Q function may be expressed as:

$$
i_Q = \sum_{x=0}^{1} |x\rangle\langle x|
$$

and

$$
not_Q = \sum_{x=0}^{1} |\overline{x}\rangle\langle x|
$$

respectively. Alternatively, we may write:

$$
i_Q = \sum_{x=0}^{1} |i_D(x)\rangle\langle x|,
$$

where $i_D(x) = x$, and

$$
not_Q = \sum_{x=0}^{1} |not_D(x)\rangle\langle x|.
$$

16.7.2 On $\mathbb{C}^2 \otimes \mathbb{C}^2$

Writing the digital Feynman gate F_D as

$$
F_D(x,y) = (F_D(x,y)_1, F_D(x,y)_2)
$$

where $F_D(x,y)_1 = x$ and $F_D(x,y)_2 = y \oplus x$, the quantum Feynman gate F_Q on the basis states $B_{\otimes^2 \mathbb{C}^2}$ can be written as

$$
\begin{aligned}
F_Q|x\rangle|y\rangle &= |x\rangle|y \oplus x\rangle \\
&= |F_D(x,y)_1\rangle|F_D(x,y)_2\rangle.
\end{aligned}
$$

Introducing the shorthand notation $|F_D(x,y)\rangle$ for $|F_D(x,y)_1\rangle|F_D(x,y)_2\rangle$, the relationship between the digital representation on $\mathbb{B}^2$ and the quantum representation on $B_{\otimes^2 \mathbb{C}^2}$ may be expressed as:

$$
F_Q|x\rangle|y\rangle = |F_D(x,y)\rangle
$$

or alternatively and more succinctly,

$$F_Q|xy\rangle = |F_D(xy)\rangle.$$

Now consider the general state of $\mathbb{C}^2 \otimes \mathbb{C}^2$:

$$|\psi\rangle = \alpha_{00}|00\rangle + \alpha_{01}|01\rangle + \alpha_{10}|10\rangle + \alpha_{11}|11\rangle.$$

Taking the inner product with the basis state $|00\rangle$ we obtain, due to the orthonormality of the basis states,

$$\langle 00|\psi\rangle = \alpha_{00}\langle 00|00\rangle + \alpha_{01}\langle 00|01\rangle + \alpha_{10}\langle 00|10\rangle + \alpha_{11}\langle 00|11\rangle$$
$$= \alpha_{00}.$$

Likewise $\langle 01|\psi\rangle = \alpha_{01}$, $\langle 10|\psi\rangle = \alpha_{10}$, $\langle 11|\psi\rangle = \alpha_{11}$.

Extending F_Q to the general state, we obtain

$$\begin{aligned}
F_Q|\psi\rangle &= \alpha_{00}F_Q|00\rangle + \alpha_{01}F_Q|01\rangle + \alpha_{10}F_Q|10\rangle + \alpha_{11}F_Q|11\rangle \\
&= \langle 00|\psi\rangle F_Q|00\rangle + \langle 01|\psi\rangle F_Q|01\rangle + \langle 10|\psi\rangle F_Q|10\rangle + \langle 11|\psi\rangle F_Q|11\rangle \\
&= (F_Q|00\rangle\langle 00| + F_Q|01\rangle\langle 01| + F_Q|10\rangle\langle 10| + F_Q|11\rangle\langle 11|)|\psi\rangle \\
&= (|F_D(00)\rangle\langle 00| + |F_D(01)\rangle\langle 01| + |F_D(10)\rangle\langle 10| + |F_D(11)\rangle\langle 11|)|\psi\rangle \\
&= \left(\sum_{xy=00}^{11} |F_D(xy)\rangle\langle xy| \right)|\psi\rangle
\end{aligned}$$

i.e., we may write the linearly extended gate as

$$\begin{aligned}
F_Q &= \sum_{xy=00}^{11} |F_D(xy)\rangle\langle xy| \\
&= \sum_{xy=00}^{11} |x\rangle|y \oplus x\rangle\langle xy|.
\end{aligned}$$

16.7.3   On $\mathbb{C}^2 \otimes \mathbb{C}^2 \otimes \mathbb{C}^2$

For the unitary extension of the quantum Toffoli gate to $\mathbb{C}^2 \otimes \mathbb{C}^2 \otimes \mathbb{C}^2$, we have:

$$\begin{aligned}
T_Q &= \sum_{xyz=000}^{111} |T_D(xyz)\rangle\langle xyz| \\
&= \sum_{xyz=000}^{111} |x\rangle|y\rangle|z \oplus (x \wedge y)\rangle\langle xyz|.
\end{aligned}$$

The unitary extension of the Peres gate may be represented similarly.

16.7.4   Notation and closing observations

Moving forward it is helpful to simplify the notation for the unitarily extended operators above to U_f where f is the invertible Boolean function being emulated. For example, we write U_i for i_Q, U_{not} for not_Q, U_{cnot} for $cnot_Q$, U_T for T_Q etc.

We note that the general form of the unitary extensions discussed above may then be expressed as:

$$
\begin{aligned}
U_f &= \sum_{s \in \mathbb{B}^n} |f(s)\rangle\langle s| \\
&= \sum_{s \in \mathbb{B}^n} |f_1(s) \cdots f_n(s)\rangle\langle s| \\
&= \sum_{s \in \mathbb{B}^n} |f_1(s)\rangle \cdots |f_n(s)\rangle\langle s|
\end{aligned}
$$

where $f : \mathbb{B}^n \to \mathbb{B}^n$ is an invertible Boolean function and $s \in \mathbb{B}^n$ denotes a string of n Boolean digits.

On a basis element $|s'\rangle$ of $\otimes^n \mathbb{C}^n$, where $s' \in \mathbb{B}^n$, we obtain

$$
U_f|s'\rangle = |f_1(s')\rangle \cdots |f_n(s')\rangle,
$$

which follows from the observation that

$$
\langle s|s'\rangle = \begin{cases} 1 & \text{if } s' = s \\ 0 & \text{if } s' \neq s. \end{cases}
$$

States such as $U_f|s'\rangle$ are not superimposed and may therefore be measured using Axiom 3.2. If $|\psi\rangle \in \otimes^n \mathbb{C}^2$ is a superimposed state then Axiom 3 applies to the measurement of $U_f|\psi\rangle$.

Exercises

16.1 Show that from the expressions for i_Q and not_Q given in Section 16.7.1, we obtain

$$
\begin{aligned}
i_Q(\alpha_0|0\rangle + \alpha_1|1\rangle) &= \alpha_0|0\rangle + \alpha_1|1\rangle, \text{ and} \\
not_Q(\alpha_0|0\rangle + \alpha_1|1\rangle) &= \alpha_0|1\rangle + \alpha_1|0\rangle.
\end{aligned}
$$

17

Quantum information processing 2: the quantum emulation of arbitrary Boolean functions

17.1 Objectives

Chapters 15 and 16 relate to the quantum emulation, and unitary extension, of a number of special invertible Boolean functions. In this chapter, we turn our attention to the quantum emulation of arbitrary Boolean functions. This extends the work of Chapters 15 and 16 in two respects:

1. to arbitrary invertible Boolean functions, and

2. to arbitrary non-invertible Boolean functions.

Below we consider them separately beginning with the case of invertible functions, which is similar to the treatment of the special cases. For the non-invertible functions we first define invertible representations of them, from which quantum emulations may be constructed. We first consider the invertible case.

17.2 Notation

We denote the set of all functions from $\mathbb{B}^n$ to $\mathbb{B}^m$ by $\mathcal{F}(\mathbb{B}^n, \mathbb{B}^m)$; i.e.,

$$\mathcal{F}(\mathbb{B}^n, \mathbb{B}^m) = \{f : f : \mathbb{B}^n \to \mathbb{B}^m\}.$$

For $m \neq n$ the functions of $\mathcal{F}(\mathbb{B}^n, \mathbb{B}^m)$ are not invertible; for $m = n$ the set $\mathcal{F}(\mathbb{B}^n, \mathbb{B}^n)$ comprises both invertible and non-invertible functions. For example $\mathcal{F}(\mathbb{B}, \mathbb{B})$ comprises four functions, two of which are invertible and two of which are not. For the special invertible Boolean functions, considered in Chapters 15 and 16, we note that $not \in \mathcal{F}(\mathbb{B}, \mathbb{B})$, $cnot \in \mathcal{F}(\mathbb{B}^2, \mathbb{B}^2)$ and the Boolean Toffoli and Peres functions are invertible elements of $\mathcal{F}(\mathbb{B}^3, \mathbb{B}^3)$.

17.3 Quantum emulation of arbitrary invertible Boolean functions

17.3.1 The quantum emulation of the invertible subset of $\mathcal{F}(\mathbb{B}, \mathbb{B})$

There are just two invertible functions in $\mathcal{F}(\mathbb{B}, \mathbb{B})$, namely the identity function $i(x) = x$ and the function $not(x) = \overline{x}$, both of which have been covered as special cases in Chapters 15 and 16 where the unitary operators U_i and U_{not} have been defined on $\mathbb{C}^2$.

DOI: 10.1201/9781003264569-17

17.3.2   The quantum emulation of the invertible subset of $\mathcal{F}(\mathbb{B}^2, \mathbb{B}^2)$

The following lemma, which we prove for the invertible functions of $\mathcal{F}(\mathbb{B}^2, \mathbb{B}^2)$, generalises easily to the invertible functions of $\mathcal{F}(\mathbb{B}^n, \mathbb{B}^n)$ for $n \geq 3$.

Lemma 17.1 *If $f : \mathbb{B}^2 \to \mathbb{B}^2$ is invertible then the operator $U_f : \mathbb{C}^2 \otimes \mathbb{C}^2 \to \mathbb{C}^2 \otimes \mathbb{C}^2$ defined by*

$$U_f = \sum_{pq=00}^{11} |f(pq)\rangle\langle pq|$$

is unitary. If i is the identity function on $\mathbb{B}^2$ then $U_i = \sum_{pq=00}^{11} |pq\rangle\langle pq|$ is the identity operator on $\mathbb{C}^2 \otimes \mathbb{C}^2$.

Proof

For $|x\rangle|y\rangle \in B_{\mathbb{C}^2 \otimes \mathbb{C}^2}$ we have:

$$U_f|x\rangle|y\rangle \;=\; |f(xy)\rangle.$$

As $f(xy) \in \mathbb{B}^2$ it follows that $|f(xy)\rangle \in B_{\mathbb{C}^2 \otimes \mathbb{C}^2}$ and because f is invertible we have $|f(xy)\rangle \neq |f(x'y')\rangle$ unless $xy = x'y'$. Hence U_f permutes the elements of the basis $B_{\mathbb{C}^2 \otimes \mathbb{C}^2}$ and is, by Lemma 16.1, unitary. On the general state

$$|\psi\rangle = \alpha_{00}|00\rangle + \alpha_{01}|01\rangle + \alpha_{10}|10\rangle + \alpha_{11}|11\rangle$$

we have

$$U_i|\psi\rangle = \alpha_{00}U_i|00\rangle + \alpha_{01}U_i|01\rangle + \alpha_{10}U_i|10\rangle + \alpha_{11}U_i|11\rangle = |\psi\rangle.$$

■

Under the conditions of the Lemma, in particular for f invertible, we now show that U_f determines a quantum emulation of f in the sense that the values of f may be computed deterministically from it. Clearly for some $f_1, f_2 \in \mathcal{F}(\mathbb{B}^2, \mathbb{B})$ we may write any $f \in \mathcal{F}(\mathbb{B}^2, \mathbb{B}^2)$ as $f(x, y) = (f_1(x, y), f_2(x, y))$ and, in the compact notation, introduced in Section 16.7.4 we have, for f invertible, the unitary operator:

$$U_f = \sum_{pq=00}^{11} |f_1(pq)\rangle|f_2(pq)\rangle\langle pq|.$$

On the basis element $|x\rangle|y\rangle \in B_{\mathbb{C}^2 \otimes \mathbb{C}^2}$ we obtain, using the result in Section 16.7.4:

$$U_f|x\rangle|y\rangle = |f_1(xy)\rangle|f_2(xy)\rangle$$

from which the values $f_1(xy)$ and $f_2(xy)$, of f, may be obtained deterministically by measuring the observables $\mathcal{A} \otimes I$ and $I \otimes \mathcal{A}$, where $\mathcal{A} = |1\rangle\langle 1|$, on the first and second qubits of the right hand side.

In the case of $\mathcal{F}(\mathbb{B}^2, \mathbb{B}^2)$ the invertible functions are readily identified and enumerated; hence their quantum emulations may be determined explicitly and we do this in Section 17.3.3. The invertible functions of $\mathcal{F}(\mathbb{B}^n, \mathbb{B}^n)$, for $n \geq 3$, are too numerous to list, and for these cases we discuss their general unitary representation and provide specific examples – see Section 17.3.4.

17.3.3 Explicit forms of the emulations of Section 17.3.2

Continuing the consideration of $\mathcal{F}(\mathbb{B}^2, \mathbb{B}^2)$; it is shown in Chapter 2 that there is a total of 24 invertible elements of $\mathcal{F}(\mathbb{B}^2, \mathbb{B}^2)$ determined by the following pairs from $\mathcal{F}(\mathbb{B}^2, \mathbb{B})$ (refer to Table 2.13):

$$(f_3, f_5), \ (f_3, f_6), \ (f_3, f_9), \ (f_3, f_{10}),$$

$$(f_5, f_6), \ (f_5, f_9), \ (f_5, f_{12}),$$

$$(f_6, f_{10}), \ (f_6, f_{12}),$$

$$(f_9, f_{10}), \ (f_9, f_{12}),$$

$$(f_{10}, f_{12})$$

and the same pairs in contrary order. Explicitly, the pairs are:

$$(x, y), \ (x, x \oplus y), \ (x, \overline{x \oplus y}), \ (x, \overline{y}),$$

$$(y, x \oplus y), \ (y, \overline{x \oplus y}), \ (y, \overline{x}),$$

$$(x \oplus y, \overline{y}), \ (x \oplus y, \overline{x}),$$

$$(\overline{x \oplus y}, \overline{y}), \ (\overline{x \oplus y}, \overline{x}),$$

$$(\overline{y}, \overline{x}).$$

Denoting the pairs by $f_{i,j}$, their quantum emulations on the basis elements $B_{\mathbb{C}^2 \otimes \mathbb{C}^2}$ are:

$$U_{f_{i,j}}|x\rangle|y\rangle = |f_{i,j}(xy)\rangle = |f_i(xy)\rangle|f_j(xy)\rangle,$$

which produce the same outputs as their Boolean equivalents when the observables $\mathcal{A} \otimes I$ and $I \otimes \mathcal{A}$ are measured on the first and second qubits. It follows from Lemma 17.1 that the linear extension of $U_{f_{i,j}}$ to $\mathbb{C}^2 \otimes \mathbb{C}^2$ is unitary.

Example 17.3.1

For $f_{3,6}(x, y) = (f_3(x, y), f_6(x, y)) = (x, x \oplus y)$ we obtain:

$$
\begin{aligned}
U_{f_{3,6}}|x\rangle|y\rangle &= |f_{3,6}(xy)\rangle \\
&= |f_3(xy)\rangle|f_6(xy)\rangle \\
&= |x\rangle|x \oplus y\rangle
\end{aligned}
$$

which is the quantum Feynman gate F_Q discussed in Chapter 15. The matrix representation of $U_{f_{3,6}}$, in the computational basis, is given in Section 16.4.

Example 17.3.2

For $f_{6,10}(x, y) = (x \oplus y, \overline{y})$ we have

$$U_{f_{6,10}}|x\rangle|y\rangle = |x \oplus y\rangle|\overline{y}\rangle$$

from which we obtain:

$$
\begin{aligned}
U_{f_{6,10}}|0\rangle|0\rangle &= |01\rangle = 0|00\rangle + 1|01\rangle + 0|10\rangle + 0|11\rangle \\
U_{f_{6,10}}|0\rangle|1\rangle &= |10\rangle = 0|00\rangle + 0|01\rangle + 1|10\rangle + 0|11\rangle \\
U_{f_{6,10}}|1\rangle|0\rangle &= |11\rangle = 0|00\rangle + 0|01\rangle + 0|10\rangle + 1|11\rangle \\
U_{f_{6,10}}|1\rangle|1\rangle &= |00\rangle = 1|00\rangle + 0|01\rangle + 0|10\rangle + 0|11\rangle.
\end{aligned}
$$

i.e., the matrix representation of $U_{f_{6,10}}$, in the computational basis, is:

$$
\begin{pmatrix}
0 & 0 & 0 & 1 \\
1 & 0 & 0 & 0 \\
0 & 1 & 0 & 0 \\
0 & 0 & 1 & 0
\end{pmatrix}.
$$

The remaining 22 invertible Boolean functions, of type $\mathcal{F}(\mathbb{B}^2, \mathbb{B}^2)$, have similar quantum emulations that may be measured in the same way.

17.3.4   The invertible subset of $\mathcal{F}(\mathbb{B}^n, \mathbb{B}^n), n \geq 3$

The case $n = 3$: we do not consider the explicit forms of all invertible functions of $\mathcal{F}(\mathbb{B}^3, \mathbb{B}^3)$, but note that the Toffoli and Peres gates are examples. For a general invertible $f \in \mathcal{F}(\mathbb{B}^3, \mathbb{B}^3)$ we define:

$$
U_f = \sum_{pqr=000}^{111} |f(pqr)\rangle\langle pqr|
$$

from which we obtain:

$$
\begin{aligned}
U_f|x\rangle|y\rangle|z\rangle &= |f(xyz)\rangle \\
&= |f_1(xyz)\rangle|f_2(xyz)\rangle|f_3(xyz)\rangle.
\end{aligned}
$$

Measuring the three qubits with the operators $\mathcal{A} \otimes I \otimes I$, $I \otimes \mathcal{A} \otimes I$ and $I \otimes I \otimes \mathcal{A}$ produces the same outputs as their digital counterparts. The linear extension of U_f defines a unitary operator on $\mathbb{C}^2 \otimes \mathbb{C}^2 \otimes \mathbb{C}^2$.

For $n > 3$, U_f is constructed from f in a similar way.

17.4   Quantum emulation of arbitrary non-invertible Boolean functions

Here we consider the quantum emulation of the arbitrary Boolean functions $\mathcal{F}(\mathbb{B}^n, \mathbb{B}^m)$. Such functions are not, in general, invertible. For every non-invertible function, f, we determine an invertible 'representation' which 'embeds' the function f and from which the values of f, at arbitrary domain points, are easily obtained. We will see that the $f_{i,j}$ functions of Section 17.3.3 embed a subset of the functions $\mathcal{F}(\mathbb{B}^2, \mathbb{B})$, and for general non-invertible functions, f, a group theory based construction produces embeddings, which we denote by $\hat{f}$; see Lemma 17.2 and its corollary below.

17.4.1   A fundamental lemma

Here we construct, for any $f \in \mathcal{F}(\mathbb{B}^n, \mathbb{B}^m)$, an invertible representation, $\hat{f} : \mathbb{B}^{n+m} \to \mathbb{B}^{n+m}$, from which a unitary quantum equivalent, on $\otimes^{n+m}\mathbb{C}^2$, is easily determined. The following lemma and corollary are fundamental to the work of this chapter and play a role both in invertible-digital and quantum computation.

Lemma 17.2 *If G and H are groups and $f : G \to H$ is an arbitrary function, then the function $\hat{f} : G \times H \to G \times H$ defined by:*

$$\hat{f}(g, h) = (g, hf(g))$$

is one-to-one and onto, with inverse

$$\hat{f}^{-1}(g, h) = (g, hf(g)^{-1}).$$

Proof

1. to show that $\hat{f}$ is one-to-one: we have if $(g, hf(g)) = (g^*, h^*f(g^*))$ then $g^* = g$ and $h^*f(g^*) = h^*f(g) = hf(g)$, i.e., $h^* = h$.

2. to show that $\hat{f}$ is onto: for arbitrary $(a, b) \in G \times H$ seek a pair $(g, hf(g))$ such that $(g, hf(g)) = (a, b)$. We have $g = a$ and $hf(a) = b$, i.e., $h = bf(a)^{-1}$.

3. It is easy to show that $\hat{f}^{-1}(\hat{f}(g, h)) = (g, h)$.

■

Corollary 1 *With $G = (\mathbb{B}^n, \oplus)$ and $H = (\mathbb{B}^m, \oplus)$ we have: if $f : \mathbb{B}^n \to \mathbb{B}^m$ then the function $\hat{f} : \mathbb{B}^{n+m} \to \mathbb{B}^{n+m}$ defined by $\hat{f}(x, y) = (x, \ y \oplus f(x))$ is one-to-one, onto and self-inverse.*

Proof

That $\hat{f}$ is one-to-one and onto is immediate from Lemma 17.2; that it is self-inverse follows from $\hat{f}(\hat{f}(x, y)) = (x, y \oplus f(x) \oplus f(x)) = (x, y)$.

■

We note that for the function $\hat{f}$ of the corollary, defined for $x \in \mathbb{B}^n$ and $y \in \mathbb{B}^m$, we obtain for $y = 0^m$

$$\hat{f}(x, 0^m) = (x, f(x))$$

i.e., the function value at x occurs in the second component of $\hat{f}(x, 0^m)$ and we therefore refer to $\hat{f}$ as an invertible 'representation of f'. It will be seen, in Chapter 18, that Corollary 1 leads to auxiliary bits/qubits being introduced in both invertible-digital and quantum computations. In the remainder of this chapter, we consider invertible representations of:

1. the non-invertible subset of $\mathcal{F}(\mathbb{B}, \mathbb{B})$,

2. the non-invertible functions $\mathcal{F}(\mathbb{B}^2, \mathbb{B})$ and

3. the non-invertible subsets of the general case $\mathcal{F}(\mathbb{B}^n, \mathbb{B}^m)$, for $n > 2$ and $m \geq 1$, beginning with the non-invertible subset of $\mathcal{F}(\mathbb{B}^n, \mathbb{B}^n)$.

17.4.2   The non-invertible subset of $\mathcal{F}(\mathbb{B}, \mathbb{B})$

The non-invertible functions of $\mathcal{F}(\mathbb{B}, \mathbb{B})$ are $f_0(x) = 0$ and $f_1(x) = 1$, for $x \in \mathbb{B}$. For each of these we apply Corollary 1 and define invertible Boolean representations of f_0 and f_1

$$
\begin{aligned}
\hat{f}_0(x, y) &= (x, y \oplus f_0(x)) \\
&= (x, y \oplus 0) \\
&= (x, y) \quad \text{(the identity function on } \mathbb{B}^2\text{)},
\end{aligned}
$$

and

$$
\begin{aligned}
\hat{f}_1(x, y) &= (x, y \oplus f_1(x)) \\
&= (x, y \oplus 1) \\
&= (x, \bar{y})
\end{aligned}
$$

from which we have

$$
\begin{aligned}
\hat{f}_0(x, 0) &= (x, f_0(x)) \\
&= (x, 0)
\end{aligned}
$$

and

$$
\begin{aligned}
\hat{f}_1(x, 0) &= (x, f_1(x)) \\
&= (x, 1).
\end{aligned}
$$

Quantum emulations of f_0 and f_1 may then be defined on the basis elements of $B_{\mathbb{C}^2 \otimes \mathbb{C}^2}$ by

$$
\begin{aligned}
U_{\hat{f}_0} |x\rangle |y\rangle &= |x\rangle |y \oplus f_0(x)\rangle \\
&= |x\rangle |y\rangle
\end{aligned}
$$

and

$$
\begin{aligned}
U_{\hat{f}_1} |x\rangle |y\rangle &= |x\rangle |y \oplus f_1(x)\rangle \\
&= |x\rangle |\bar{y}\rangle
\end{aligned}
$$

which, by linear extension, determine the unitary operators (also denoted $U_{\hat{f}_0}$ and $U_{\hat{f}_1}$):

$$
U_{\hat{f}_0} = \sum_{pq=00}^{11} |pq\rangle \langle pq|
$$

$$
U_{\hat{f}_1} = \sum_{pq=00}^{11} |pq\rangle \langle p\bar{q}|
$$

on $\mathbb{C}^2 \otimes \mathbb{C}^2$. We have

$$
U_{\hat{f}_0} |x\rangle |0\rangle = |x\rangle |0\rangle \quad \text{and} \quad U_{\hat{f}_1} |x\rangle |0\rangle = |x\rangle |1\rangle
$$

which produce the same outputs as their Boolean equivalents, for $x \in \mathbb{B}$, when measured with the operator $I \otimes \mathcal{A}$. We note, finally, that the identity function $\hat{f}_0$ on $\mathbb{B}^2$ is emulated by the identity operator $U_{\hat{f}_0}$ on $\mathbb{C}^2 \otimes \mathbb{C}^2$.

17.4.3 The non-invertible functions $\mathcal{F}(\mathbb{B}^2, \mathbb{B})$

The functions of $\mathcal{F}(\mathbb{B}^2, \mathbb{B})$, which we denote by $f_0, \ldots, f_{15}$, are clearly not invertible (see Table 2.13). For the purposes of constructing quantum emulations we consider the two following subsets separately.

$$\mathcal{F}_1(\mathbb{B}^2, \mathbb{B}) = \{f_5, f_6, f_9, f_{10}\}$$

and

$$\mathcal{F}_2(\mathbb{B}^2, \mathbb{B}) = \{f_0, f_1, f_2, f_3, f_4, f_7, f_8, f_{11}, f_{12}, f_{13}, f_{14}, f_{15}\}.$$

I. The subset $\mathcal{F}_1(\mathbb{B}^2, \mathbb{B})$ of $\mathcal{F}(\mathbb{B}^2, \mathbb{B})$:

Although Corollary 1 may be invoked to construct invertible representations of these functions, it produces representations on $\mathbb{B}^3$. Instead representations on $\mathbb{B}^2$ are possible and, as shown below, we have already constructed suitable quantum emulations for them in Section 17.3.3.

Specifically, we have $f_5(x, y) = y$, $f_6(x, y) = x \oplus y$, $f_9(x, y) = \overline{x \oplus y}$, $f_{10}(x, y) = \overline{y}$, and note that, from Section 17.3.3, the functions $f_{3,5}, f_{3,6}, f_{3,9}$ and $f_{3,10}$ are invertible and take the form:

$$
\begin{aligned}
f_{3,5}(x, y) &= (x, f_5(x, y)) \\
f_{3,6}(x, y) &= (x, f_6(x, y)) \\
f_{3,9}(x, y) &= (x, f_9(x, y)) \\
f_{3,10}(x, y) &= (x, f_{10}(x, y))
\end{aligned}
$$

with the quantum emulations (again see Section 17.3.3):

$$
\begin{aligned}
U_{f_{3,5}} |x\rangle |y\rangle &= |x\rangle |f_5(x, y)\rangle \\
U_{f_{3,6}} |x\rangle |y\rangle &= |x\rangle |f_6(x, y)\rangle \\
U_{f_{3,9}} |x\rangle |y\rangle &= |x\rangle |f_9(x, y)\rangle \\
U_{f_{3,10}} |x\rangle |y\rangle &= |x\rangle |f_{10}(x, y)\rangle
\end{aligned}
$$

on the basis states of $\mathbb{C}^2 \otimes \mathbb{C}^2$. Measuring the 2nd qubit, in each case provides, deterministically, the value of the associated non-invertible Boolean function of $\mathcal{F}_1(\mathbb{B}^2, \mathbb{B})$ on (x, y).

The extended operators $U_{f_{3,i}}$, on $\mathbb{C}^2 \otimes \mathbb{C}^2$, may be expressed as

$$U_{f_{3,i}} = \sum_{pq=00}^{11} |p\rangle |f_i(pq)\rangle \langle p| \langle q|.$$

II. The subset $\mathcal{F}_2(\mathbb{B}^2, \mathbb{B})$ of $\mathcal{F}(\mathbb{B}^2, \mathbb{B})$:

It is easy to show that the above constructions, i.e., $(x, f_i(x, y))$, are not invertible for the functions of $\mathcal{F}_2(\mathbb{B}^2, \mathbb{B})$ (see Exercise 17.1 below). It follows from the explicit forms for the invertible functions of $\mathcal{F}(\mathbb{B}^2, \mathbb{B}^2)$, identified in Section 17.3.3, that no invertible representations from $\mathbb{B}^2$ to $\mathbb{B}^2$ are possible for the functions of $\mathcal{F}_2(\mathbb{B}^2, \mathbb{B})$. Instead we apply Corollary 1 and define invertible representations with domain and range $\mathbb{B}^3$; i.e., for $f_i \in \mathcal{F}_2(\mathbb{B}^2, \mathbb{B})$ we define

$$\hat{f}_i(x, y, z) = (x, y, z \oplus f_i(x, y))$$

and note that:

- the process, applied to f_1, i.e., $f_1(x, y) = x \wedge y$, yields the Toffoli gate:

$$\hat{f}_1(x, y, z) = (x, y, z \oplus (x \wedge y)).$$

which is an invertible form of the $\wedge$ function. There exist similar representations for all the functions of $\mathcal{F}_2(\mathbb{B}^2, \mathbb{B})$.

From Corollary 1 we have invertible representations $\hat{f}_i : \mathbb{B}^3 \to \mathbb{B}^3$, of $f_i \in \mathcal{F}_2(\mathbb{B}^2, \mathbb{B})$, defined by:

$$\hat{f}_i(x, y, z) = (x, y, z \oplus f_i(x, y)),$$

and the quantum operators $U_{\hat{f}_i}$ on the basis states $B_{\otimes^3 \mathbb{C}^2}$

$$U_{\hat{f}_i}|x\rangle|y\rangle|z\rangle = |x\rangle|y\rangle|z \oplus f_i(x, y)\rangle$$

from which we note that:

$$U_{\hat{f}_i}|x\rangle|y\rangle|0\rangle = |x\rangle|y\rangle|f_i(x, y)\rangle.$$

The function value $f_i(x, y) \in \mathbb{B}$ can be obtained by measuring the 3rd qubit of $U_{\hat{f}_i}|x\rangle|y\rangle|0\rangle$ with the operator $I \otimes I \otimes \mathcal{A}$.

The extended operator $U_{\hat{f}_i}$, on $\mathbb{C}^2 \otimes \mathbb{C}^2 \otimes \mathbb{C}^2$, may be expressed as:

$$U_{\hat{f}_i} = \sum_{pqr=000}^{111} |p\rangle|q\rangle|r \oplus f_i(pq)\rangle\langle p|\langle q|\langle r|.$$

17.4.4   The non-invertible subset of $\mathcal{F}(\mathbb{B}^n, \mathbb{B}^n)$

From Corollary 1 it follows that for any non-invertible $f : \mathbb{B}^n \to \mathbb{B}^n$, the invertible representation $\hat{f} : \mathbb{B}^{2n} \to \mathbb{B}^{2n}$ is defined by

$$
\begin{aligned}
\hat{f}(x, y) &= (x, y \oplus f(x)) \\
&\equiv (x_1, \ldots, x_n, \ y_1 \oplus f_1(x), \ldots, y_n \oplus f_n(x))
\end{aligned}
$$

where $x = x_1, \ldots, x_n \in \mathbb{B}^n$ and $y = y_1, \ldots, y_n$. For $\hat{f} : \mathbb{B}^{2n} \to \mathbb{B}^{2n}$ we have:

$$\hat{f}(x, 0) = (x, f(x)),$$

or

$$\hat{f}(x_1, \ldots, x_n, 0) = (x_1, \ldots x_n, \ f_1(x_1, \ldots, x_n), \ldots, f_n(x_1, \ldots, x_n))$$

i.e., the components of $f(x)$ comprise the $n + 1, \ldots, 2n$ components of $\hat{f}(x, 0)$.

The quantum emulations of these functions correspond to the special case of $m = n$ in the section, immediately below, which discusses the more general functions $\mathcal{F}(\mathbb{B}^n, \mathbb{B}^m)$.

17.4.5   The general functions $\mathcal{F}(\mathbb{B}^n, \mathbb{B}^m)$

Again, following Corollary 1 we have for any $f : \mathbb{B}^n \to \mathbb{B}^m$, the invertible representation $\hat{f} : \mathbb{B}^{n+m} \to \mathbb{B}^{n+m}$ defined by:

$$
\begin{aligned}
\hat{f}(x, y) &= (x, y \oplus f(x)) \\
&\equiv (x_1, \ldots, x_n, \ y_1 \oplus f_1(x), \ldots, y_m \oplus f_m(x))
\end{aligned}
$$

where $x = x_1, \ldots, x_n \in \mathbb{B}^n$ and $y = y_1, \ldots, y_m \in \mathbb{B}^m$. We have:

$$\hat{f}(x_1, \ldots, x_n, 0) = (x_1, \ldots, x_n, \ f_1(x_1, \ldots, x_n), \ldots, f_m(x_1, \ldots, x_n))$$

i.e., the components of $f(x)$ comprise the $n + 1, \ldots, n + m$ components of $\hat{f}(x, 0)$. For $f : \mathbb{B}^n \to \mathbb{B}^m$ we define U_f on the basis states $B_{\otimes^{n+m}\mathbb{C}^2}$ by:

$$U_{\hat{f}}|x\rangle|y\rangle \ = \ |x\rangle|y \oplus f(x)\rangle$$

where $|x\rangle \in \otimes^n \mathbb{C}^2$, $|y\rangle \in \otimes^m \mathbb{C}^2$. From this we obtain:

$$
\begin{aligned}
U_{\hat{f}}|x\rangle|0\rangle \ &= \ |x\rangle|f(x)\rangle \\
&= \ |x_1\rangle \cdots |x_n\rangle|f_1(x)\rangle|f_2(x)\rangle \cdots |f_m(x)\rangle
\end{aligned}
$$

and the Boolean values $f_1(x), \ldots, f_m(x)$ may be obtained by measuring the state $U_{\hat{f}}|x\rangle|0\rangle$ with the operators:

$$I^{\otimes n} \otimes \mathcal{A} \otimes I^{\otimes(m-1)}$$

$$\vdots$$

$$I^{\otimes n} \otimes I^{\otimes(m-1)} \otimes \mathcal{A}$$

respectively. We note that the linear extension of $U_{\hat{f}}$ to $\otimes^{n+m}\mathbb{C}^2$, also denoted $U_{\hat{f}}$, may be expressed as:

$$U_{\hat{f}} = \sum_{p,q} |p\rangle|q \oplus f(p)\rangle|\langle p|\langle q|$$

where $p \in \mathbb{B}^n$ and $q \in \mathbb{B}^m$.

17.5   Black-box representations of Boolean functions and their quantum emulations

It is common, in algorithms, to use black-box (or 'oracle') representations of invertible Boolean functions and their corresponding quantum operators. The examples in the figures below are for:

1. invertible functions of $\mathcal{F}(\mathbb{B}^2, \mathbb{B}^2)$ and their quantum emulations – see Section 17.3.2 and Figures 17.1 and 17.2,

2. invertible representations of the functions $\mathcal{F}_1(\mathbb{B}^2, \mathbb{B})$ and their quantum emulations – see Section 17.4.3 and Figures 17.3 and 17.4,

3. invertible representations of the functions $\mathcal{F}_2(\mathbb{B}^2, \mathbb{B})$ and their quantum emulations – see Section 17.4.3 and Figures 17.5 and 17.6.

Oracles, in general, show what is to be computed on the input specified without details of how the computations are done or even what f is. Often, in the design of algorithms, we only have query access to an oracle for a function – from which we may investigate its properties.

We note that for the quantum operators the general states of $\mathbb{C}^2 \otimes \mathbb{C}^2$ are also valid input.

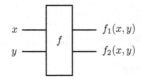

FIGURE 17.1
Black-box representation of an invertible function $f : \mathbb{B}^2 \to \mathbb{B}^2$, where $f(x,y) = (f_1(x,y), f_2(x,y))$ on $\mathbb{B}^2$.

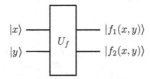

FIGURE 17.2
The black-box representation of the unitary operator U_f, on the basis states $B_{\otimes^2 \mathbb{C}^2}$, for f of Figure 17.1.

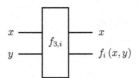

FIGURE 17.3
Black-box representation of the non-invertible function $f_i \in \mathcal{F}_1(\mathbb{B}^2, \mathbb{B})$ embedded in $f_{3,i}$.

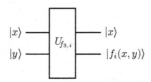

FIGURE 17.4
Black-box representation of the quantum operator $U_{f_{3,i}}$ on $B_{\otimes^2 \mathbb{C}^2}$.

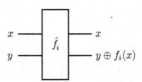

FIGURE 17.5
Black-box representation of a non-invertible function $f_i \in \mathcal{F}_2(\mathbb{B}^2, \mathbb{B})$.

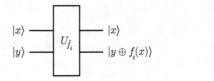

FIGURE 17.6
Black-box representation of the quantum operator $U_{\hat{f}_i}$ on $B_{\otimes^2 \mathbb{C}^2}$ for $f_i \in \mathcal{F}_2(\mathbb{B}^2, \mathbb{B})$.

Exercises

17.1 Show that the functions

(a) $(x, y) \rightarrow (x, x \wedge y)$ and
(b) $(x, y) \rightarrow (x, x \vee y)$,

from $\mathbb{B}^2$ to $\mathbb{B}^2$, are not invertible.

17.2 Show that the function

$$
\begin{aligned}
\hat{f}_7(x, y, z) &= (x, y, z \oplus f_7(x, y)) \\
&= (x, y, z \oplus (x \vee y))
\end{aligned}
$$

is invertible.

17.3 If $f : \mathbb{B}^2 \rightarrow \mathbb{B}$ is defined by $f(x_1, x_2) = x_1 \wedge x_2$, show that the associated $\hat{f}$ function is the Toffoli gate.

17.4 Show that the invertible subset of $\mathcal{F}(\mathbb{B}^n, \mathbb{B}^n)$, which we can denote by $G(n, \mathbb{B})$, determines a group under the functional composition operator $\circ$.

17.5 Show that the functions of $\mathcal{F}(\mathbb{B}^n, \mathbb{B}^m)$ determine a group under the $\oplus$ operator.

17.6 If $f \in \mathcal{F}(\mathbb{B}^n, \mathbb{B}^m)$ and $\hat{\mathcal{F}}_{n,m}$ comprises the functions $\hat{f}$, where $\hat{f}(x, y) = (x, y \oplus f(x))$ for $x \in \mathbb{B}^n$ and $y \in \mathbb{B}^m$, show that $\hat{\mathcal{F}}_{n,m}$ determines a group under functional composition.

17.7 Show that the groups $(\mathcal{F}(\mathbb{B}^n, \mathbb{B}^m), \oplus)$ and $(\hat{\mathcal{F}}_{n,m}, \circ)$ are isomorphic.

17.8 Show that for all $g, g^* \in G(n, \mathbb{B})$ the quantum emulation, $g \rightarrow U_g$ of g, defined by $U_g |x\rangle = |g(x)\rangle$, satisfies

$$
U_{gg^*} = U_g U_{g^*}.
$$

Under these conditions we say that $g \rightarrow U_g$ is a representation of the group $G(n, \mathbb{B})$ in the unitary group of $\otimes^n \mathbb{C}^2$.

17.9 Show that the group $\hat{\mathcal{F}}_{n,m}$ has a representation in the unitary group of $\otimes^{n+m} \mathbb{C}^2$.

18

Invertible digital circuits and their quantum emulations

18.1 Objectives

Black box representations of functions, by definition, do not provide information on how a function, so represented, is computed. By a 'circuit' for the computation of a function we mean a sequence of 'basic' gates that implement the function. It can be shown that, given a classical circuit for the Boolean function $\hat{f}$, there is a quantum circuit of equal 'efficiency' for the implementation of the unitary operator $U_{\hat{f}}$ and hence the quantum computation of f. In this chapter we consider examples of invertible digital functions and their evaluation using basic invertible Boolean gates; specifically the Toffoli and Feynman gates T_D and F_D. We recall that the Toffoli gate is complete for invertible digital computation, however, the inclusion of the Feynman gate, F_D, simplifies the circuits. We proceed by example to demonstrate the process and consequential side-effects of such implementations and their quantum emulations. It is seen that the cost of using only Toffoli gates is that additional input bits/qubits may be required, and the computation of intermediate results (usually referred to as garbage or junk) may be necessary. We also consider a general circuit for re-setting registers containing 'junk', this is helpful to avoid problems with downstream computation.

18.2 Invertible digital circuits

We begin with a general Boolean function $f : \mathbb{B}^n \to \mathbb{B}^m$ and the general invertible representation $\hat{f} : \mathbb{B}^{n+m} \to \mathbb{B}^{n+m}$, of f, discussed earlier and defined by:

$$\hat{f}(x,y) = (x, y \oplus f(x)) \text{ for } x \in \mathbb{B}^n \text{ and } y \in \mathbb{B}^m.$$

We note that invertible representations need not be of this form and representations with fewer than $n+m$ variables may be possible (for example, the Peres gate); but what follows is based on the $\hat{f}$ representation, which works for any f. We have:

$$\hat{f}(x, 0^m) = (x_1, \ldots, x_n, f_1(x), \ldots, f_m(x)).$$

The n-input Toffoli gate is defined by:

$$T^{(n)}(x_1, x_2, \ldots, x_{n-1}, y) = (x_1, x_2, \ldots, x_{n-1}, y \oplus (x_1 \wedge x_2 \wedge \ldots \wedge x_{n-1})).$$

For $n = 3$ we have the 'usual' Toffoli gate $T^{(3)}(x_1, x_2, y) = (x_1, x_2, y \oplus (x_1 \wedge x_2))$ with the circuit shown in Figure 18.1.

DOI: 10.1201/9781003264569-18

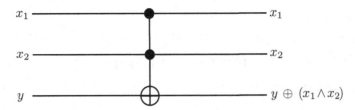

FIGURE 18.1
The Boolean Toffoli, or *ccnot*, 3-bit invertible gate.

In the 4-input case we have $f : \mathbb{B}^3 \to \mathbb{B}$ and $\hat{f} : \mathbb{B}^4 \to \mathbb{B}^4$ defined by

$$\begin{aligned} f(x_1, x_2, x_3) &= x_1 \wedge x_2 \wedge x_3 \\ \hat{f}(x_1, x_2, x_3, y) &= (x_1, x_2, x_3, y \oplus (x_1 \wedge x_2 \wedge x_3)). \end{aligned}$$

Clearly, more than one 3-input Toffoli gate is necessary to implement the 4-input, invertible Toffoli gate $\hat{f}$. It may, however, be implemented using two 3-input Toffoli gates – see Figure 18.2. We note that a workspace register, y_1, is required to hold the intermediate result $x_1 \wedge x_2$, the y_2 register holds the required output of the circuit when y_1 and y_2 are suitably initialised.

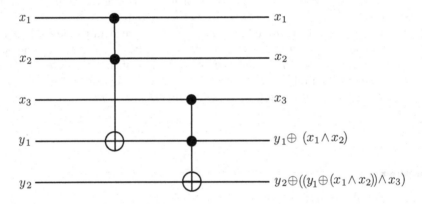

FIGURE 18.2
A realisation of the Toffoli 4-bit invertible gate using two Toffoli 3-input gates.

The circuit outputs $(x_1, x_2, x_3, x_1 \wedge x_2, x_1 \wedge x_2 \wedge x_3)$ on the input of $(x_1, x_2, x_3, 0, 0)$. (Figure 18.3). In addition to the function value $f(x_1, x_2, x_3)$ the value $x_1 \wedge x_2$ is also output. This is a direct consequence of computing a non-invertible function using an invertible representation. Unwanted outputs, $x_1 \wedge x_2$, in this case, generated by networks such as $C_{\hat{f}}$ are often referred to as 'junk'.

Alternatively, to re-order the outputs, consider Figure 18.4 from which we obtain, by setting $y_1 = y_2 = 0$, the network in Figure 18.5.

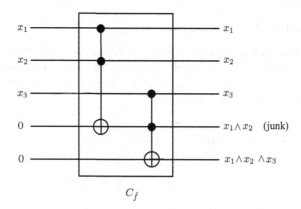

FIGURE 18.3
A network $C_{\hat{f}}$ for the Toffoli 4-bit invertible gate $\hat{f}$.

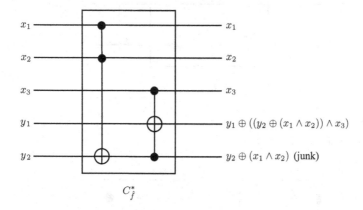

FIGURE 18.4
A modification of $C_{\hat{f}}$ to re-order the output.

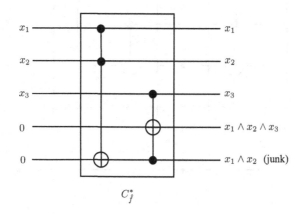

FIGURE 18.5
The network $C_{\hat{f}}^{*}$ with $y_1 = y_2 = 0$.

18.3   Junk removal

Registers holding junk can be reset to their initial values – this is sometimes referred to as 'un-computing' the junk. For example, in the case above, this can be done using an additional Toffoli gate as shown in Figure 18.6. A more general structure for resetting registers holding junk is illustrated in Figure 18.7.

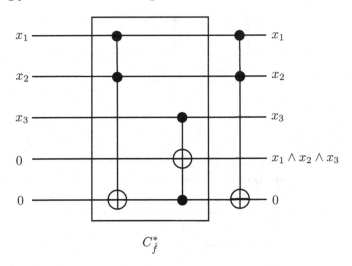

FIGURE 18.6
A junk-removed network for $'\hat{f}$.

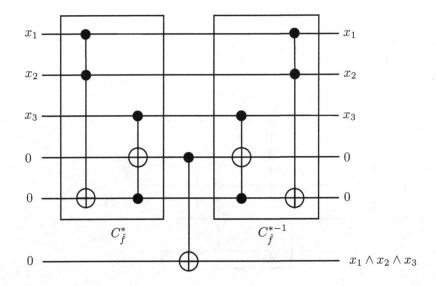

FIGURE 18.7
An alternative network for junk removal from $\hat{f}$.

The circuit of Figure 18.7 introduces an additional input register and further gates, however, it is representative of a general network structure for the removal of junk in invertible digital computation – see Figure 18.8. The inputs, other than $x_1, \ldots, x_n$, to the network above are often called 'auxiliary' or 'ancilla' inputs necessary for the 'junk-removed' computation of $f(x)$ using an invertible circuit.

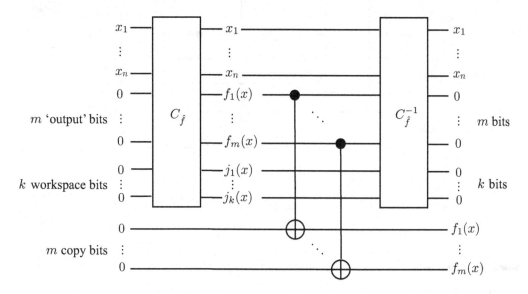

FIGURE 18.8
A general network structure for the computation of any $f : \mathbb{B}^n \rightarrow \mathbb{B}^m$ with junk $j_1(x), \ldots j_k(x)$ removed – i.e., the workspace registers reset to 0.

18.4 Quantum emulation

Figure 18.9 shows a quantum network for the implementation of $|f(x)\rangle$ with junk removed. Mathematically, for the network of Figure 18.9 we have:

$$
\begin{aligned}
(U_f^{-1} \otimes I)cnot(U_f \otimes I)|x\rangle|0^m\rangle|0^k\rangle|0^m\rangle &= (U_f^{-1} \otimes I)cnot|x\rangle|f(x)\rangle|j\rangle|0^m\rangle \\
&= (U_f^{-1} \otimes I)(|x\rangle|f(x)\rangle|j\rangle|f(x)\rangle) \\
&= |x\rangle|0^m\rangle|0^k\rangle|f(x)\rangle.
\end{aligned}
$$

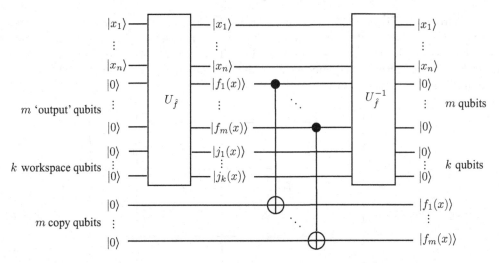

FIGURE 18.9
A quantum network structure for the computation of $|f(x)\rangle$ with junk $|j_1(x)\rangle, \ldots |j_k(x)\rangle$ removed.

Exercises

18.1 Show that the digital gate

$$T_D^\vee(x, y, z) = (x, y, z \oplus (y \vee x))$$

is invertible.

18.2 Design an invertible digital circuit to output the pair $(x_1, x_2) \in \mathbb{B}^2$ in descending order. (Hint: note that $\max(x_1, x_2) = x_1 \vee x_2$ and $\min(x_1, x_2) = x_1 \wedge x_2$).

19

Quantum measurement 2: general pure states, Bell states

19.1 Objectives

The relatively simple examples of quantum computation (emulating digital computation) discussed in Chapters 15, 17 and 18 involved only the simplest quantum inputs and produced, deterministically, outputs identical to their digital counterparts. The simplified statement of the von Neumann-Lüders measurement postulate, discussed in Chapter 14, was applicable to these computations. In this chapter we introduce the processing of more general, i.e., super-imposed, quantum states. This is divergent from the digital case and, as will be clear later, imparts quantum computation with extended power.

19.2 The measurement of super-imposed quantum states

In moving forward to consider these generalisations, it is necessary to invoke the general von Neumann-Lüders measurement model (i.e., Axiom 3 of Section 13.6) which we restate below and apply it to an example with which we already have some familiarity.

Axiom 3
Following the measurement of an observable $\mathcal{O}$ on the state $|\psi\rangle \in \otimes^n \mathbb{C}^2$, the post-measurement state is the normalisation of the vector $P|\psi\rangle$, where P is the orthogonal projection of $|\psi\rangle$ onto the subspace of $\otimes^n \mathbb{C}^2$ generated by the eigenstates of $\mathcal{O}$ that

1. occur in the linear-superposition of $|\psi\rangle$, and

2. are compatible with the outcome of the measurement of $\mathcal{O}$.

The probability of the outcome obtained is $\|P|\psi\rangle\|^2$.

Example 19.2.1 An application of the general measurement postulate (Axiom 3)

Consider the quantum state

$$|\psi\rangle = \alpha_{00}|0\rangle|0\rangle + \alpha_{01}|0\rangle|1\rangle + \alpha_{10}|1\rangle|0\rangle + \alpha_{11}|1\rangle|1\rangle \in \mathbb{C}^2 \otimes \mathbb{C}^2$$

where $|\alpha_{00}|^2 + |\alpha_{01}|^2 + |\alpha_{10}|^2 + |\alpha_{11}|^2 = 1$. If $\mathcal{A}$ is the observable with matrix

$$\begin{pmatrix} 0 & 0 \\ 0 & 1 \end{pmatrix}$$

DOI: 10.1201/9781003264569-19

in the computational basis, then the vectors $\{|0\rangle|0\rangle, |0\rangle|1\rangle, |1\rangle|0\rangle, |1\rangle|1\rangle\}$ are eigenvectors of the observable $\mathcal{A} \otimes I$ and they determine an orthonormal basis of $\mathbb{C}^2 \otimes \mathbb{C}^2$ (see Example 11.8.6). The eigenvalues of $\mathcal{A} \otimes I$ are 0 and 1 and they are both degenerate. Specifically, we have

$$\mathcal{A} \otimes I|0\rangle|0\rangle = 0|0\rangle|0\rangle, \quad \mathcal{A} \otimes I|0\rangle|1\rangle = 0|0\rangle|1\rangle$$

and

$$\mathcal{A} \otimes I|1\rangle|0\rangle = 1|1\rangle|0\rangle, \quad \mathcal{A} \otimes I|1\rangle|1\rangle = 1|1\rangle|1\rangle.$$

Hence if 0 is measured for $\mathcal{A} \otimes I$ on $|\psi\rangle$, we have

1.
$$P|\psi\rangle = \alpha_{00}|0\rangle|0\rangle + \alpha_{01}|0\rangle|1\rangle$$

which happens with probability $\|P|\psi\rangle\|^2 = |\alpha_{00}|^2 + |\alpha_{01}|^2$, and

2. the post-measurement state

$$\frac{\alpha_{00}|0\rangle|0\rangle + \alpha_{01}|0\rangle|1\rangle}{\sqrt{|\alpha_{00}|^2 + |\alpha_{01}|^2}}.$$

Similarly if 1 is measured for $\mathcal{A} \otimes I$ on $|\psi\rangle$, we have:

1.
$$P|\psi\rangle = \alpha_{10}|1\rangle|0\rangle + \alpha_{11}|1\rangle|1\rangle$$

which happens with probability $\|P|\psi\rangle\|^2 = |\alpha_{10}|^2 + |\alpha_{11}|^2$, and

2. the post-measurement state

$$\frac{\alpha_{10}|1\rangle|0\rangle + \alpha_{11}|1\rangle|1\rangle}{\sqrt{|\alpha_{10}|^2 + |\alpha_{11}|^2}}.$$

19.3   Measuring the EPR-Bell state $\frac{1}{\sqrt{2}}(|0\rangle|0\rangle + |1\rangle|1\rangle)$

From the above we also note that if 0 is obtained when measuring the first qubit of the Bell state

$$|B_1\rangle = \frac{1}{\sqrt{2}}(|0\rangle|0\rangle + |1\rangle|1\rangle)$$

with the operator $\mathcal{A} \otimes I$, the post-measurement state is

$$|0\rangle|0\rangle$$

and if 1 is measured then the post-measurement state is

$$|1\rangle|1\rangle$$

these being the states that are

1. compatible with the measured value of the observable and

2. occur in the representation of $|B_1\rangle$ as a superposition of the eigenstates of the observable $\mathcal{A} \otimes I$ (the von Neumann-Lüders postulate).

This is a striking result in that the post-measurement state of the second qubit, although nothing is actually done to it (it being 'trivially' measured by the identity operator), is identical to that of the first qubit. The physical implications of this result are particularly interesting. It is possible to design an experiment that outputs two photons moving in opposite directions (say left and right) with the same, but unknown helicity. The helicity of a photon is a quantum observable that takes just two values which are usually denoted by v and h. It follows that the state of the combined photon pair is the entangled tensor:

$$\frac{1}{\sqrt{2}}(|v\rangle|v\rangle + |h\rangle|h\rangle).$$

If the helicity of the left-travelling photon is measured with result v, then the post-measurement state of the pair is

$$|v\rangle|v\rangle$$

i.e., the right travelling photon, the state of which has not been measured (and which may be millions of miles separated from the left-travelling photon at the time when the helicity of the left-travelling photon is measured), has the same post-measurement state as that of the left-travelling photon. It is as though the right-travelling photon has also been measured. Of course, if the helicity of the left travelling photon was measured to be h then the post-measurement state of the pair would be $|h\rangle|h\rangle$. The state $\frac{1}{\sqrt{2}}(|v\rangle|v\rangle + |h\rangle|h\rangle)$ is known as the Einstein-Podolsky-Rosen (EPR) state, these being the physicists to first identify this counter-intuitive quantum phenomenon. This theoretically-predicted behaviour was initially regarded as paradoxical and referred to, by Einstein, as 'spooky action at a distance'. In quantum mechanics the EPR and Bell states are said to be entangled – a term defined formally in Chapter 10. More recent research, particularly due to the Irish physicist John Bell, and experiments by various institutions have supported the view that there is no paradox and that quantum mechanics may be 'non-local'.

Another view: Alice and Bob work in physics laboratories in locations separated by some distance. They each have a 2-state quantum particle and their particles are entangled; i.e., the state of the pair is $\frac{1}{\sqrt{2}}(|0\rangle|0\rangle + |1\rangle|1\rangle)$. If Alice measures 0 for her particle then the post-measurement state of the pair is $|0\rangle|0\rangle$; Bob will now measure 0 for his particle with probability 1, and Alice knows this in advance without any communication between the two. Prior to Alice's measurement, Bob would have measured 0 with probability $\frac{1}{2}$ and 1 with probability $\frac{1}{2}$ for his particle.

This was regarded as paradoxical in the sense that faster-than-light 'communication' appears to be taking place and this is incompatible with relativity theory. However, unless Alice tells Bob (by classical means e.g., telephone, carrier pigeon etc.) that she has measured her particle and dis-entangled the pair, Bob wouldn't know that his particle was dis-entangled – and he couldn't determine this by measurement. Because if the dis-entangled state were $|0\rangle|0\rangle$ he would measure 0 with probability 1, but he couldn't conclude that the particles were dis-entangled prior to his measurement – because Bob's probability of measuring 0 for his particle was $\frac{1}{2}$ whilst they were entangled.

19.4   More general measurements of 2-qubit states in the computational basis

Again we consider measurement of the observables $\mathcal{A} \otimes I : \mathbb{C}^2 \otimes \mathbb{C}^2 \to \mathbb{C}^2 \otimes \mathbb{C}^2$ and $I \otimes \mathcal{A} : \mathbb{C}^2 \otimes \mathbb{C}^2 \to \mathbb{C}^2 \otimes \mathbb{C}^2$ on the general state $|\psi\rangle \in \mathbb{C}^2 \otimes \mathbb{C}^2$ defined by

$$|\psi\rangle = \alpha_{00}|0\rangle|0\rangle + \alpha_{01}|0\rangle|1\rangle + \alpha_{10}|1\rangle|0\rangle + \alpha_{11}|1\rangle|1\rangle \in \mathbb{C}^2 \otimes \mathbb{C}^2$$

where $|\alpha_{00}|^2 + |\alpha_{01}|^2 + |\alpha_{10}|^2 + |\alpha_{11}|^2 = 1$. The observables have the matrix form

$$A \otimes I = \begin{pmatrix} 0 & 0 & 0 & 0 \\ 0 & 0 & 0 & 0 \\ 0 & 0 & 1 & 0 \\ 0 & 0 & 0 & 1 \end{pmatrix}$$

and

$$I \otimes A = \begin{pmatrix} 0 & 0 & 0 & 0 \\ 0 & 1 & 0 & 0 \\ 0 & 0 & 0 & 0 \\ 0 & 0 & 0 & 1 \end{pmatrix}.$$

Measuring $\mathcal{A} \otimes I$ on $|\psi\rangle$ we have:

1. if 0 is measured then the post-measurement state is, as we have seen earlier, given by

$$|\psi_1\rangle = \frac{\alpha_{00}|0\rangle|0\rangle + \alpha_{01}|0\rangle|1\rangle}{\sqrt{|\alpha_{00}|^2 + |\alpha_{01}|^2}}$$

1.1 Measuring $|\psi_1\rangle$ with $I \otimes \mathcal{A}$ with the result 0 leads to the post-measurement state

$$|0\rangle|0\rangle$$

1.2 Measuring $|\psi_1\rangle$ with $I \otimes \mathcal{A}$ with the result 1 leads to the post-measurement state

$$|0\rangle|1\rangle$$

2. if 1 is measured for $\mathcal{A} \otimes I$ then the post-measurement state is, as we have seen earlier, given by:

$$|\psi_2\rangle = \frac{\alpha_{10}|1\rangle|0\rangle + \alpha_{11}|1\rangle|1\rangle}{\sqrt{|\alpha_{10}|^2 + |\alpha_{11}|^2}}$$

2.1 Measuring $|\psi_2\rangle$ with $I \otimes \mathcal{A}$ with the result 0 leads to the post-measurement state

$$|1\rangle|0\rangle$$

2.2 Measuring $|\psi_2\rangle$ with $I \otimes \mathcal{A}$ with the result 1 leads to the post-measurement state

$$|1\rangle|1\rangle,$$

i.e., we have for consecutive measurements of the observables, $\mathcal{A} \otimes I$ and $I \otimes \mathcal{A}$, the four possible outcomes:

- for measurements $(0, 0)$ the post-measurement state will be $|0\rangle|0\rangle$,

- for measurements $(0, 1)$ the post-measurement state will be $|0\rangle|1\rangle$,

- for measurements $(1, 0)$ the post-measurement state will be $|1\rangle|0\rangle$,

- for measurements $(1, 1)$ the post-measurement state will be $|1\rangle|1\rangle$.

19.5 Measuring 2-qubit states in the Bell basis

To this point we have only considered quantum measurement in 'computational' bases, i.e., $\{|0\rangle, |1\rangle\}$ and the tensor product bases constructed therefrom. Here, we consider measurement in the orthonormal (and entangled) Bell basis:

$$|B_1\rangle = \frac{1}{\sqrt{2}}(|0\rangle|0\rangle + |1\rangle|1\rangle)$$

$$|B_2\rangle = \frac{1}{\sqrt{2}}(|0\rangle|1\rangle + |1\rangle|0\rangle)$$

$$|B_3\rangle = \frac{1}{\sqrt{2}}(|0\rangle|0\rangle - |1\rangle|1\rangle)$$

$$|B_4\rangle = \frac{1}{\sqrt{2}}(|0\rangle|1\rangle - |1\rangle|0\rangle)$$

of $\mathbb{C}^2 \otimes \mathbb{C}^2$. For any normalised state $|\psi\rangle \in \mathbb{C}^2 \otimes \mathbb{C}^2$ there exist $\beta_1, \beta_2, \beta_3, \beta_4 \in \mathbb{C}$ with $\sum_{i=1}^{4} |\beta_i|^2 = 1$ and

$$|\psi\rangle = \beta_1|B_1\rangle + \beta_2|B_2\rangle + \beta_3|B_3\rangle + \beta_4|B_4\rangle.$$

As we have seen, the von Neumann-Lüders postulate relates measurement to self-adjoint operators having eigenvectors corresponding to the basis chosen to represent the state – in this case the Bell vectors $|B_1\rangle, |B_2\rangle, |B_3\rangle, |B_4\rangle$. The following observables S_1^2, S_2^2, S_3^2, equivalently self-adjoint operators on $\mathbb{C}^2 \otimes \mathbb{C}^2$, each have the Bell states as eigenstates.

$$S_1^2 = \frac{1}{2}\begin{pmatrix} 1 & 0 & 0 & 1 \\ 0 & 1 & 1 & 0 \\ 0 & 1 & 1 & 0 \\ 1 & 0 & 0 & 1 \end{pmatrix}, \quad S_2^2 = \frac{1}{2}\begin{pmatrix} 1 & 0 & 0 & -1 \\ 0 & 1 & 1 & 0 \\ 0 & 1 & 1 & 0 \\ -1 & 0 & 0 & 1 \end{pmatrix}, \quad S_3^2 = \begin{pmatrix} 1 & 0 & 0 & 0 \\ 0 & 0 & 0 & 0 \\ 0 & 0 & 0 & 0 \\ 0 & 0 & 0 & 1 \end{pmatrix}.$$

Lemma 19.1 *The Bell states $|B_1\rangle, |B_2\rangle, |B_3\rangle, |B_4\rangle$ are eigenvectors of each of the observables S_1^2, S_2^2 and S_3^2. The eigenvalues are 0 and 1, for each of the observables, and the spectral properties are as indicated in the following table:*

	S_1^2	S_2^2	S_3^2
$|B_1\rangle$	1	0	1
$|B_2\rangle$	1	1	0
$|B_3\rangle$	0	1	1
$|B_4\rangle$	0	0	0

Proof
Left as an exercise.

19.5.1 Measuring the observables S_3^2 and S_1^2

1. If measuring the state $|\psi\rangle$ with the observable S_3^2 results in the outcome 1, then the post-measurement state is:

$$|\psi_1\rangle = \frac{\beta_1|B_1\rangle + \beta_3|B_3\rangle}{\sqrt{|\beta_1|^2 + |\beta_3|^2}}.$$

Then if a measurement of the observable S_1^2 on the state $|\psi_1\rangle$ results in the outcome 1 then the resulting state is:

$$|B_1\rangle.$$

If a measurement of the observable S_1^2 on the state $|\psi_1\rangle$ results in the outcome 0 then the resulting state is:

$$|B_3\rangle.$$

2. If measuring the state $|\psi\rangle$ with the observable S_3^2 results in the outcome 0 then the post-measurement state is:

$$|\psi_2\rangle = \frac{\beta_2|B_2\rangle + \beta_4|B_4\rangle}{\sqrt{|\beta_2|^2 + |\beta_4|^2}}.$$

Then if a measurement of the observable S_1^2 on the state $|\psi_2\rangle$ results in the outcome 1 then the resulting state is:

$$|B_2\rangle.$$

If a measurement of the observable S_1^2 on the state $|\psi_2\rangle$ results in the outcome 0 then the resulting state is:

$$|B_4\rangle.$$

That is, when measuring the observables S_3^2 and S_1^2 on the state $|\psi\rangle$ we have:

- if $(1, 1)$ is measured then the post-measurement state will be $|B_1\rangle$,

- if $(1, 0)$ is measured then the post-measurement state will be $|B_3\rangle$,

- if $(0, 1)$ is measured then the post-measurement state will be $|B_2\rangle$,

- if $(0, 0)$ is measured then the post-measurement state will be $|B_4\rangle$.

Exercises

19.1 Following the calculations above to measure the observables S_3^2 and S_1^2 on the state $|\psi\rangle$:

(a) perform the calculations to measure the observables S_1^2 and S_2^2 on the state $|\psi\rangle$.

(b) perform the calculations to measure the observables S_2^2 and S_3^2 on the state $|\psi\rangle$.

20

Quantum information processing 3

20.1 Objectives

We have seen in earlier chapters how digital computations can be emulated quantum mechanically. In this chapter, we cover further aspects of processing, and highlight some significant differences that occur between the digital and quantum cases. We first introduce the concepts of quantum parallelism and quantum swapping both of which have analogs in the digital case. We demonstrate that copying, which is a trivial operation in digital computing, is not generally achievable in the quantum case. The quantum concept of teleportation, which has no analog in digital computing, is also introduced.

20.2 Quantum parallelism

For $f : \mathbb{B} \to \mathbb{B}$, consider the circuit for $U_f|x\rangle|y\rangle = |x\rangle|y \oplus f(x)\rangle$, extended linearly to the whole of $\mathbb{C}^2 \otimes \mathbb{C}^2$, with the inputs shown in Figure 20.1.

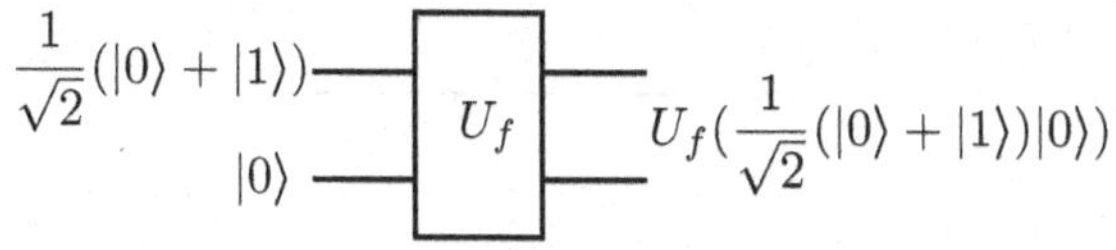

$$\frac{1}{\sqrt{2}}(|0\rangle + |1\rangle) \longrightarrow \boxed{U_f} \longrightarrow U_f(\frac{1}{\sqrt{2}}(|0\rangle + |1\rangle)|0\rangle)$$

$$|0\rangle \longrightarrow$$

FIGURE 20.1
Quantum parallelism.

$$
\begin{aligned}
U_f(\frac{1}{\sqrt{2}}(|0\rangle + |1\rangle)|0\rangle) &= \frac{1}{\sqrt{2}}(U_f|0\rangle|0\rangle + U_f|1\rangle|0\rangle) \\
&= \frac{1}{\sqrt{2}}(|0\rangle|f(0)\rangle + |1\rangle|f(1)\rangle).
\end{aligned}
$$

That is, the input of a single state $\frac{1}{\sqrt{2}}(|0\rangle + |1\rangle) \otimes |0\rangle \in \mathbb{C}^2 \otimes \mathbb{C}^2$ to the 'gate' to evaluate f has given rise to two evaluations, $f(0)$ and $f(1)$, of f. This is known as **quantum parallelism** and occurs because superpositions of states are also valid states in quantum mechanics. The two input registers hold $|0\rangle \in \mathbb{C}^2$ and $\frac{1}{\sqrt{2}}(|0\rangle + |1\rangle) \in \mathbb{C}^2$; the latter could be generated, for example, by the Hadamard gate (see Section 21.4). We should also note that if $f(0) \neq f(1)$ then the output state is entangled (see Section 10.3 on tensor products).

DOI: 10.1201/9781003264569-20

20.3   Qubit swapping

In certain quantum algorithms, e.g., the quantum Fourier transform (which is beyond the scope of this book), it is necessary to swap qubits. In particular, a unitary operator that maps the state

$$|\phi\rangle|\psi\rangle \equiv (\alpha_1|0\rangle + \beta_1|1\rangle) \otimes (\alpha_2|0\rangle + \beta_2|1\rangle)$$

to the state

$$|\psi\rangle|\phi\rangle \equiv (\alpha_2|0\rangle + \beta_2|1\rangle) \otimes (\alpha_1|0\rangle + \beta_1|1\rangle)$$

is required. Much as in the digital case, we define a function S on the basis states $B_{\otimes^2\mathbb{C}^2}$ of $\mathbb{C}^2 \otimes \mathbb{C}^2$ by

$$S|x\rangle|y\rangle = |y\rangle|x\rangle \quad \text{where } x, y \in \mathbb{B},$$

the linear extension of which to $\mathbb{C}^2 \otimes \mathbb{C}^2$, also denoted S and referred to as the swap operator, is unitary and has the matrix representation:

$$\begin{pmatrix} 1 & 0 & 0 & 0 \\ 0 & 0 & 1 & 0 \\ 0 & 1 & 0 & 0 \\ 0 & 0 & 0 & 1 \end{pmatrix}.$$

It is straightforward to show that for a general qubit pair, i.e., $|\phi\rangle$ and $|\psi\rangle$, as defined above, we obtain:

$$S|\phi\rangle|\psi\rangle = |\psi\rangle|\phi\rangle,$$

i.e., S swaps the qubits as required. Figure 20.2 illustrates S on the basis states $B_{\otimes^2\mathbb{C}^2}$; and the linear extension of S, to $\mathbb{C}^2 \otimes \mathbb{C}^2$, showing the action of S in the general case is represented in Figure 20.3.

$$|x\rangle \quad\quad\quad\quad\quad |y\rangle$$
$$|y\rangle \quad\quad\quad\quad\quad |x\rangle$$

FIGURE 20.2
The swap operator on $B_{\otimes^2\mathbb{C}^2}$.

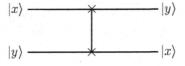

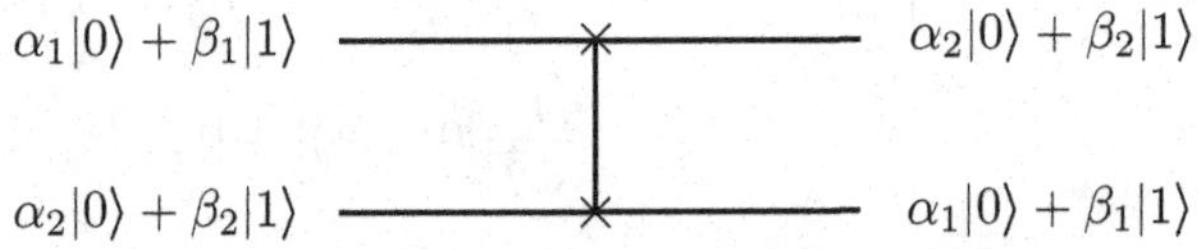

$$\alpha_1|0\rangle + \beta_1|1\rangle \quad\quad\quad \alpha_2|0\rangle + \beta_2|1\rangle$$
$$\alpha_2|0\rangle + \beta_2|1\rangle \quad\quad\quad \alpha_1|0\rangle + \beta_1|1\rangle$$

FIGURE 20.3
The linearly extended swap operator on $\mathbb{C}^2 \otimes \mathbb{C}^2$.

Suitable generalisations of S to swap pairs of qubits in the general qubit space $\otimes^n\mathbb{C}^2$ are easily defined.

20.4 Quantum copying – the no-cloning theorem

We compare the quantum case with the functions, introduced earlier, to copy in the digital domain.

The quantum equivalent of the classical *dupe* function $(i, i) : \mathbb{B} \to \mathbb{B}^2$ defined by $(i, i)(x) = (x, x)$ is the function $U : \{|0\rangle, |1\rangle\} \to \mathbb{C}^2 \otimes \mathbb{C}^2$ defined by

$$U|x\rangle = |x\rangle|x\rangle \qquad \text{for } x \in \mathbb{B}.$$

That is, U is such that

$$U|0\rangle = |0\rangle|0\rangle$$

and

$$U|1\rangle = |1\rangle|1\rangle.$$

Clearly U copies the states $|0\rangle$ and $|1\rangle$. However, the following is true:

Lemma 20.1 *The linear extension of U to $\mathbb{C}^2$, defined by*

$$U(\alpha_0|0\rangle + \alpha_1|1\rangle) \equiv \alpha_0 U|0\rangle + \alpha_1 U|1\rangle,$$

does not copy the general quantum state, $\alpha_0|0\rangle + \alpha_1|1\rangle$, of $\mathbb{C}^2$.

Proof

In general $\alpha_0 U|0\rangle + \alpha_1 U|1\rangle \neq (\alpha_0|0\rangle + \alpha_1|1\rangle) \otimes (\alpha_0|0\rangle + \alpha_1|1\rangle)$. Equality only occurs in the cases $\alpha_0 = 1, \alpha_1 = 0$ and $\alpha_0 = 0, \alpha_1 = 1$ for which we have $U|0\rangle = |0\rangle|0\rangle$ and $U|1\rangle = |1\rangle|1\rangle$.

∎

We conclude that while the classical bit-copying function $(i, i)(x) = (x, x)$ is easily extended to copy a general digital register (see Figure 2.28) this is not the case for the quantum equivalent.

We have also seen that the digital *cnot* function enables arbitrary Boolean strings to be copied – see Figure 2.35. The quantum *cnot* operator, $U_{cnot}|x\rangle|y\rangle = |x\rangle|y \oplus x\rangle$ copies the states $|0\rangle$ and $|1\rangle$; i.e., we have $U_{cnot}|0\rangle|0\rangle = |0\rangle|0\rangle$ and $U_{cnot}|1\rangle|0\rangle = |1\rangle|1\rangle$. However, the linear extension of U_{cnot} to $\mathbb{C}^2 \otimes \mathbb{C}^2$ gives:

$$
\begin{aligned}
U_{cnot}(\alpha_0|0\rangle + \alpha_1|1\rangle)|0\rangle &= \alpha_0 U_{cnot}|0\rangle|0\rangle + \alpha_1 U_{cnot}|1\rangle|0\rangle \\
&= \alpha_0|0\rangle|0\rangle + \alpha_1|1\rangle|1\rangle \\
&\neq (\alpha_0|0\rangle + \alpha_1|1\rangle) \otimes (\alpha_0|0\rangle + \alpha_1|1\rangle)
\end{aligned}
$$

and the input state $\alpha_0|0\rangle + \alpha_1|1\rangle$ is not duplicated. The following Lemma explains why our attempts to clone general quantum states has not been successful.

Lemma 20.2 *Let V be a finite dimensional vector space, then*

1. the mapping $C_1 : V \to V \otimes V$ defined by $C_1|\phi\rangle = |\phi\rangle|\phi\rangle$ is non-linear,

2. the mapping $C_2 : V \otimes V \to V \otimes V$ defined by $C_2|\phi\rangle|\phi_0\rangle = |\phi\rangle|\phi\rangle$, for any constant $|\phi_0\rangle \in V$, is non-linear.

Proof

1. We have $C_1(|\phi\rangle + |\chi\rangle) = (|\phi\rangle + |\chi\rangle)(|\phi\rangle + |\chi\rangle) = |\phi\rangle|\phi\rangle + |\chi\rangle|\chi\rangle + |\phi\rangle|\chi\rangle + |\chi\rangle|\phi\rangle$ and $C_1|\phi\rangle + C_1|\chi\rangle = |\phi\rangle|\phi\rangle + |\chi\rangle|\chi\rangle$. Hence $C_1(|\phi\rangle + |\chi\rangle) \neq C_1|\phi\rangle + C_1|\chi\rangle$ and C_1 is not linear.

2. Similarly $C_2(|\phi\rangle + |\chi\rangle)|\phi_0\rangle \neq C_2(|\phi\rangle|\phi_0\rangle) + C_2(|\chi\rangle|\phi_0\rangle)$.

∎

As quantum processes are unitary (and hence linear) the Lemma tells us that a general quantum cloning operator does not exist. In compensation for not being able to copy quantum states (the no-cloning theorem), quantum theory enables us to reproduce a state wherever we like, at the expense of losing the original. This is what has become known as quantum teleportation. The process depends, fundamentally, on quantum entanglement. Teleportation enables us to demonstrate measurement in both the computational and Bell bases, to achieve the same outcome.

20.5   Quantum teleportation 1, computational-basis measurement

Alice has a qubit $|\psi\rangle = \alpha|0\rangle_c + \beta|1\rangle_c$ in her possession that she wishes to convey to Bob. For a classical state Alice could measure its properties and send them to Bob, who could then reproduce it. However, if Alice measures the quantum state $|\psi\rangle$, the state may change and a different approach is required.

The objective of teleportation is to transmit $|\psi\rangle$ to Bob using only classical information; i.e., bits. The teleportation process requires that Alice and Bob share an entangled state, which is fixed in advance by agreement between them. We assume this state to be:

$$|\phi\rangle = \frac{1}{\sqrt{2}}(|0\rangle_a|0\rangle_b + |1\rangle_a|1\rangle_b)$$

of which each has a qubit which we distinguish by the subscripts a and b. This can be implemented by preparing the particles together and firing them to Alice and Bob from the same source. The subscripts a and b in the entangled state refer to Alice's and Bob's particles. The combined state of the particles held by Alice is therefore

$$
\begin{aligned}
|\psi\rangle \otimes |\phi\rangle &= \frac{1}{\sqrt{2}}(\alpha|0\rangle_c + \beta|1\rangle_c) \otimes (|0\rangle_a|0\rangle_b + |1\rangle_a|1\rangle_b) \\
&= \frac{1}{\sqrt{2}}(\alpha|0\rangle_c|0\rangle_a|0\rangle_b + \alpha|0\rangle_c|1\rangle_a|1\rangle_b + \beta|1\rangle_c|0\rangle_a|0\rangle_b + \beta|1\rangle_c|1\rangle_a|1\rangle_b)
\end{aligned}
$$

to which she applies the unitary operators $cnot \otimes I$ and $H \otimes I \otimes I$ to obtain:

$$(H \otimes I \otimes I)(cnot \otimes I)\frac{1}{\sqrt{2}}(\alpha|0\rangle_c|0\rangle_a|0\rangle_b + \alpha|0\rangle_c|1\rangle_a|1\rangle_b + \beta|1\rangle_c|0\rangle_a|0\rangle_b + \beta|1\rangle_c|1\rangle_a|1\rangle_b)$$

$$
\begin{aligned}
&= (H \otimes I \otimes I)\frac{1}{\sqrt{2}}(\alpha|0\rangle_c|0\rangle_a|0\rangle_b + \alpha|0\rangle_c|1\rangle_a|1\rangle_b + \beta|1\rangle_c|1\rangle_a|0\rangle_b + \beta|1\rangle_c|0\rangle_a|1\rangle_b) \\
&= \frac{1}{\sqrt{2}}(\alpha H(|0\rangle_c)|0\rangle_a|0\rangle_b + \alpha H(|0\rangle_c)|1\rangle_a|1\rangle_b + \beta H(|1\rangle_c)|1\rangle_a|0\rangle_b + \beta H(|1\rangle_c)|0\rangle_a|1\rangle_b) \\
&= \frac{1}{\sqrt{2}}\left(\alpha H(|0\rangle_c)[|0\rangle_a|0\rangle_b + |1\rangle_a|1\rangle_b] + \beta H(|1\rangle_c)[|1\rangle_a|0\rangle_b + |0\rangle_a|1\rangle_b]\right) \\
&= \frac{1}{2}\left(\alpha(|0\rangle_c + |1\rangle_c)[|0\rangle_a|0\rangle_b + |1\rangle_a|1\rangle_b] + \beta(|0\rangle_c - |1\rangle_c)[|1\rangle_a|0\rangle_b + |0\rangle_a|1\rangle_b]\right) \\
&= \frac{1}{2}\big(\alpha|0\rangle_c|0\rangle_a|0\rangle_b + \alpha|0\rangle_c|1\rangle_a|1\rangle_b + \alpha|1\rangle_c|0\rangle_a|0\rangle_b + \alpha|1\rangle_c|1\rangle_a|1\rangle_b \\
&\qquad + \beta|0\rangle_c|1\rangle_a|0\rangle_b + \beta|0\rangle_c|0\rangle_a|1\rangle_b - \beta|1\rangle_c|1\rangle_a|0\rangle_b - \beta|1\rangle_c|0\rangle_a|1\rangle_b\big) \\
&= \frac{1}{2}\big(|0\rangle_c|0\rangle_a[\alpha|0\rangle_b + \beta|1\rangle_b] + |0\rangle_c|1\rangle_a[\alpha|1\rangle_b + \beta|0\rangle_b] \\
&\qquad + |1\rangle_c|0\rangle_a[\alpha|0\rangle_b - \beta|1\rangle_b] + |1\rangle_c|1\rangle_a[\alpha|1\rangle_b - \beta|0\rangle_b]\big).
\end{aligned}
$$

Alice now measures the observable $A \otimes I$ on the first qubit and the observable $I \otimes A$ on the second qubit of the transformed state. Here A has the matrix representation

$$A = \begin{pmatrix} 0 & 0 \\ 0 & 1 \end{pmatrix}.$$

As we have seen earlier A has eigenvector $|0\rangle$ with eigenvalue 0 and eigenvector $|1\rangle$ with eigenvalue 1.

Alice communicates, using a classical channel (e.g., text message, Morse code signal), the results of her measurements; i.e., the two classical bits of information $((0,0), (0,1), (1,0)$, or $(1,1))$ obtained from her measurements. It follows that if Alice's measurements of A result in:

- $(1,1)$, then the post-measurement state of Bob's particle is $\alpha|1\rangle_b - \beta|0\rangle_b$ and he applies the gate:

$$V_{11} = \begin{pmatrix} 0 & 1 \\ -1 & 0 \end{pmatrix}$$

  to transform his particle to the one Alice wished to transmit,

- $(1,0)$, then the post-measurement state of Bob's particle is $\alpha|0\rangle_b - \beta|1\rangle_b$ and he applies the gate:

$$V_{10} = \begin{pmatrix} 1 & 0 \\ 0 & -1 \end{pmatrix},$$

- $(0,1)$, then the post-measurement state of Bob's particle is $\alpha|1\rangle_b + \beta|0\rangle_b$ and he applies the gate:

$$V_{01} = \begin{pmatrix} 0 & 1 \\ 1 & 0 \end{pmatrix},$$

- $(0,0)$, then the post-measurement state of Bob's particle is $\alpha|0\rangle_b + \beta|1\rangle_b$ and no transformation of his particle is necessary – equivalently V_{00} is the 2×2 identity matrix.

Following the measurements Alice is left with one of the states $|11\rangle, |10\rangle, |01\rangle, |00\rangle$ as shown in Figure 20.4 – where vertical bars are used to denote that the state $|\phi\rangle$ is entangled.

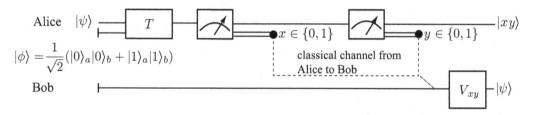

FIGURE 20.4
A quantum circuit to teleport the state $|\psi\rangle = \alpha|0\rangle + \beta|1\rangle$ from Alice to Bob, where $T = (H \otimes I \otimes I)(cnot \otimes I)$.

For each qubit teleported, Alice sends Bob two classical bits of information. If a hacker intercepts the two bits, they know what Bob needs to do to 'receive' his state. However, this information is useless as the hacker cannot interact with Bob's particle.

20.6   Quantum teleportation 2, Bell-basis measurement

Again Alice wishes to send the qubit state $|\psi\rangle = \alpha|0\rangle_c + \beta|1\rangle_c$ to Bob. As before the subscript c is used to distinguish it from states labelled a and b below, and we assume that Alice and Bob share the first of the following entangled states:

$$|B_1\rangle_{ab} = \frac{1}{\sqrt{2}}(|0\rangle_a|0\rangle_b + |1\rangle_a|1\rangle_b)$$

$$|B_2\rangle_{ab} = \frac{1}{\sqrt{2}}(|0\rangle_a|1\rangle_b + |1\rangle_a|0\rangle_b)$$

$$|B_3\rangle_{ab} = \frac{1}{\sqrt{2}}(|0\rangle_a|0\rangle_b - |1\rangle_a|1\rangle_b)$$

$$|B_4\rangle_{ab} = \frac{1}{\sqrt{2}}(|0\rangle_a|1\rangle_b - |1\rangle_a|0\rangle_b).$$

Alice has one of the particles of $|B_1\rangle_{ab}$ (labelled a), Bob has the other (labelled b). Hence at this point Alice has two particles, i.e., c (the one she wishes to teleport to Bob) and a (one of the entangled pair a, b). Bob has one particle, namely b. The state vector for the compound system is therefore:

$$|\psi\rangle \otimes |B_1\rangle_{ab} = (\alpha|0\rangle_c + \beta|1\rangle_c) \otimes \frac{1}{\sqrt{2}}(|0\rangle_a|0\rangle_b + |1\rangle_a|1\rangle_b)$$

and Alice will make a local measurement, in the Bell basis, on the two particles in her possession (i.e., c and a).

To assist the interpretation of Alice's measurements, we write her qubits as superpositions of the Bell states. We have:

$$|\psi\rangle \otimes |B_1\rangle_{ab} = (\alpha|0\rangle_c + \beta|1\rangle_c) \otimes \frac{1}{\sqrt{2}}(|0\rangle_a \otimes |0\rangle_b + |1\rangle_a \otimes |1\rangle_b)$$

$$= \frac{1}{\sqrt{2}}[\alpha(|0\rangle_c \otimes |0\rangle_a) \otimes |0\rangle_b + \alpha(|0\rangle_c \otimes |1\rangle_a) \otimes |1\rangle_b$$

$$+ \beta(|1\rangle_c \otimes |0\rangle_a) \otimes |0\rangle_b + \beta(|1\rangle_c \otimes |1\rangle_a) \otimes |1\rangle_b]$$

and the following identities can be substituted, with subscripts c and a, for the terms in emboldened braces

$$|0\rangle \otimes |0\rangle = \frac{1}{\sqrt{2}}(|B_1\rangle + |B_3\rangle), \quad |0\rangle \otimes |1\rangle = \frac{1}{\sqrt{2}}(|B_2\rangle + |B_4\rangle),$$

$$|1\rangle \otimes |0\rangle = \frac{1}{\sqrt{2}}(|B_2\rangle - |B_4\rangle), \quad |1\rangle \otimes |1\rangle = \frac{1}{\sqrt{2}}(|B_1\rangle - |B_3\rangle).$$

We obtain

$$|\psi\rangle \otimes |B_1\rangle_{ab} = \frac{1}{2}|B_1\rangle_{ca} \otimes (\alpha|0\rangle_b + \beta|1\rangle_b) + \frac{1}{2}|B_2\rangle_{ca} \otimes (\alpha|1\rangle_b + \beta|0\rangle_b)$$

$$+ \frac{1}{2}|B_3\rangle_{ca} \otimes (\alpha|0\rangle_b - \beta|1\rangle_b) + \frac{1}{2}|B_4\rangle_{ca} \otimes (\alpha|1\rangle_b - \beta|0\rangle_b).$$

We note that the above is just an alternative expression for $|\psi\rangle \otimes |B_1\rangle_{ab}$, achieved by a change-of-basis on Alice's part of the system. No operations have been performed and the three particles are still in the same total state. Note that Alice's particles, c and a, are now entangled in Bell states, and the entanglement initially shared by Alice's and Bob's particles a and b is broken.

Teleportation takes place when Alice measures the observables S_1^2 and S_3^2 on her two qubits, c and a, in the Bell basis

$$|B_1\rangle_{ca}, |B_2\rangle_{ca}, |B_3\rangle_{ca}, |B_4\rangle_{ca}.$$

This process is described in Section 19.5 and now applied to the first two qubits of $|\psi\rangle \otimes |B_1\rangle_{ab}$. The result is that $|\psi\rangle \otimes |B_1\rangle_{ab}$ collapses to one of the following four states:

- $|B_1\rangle_{ca} \otimes (\alpha|0\rangle_b + \beta|1\rangle_b)$ if Alice measures $(1,1)$, resulting in final state $|B_1\rangle$,

- $|B_3\rangle_{ca} \otimes (\alpha|0\rangle_b - \beta|1\rangle_b)$ if Alice measures $(1,0)$, resulting in final state $|B_3\rangle$,

- $|B_2\rangle_{ca} \otimes (\alpha|1\rangle_b + \beta|0\rangle_b)$ if Alice measures $(0,1)$, resulting in final state $|B_2\rangle$,

- $|B_4\rangle_{ca} \otimes (\alpha|1\rangle_b - \beta|0\rangle_b)$ if Alice measures $(0,0)$, resulting in final state $|B_4\rangle$.

In the above, each of the four possible states of Bob's qubit are related to the state Alice wishes to teleport by the following unitary operators:

$$\begin{pmatrix} 1 & 0 \\ 0 & 1 \end{pmatrix}, \begin{pmatrix} 1 & 0 \\ 0 & -1 \end{pmatrix}, \begin{pmatrix} 0 & 1 \\ 1 & 0 \end{pmatrix}, \begin{pmatrix} 0 & 1 \\ -1 & 0 \end{pmatrix}.$$

If Bob receives:

- $(1,1)$, he knows his qubit is precisely the qubit Alice wished to teleport and he need do nothing.

- $(1,0)$, he applies the gate

$$U_{10} = \begin{pmatrix} 1 & 0 \\ 0 & -1 \end{pmatrix}$$

to his qubit, which is then the state Alice wished to teleport.

- $(0,1)$, he applies the gate

$$U_{01} = \begin{pmatrix} 0 & 1 \\ 1 & 0 \end{pmatrix}$$

to his qubit.

- $(0,0)$, he applies the gate

$$U_{00} = \begin{pmatrix} 0 & 1 \\ -1 & 0 \end{pmatrix}$$

to his qubit.

Bob now has the state $|\psi\rangle$ Alice wished to teleport to him. Alice is left with one of $|B_1\rangle_{ca}, |B_2\rangle_{ca}, |B_3\rangle_{ca}, |B_4\rangle_{ca}$ from which the state $|\psi\rangle$ cannot be reconstructed.

A circuit for the teleportation process is shown in Figure 20.5. The B in the measurement graphic indicates measurement of the first two qubits of the input state

$$|\psi\rangle \otimes \frac{1}{\sqrt{2}}(|0\rangle_a|0\rangle_b + |1\rangle_a|1\rangle_b)$$

in the Bell basis. We note, in Figure 20.5, that the subscript $q = 1$ if $xy = 11$, $q = 3$ if $xy = 10$, $q = 2$ if $xy = 01$ and $q = 4$ if $xy = 00$.

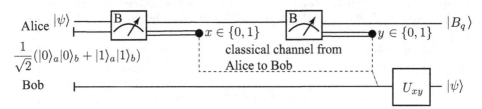

FIGURE 20.5

A quantum circuit to teleport the state $|\psi\rangle = \alpha|0\rangle + \beta|1\rangle$ from Alice to Bob.

21

More on quantum gates and circuits: those without digital equivalents

21.1 Objectives

To this point only quantum gates derived from invertible digital counterparts or general invertible Boolean functions have been considered. Here we introduce a number of quantum gates without digital equivalents. In particular, we consider the Pauli gates, the $\sqrt{not}$ gate, the phase-shift gate and the Hadamard gate, some of which occur in quantum algorithms introduced later. These gates impart additional power to quantum computation.

21.2 General 1-qubit quantum gates

Definition 21.1 1-qubit gate
A 1-qubit gate is a transformation $U : \mathbb{C}^2 \to \mathbb{C}^2$ for $U \in U(2)$, where $U(2)$ is the unitary group on $\mathbb{C}^2$.

It follows from the definition that all 1-qubit gates are represented by matrices of the form

$$U = \begin{pmatrix} u_{00} & u_{01} \\ u_{10} & u_{11} \end{pmatrix}$$

where $u_{ij} \in \mathbb{C}$ and $UU^\dagger = I_2$, the 2×2 identity matrix. U maps $|0\rangle$ which is represented by

$$\begin{pmatrix} 1 \\ 0 \end{pmatrix}$$

to $u_{00}|0\rangle + u_{10}|1\rangle$. Similarly $|1\rangle$, which is represented by

$$\begin{pmatrix} 0 \\ 1 \end{pmatrix}$$

is mapped by U to $u_{01}|0\rangle + u_{11}|1\rangle$. The general 1-qubit $|\psi\rangle = c_0|0\rangle + c_1|1\rangle$ therefore transforms, under U, to:

$$\begin{aligned} U|\psi\rangle &= c_0(u_{00}|0\rangle + u_{10}|1\rangle) + c_1(u_{01}|0\rangle + u_{11}|1\rangle) \\ &= (c_0 u_{00} + c_1 u_{01})|0\rangle + (c_0 u_{10} + c_1 u_{11})|1\rangle. \end{aligned}$$

The specific 1-qubit gates, discussed below, are special cases and their actions on qubits follow from this general expression. The matrix representations, as linear combinations of the Pauli matrices and the identity matrix, follow from the general expressions discussed earlier.

DOI: 10.1201/9781003264569-21

Following the digital case, quantum gates may be represented graphically. A 1-qubit gate is depicted as shown in Figure 21.1, where $x \in \{0, 1\}$ and the time-axis runs horizontally.

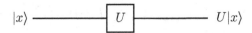

$$|x\rangle \underline{\hspace{2cm}} \boxed{U} \underline{\hspace{2cm}} U|x\rangle$$

FIGURE 21.1
A generic 1-qubit quantum gate.

21.3   The Pauli and the $\sqrt{not}$ 1-qubit gates

1. The Pauli X gate, or the quantum *not* gate:

   We have met this gate earlier. The Pauli gate has a parallel in the digital case and is included again here for two reasons:

   (a) for completeness with the other two Pauli gates,
   (b) the square root, written $\sqrt{not}$, of the *not* gate exists as a useful quantum gate.

   The other two Pauli gates and the quantum $\sqrt{not}$ have no parallels in digital computation. The matrix of the *not* is the first Pauli matrix σ_1, sometimes also denoted X, i.e.,

   $$\begin{pmatrix} 0 & 1 \\ 1 & 0 \end{pmatrix} = \sigma_1.$$

   The gate takes $|0\rangle$ to $|1\rangle$ and $|1\rangle$ to $|0\rangle$.

2. The Pauli Y gate:

   $$\begin{pmatrix} 0 & -i \\ i & 0 \end{pmatrix} = \sigma_2.$$

   The gate takes $|0\rangle$ to $i|1\rangle$ and $|1\rangle$ to $-i|0\rangle$.

3. The Pauli Z gate:

   $$\begin{pmatrix} 1 & 0 \\ 0 & -1 \end{pmatrix} = \sigma_3.$$

   The gate takes $|0\rangle$ to $|0\rangle$ and $|1\rangle$ to $-|1\rangle$.

4. The $\sqrt{not}$ gate: as a further divergence from digital computation we note that

   $$[\frac{1}{2}(1+i)(I_2 - i\sigma_1)]^2 = \sigma_1,$$

   and hence the quantum *not* gate has a square root, which is unitary and hence a quantum gate. That is, it is an element of the group $U(2)$, denoted by $\sqrt{not}$. From the above, the matrix of $\sqrt{not}$ is:

   $$\frac{1}{2}(1+i)(I_2 - i\sigma_1) = \frac{1}{2}(1+i)\begin{pmatrix} 1 & -i \\ -i & 1 \end{pmatrix}.$$

   The quantum $\sqrt{not}$ gate is depicted as shown in Figure 21.2.

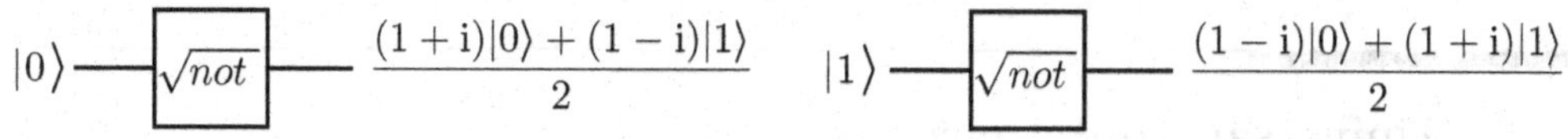

$$|0\rangle - \boxed{\sqrt{not}} - \frac{(1+i)|0\rangle + (1-i)|1\rangle}{2} \qquad |1\rangle - \boxed{\sqrt{not}} - \frac{(1-i)|0\rangle + (1+i)|1\rangle}{2}$$

FIGURE 21.2
The quantum $\sqrt{not}$ gate.

21.4 Further 1-qubit gates: phase-shift and Hadamard

1. The phase-shift gate: a phase-shift gate fixes $|0\rangle$ but shifts the phase of $|1\rangle$ by $e^{i\phi}$ and may be represented by the matrix

$$R_\phi = \begin{pmatrix} 1 & 0 \\ 0 & e^{i\phi} \end{pmatrix}.$$

We note that R_ϕ may be expressed in terms of the Pauli matrices:

$$\begin{pmatrix} 1 & 0 \\ 0 & e^{i\phi} \end{pmatrix} = e^{i\frac{1}{2}\phi} \begin{pmatrix} e^{-i\frac{1}{2}\phi} & 0 \\ 0 & e^{i\frac{1}{2}\phi} \end{pmatrix} = \frac{1}{2}(1 + e^{i\phi})I_2 + \frac{1}{2}(1 - e^{i\phi})\sigma_3.$$

and is also readily implemented physically. The gate is represented diagrammatically in Figure 21.3.

$$|0\rangle - \boxed{R_\phi} - |0\rangle \qquad |1\rangle - \boxed{R_\phi} - e^{i\phi}|1\rangle$$

FIGURE 21.3
The phase-shift gate.

2. Another useful, and easily implemented, gate is the Hadamard gate with matrix

$$H = \frac{1}{\sqrt{2}} \begin{pmatrix} 1 & 1 \\ 1 & -1 \end{pmatrix}.$$

H is the 1-qubit Fourier transform gate (Figure 21.4). We note that this can be expressed in terms of the Pauli matrices as

$$H = \frac{1}{\sqrt{2}} \begin{pmatrix} 1 & 1 \\ 1 & -1 \end{pmatrix} = \frac{1}{\sqrt{2}}(\sigma_1 + \sigma_3).$$

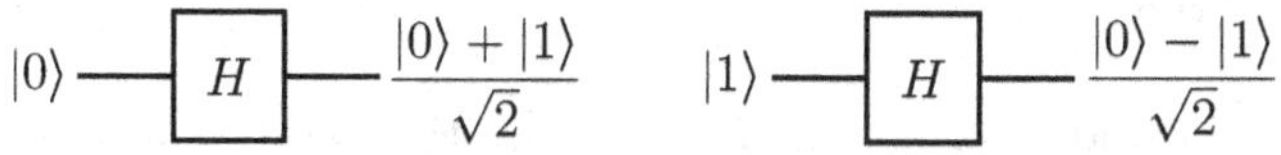

$$|0\rangle - \boxed{H} - \frac{|0\rangle + |1\rangle}{\sqrt{2}} \qquad |1\rangle - \boxed{H} - \frac{|0\rangle - |1\rangle}{\sqrt{2}}$$

FIGURE 21.4
The Hadamard gate.

21.5 Universal 1-qubit circuits

A set of 1-qubit gates is said to be universal if they generate all the unitary operators of $\mathbb{C}^2$. Equivalently, the set is universal if an arbitrary 1-qubit state $|\psi\rangle$ may be generated from the set, starting from $|0\rangle$. That is, for any $|\psi\rangle \in \mathbb{C}^2$ there is a circuit, C_ψ, of gates from the set which can be used to generate it, as shown in Figure 21.5.

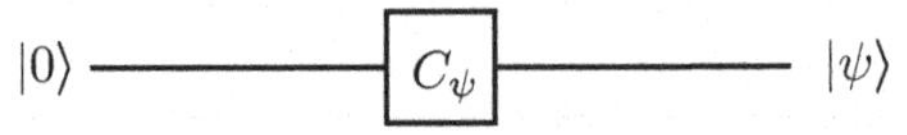

$$|0\rangle \longrightarrow \boxed{C_\psi} \longrightarrow |\psi\rangle$$

FIGURE 21.5
A universal 1-qubit gate.

Lemma 21.1 *The Hadamard and phase-shift gates are universal as shown in Figure 21.6, where*

$$|\psi\rangle = \cos\theta|0\rangle + e^{i\phi}\sin\theta|1\rangle$$

up to a global phase factor.

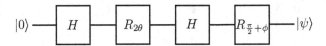

FIGURE 21.6
The universality of H and R_ϕ.

Proof
It is easy to show that the circuit, which is equivalent to the product, $R_{\frac{\pi}{2}+\phi}HR_{2\theta}H$ of unitary matrices, generates

$$\frac{1}{2}(1 + e^{2\theta i})|0\rangle + \frac{1}{2}ie^{i\phi}(1 - e^{2\theta i})|1\rangle$$

from $|0\rangle$. We have $e^{2\theta i} = e^{\theta i}e^{\theta i} = (\cos\theta + i\sin\theta)^2$ from which we obtain

$$1 + e^{2\theta i} = 2(\cos^2\theta + i\sin\theta\cos\theta),$$
$$1 - e^{2\theta i} = 2(\sin^2\theta - i\sin\theta\cos\theta).$$

A little algebra then gives

$$(1 + e^{2\theta i})|0\rangle + ie^{i\phi}(1 - e^{2\theta i})|1\rangle = e^{i\theta}(\cos\theta|0\rangle + e^{i\phi}\sin\theta|1\rangle)$$

and the Lemma is true.

∎

Equivalently, the vector $|0\rangle$ may be moved to any position on the surface of the Bloch sphere (see Chapter 13) using only Hadamard and phase-shift gates.

21.6 Some 2-qubit quantum gates

We note that the complete set of 1-qubit gates (discussed above), acting on both qubits, augmented by the quantum *cnot* gate on $\mathbb{C}^2 \otimes \mathbb{C}^2$ determine a universal set for 2-qubit quantum gates. However, the proof of this is beyond the scope of this text; instead we concentrate on 2-qubit gates that occur in the quantum algorithms we present later.

21.6.1 Tensor product gates

A 1-qubit gate $U \in U(2)$ defines the 2-qubit gate $U \otimes U : \mathbb{C}^2 \otimes \mathbb{C}^2 \to \mathbb{C}^2 \otimes \mathbb{C}^2$ – see Figure 21.7. It follows that if A_U is a matrix representation of U then $A_U \otimes A_U$ is a matrix representation of $U \otimes U$.

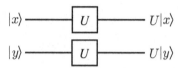

$$
|x\rangle \quad\longrightarrow\quad not_Q \quad\longrightarrow\quad |\bar{x}\rangle = |1 \oplus x\rangle
$$

e.g. for $U = not_Q$

$$
|y\rangle \quad\longrightarrow\quad not_Q \quad\longrightarrow\quad |\bar{y}\rangle = |1 \oplus y\rangle
$$

FIGURE 21.7
2-qubit gates $U \otimes U$ and $not_Q \otimes not_Q$.

More generally if U and V are distinct 1-qubit gates then gates $U \otimes V$ and $V \otimes U$ determine 2-qubit gates. However, not all 2-qubit gates may be expressed as tensor products of 1-qubit gates. An example of such is the *cnot* gate (see Exercises).

21.6.2 The Hadamard-cnot circuit

The Hadamard-cnot circuit is the product of the quantum $cnot_Q$ gate, written here as *cnot*, and the tensor product gate $H \otimes I$. H is the Hadamard gate discussed above:

$$
H|0\rangle = \frac{1}{\sqrt{2}}(|0\rangle + |1\rangle) \qquad \text{and} \qquad H|1\rangle = \frac{1}{\sqrt{2}}(|0\rangle - |1\rangle),
$$

and $cnot : \mathbb{C}^2 \otimes \mathbb{C}^2 \to \mathbb{C}^2 \otimes \mathbb{C}^2$ is the 2-qubit operator defined by:

$$
\begin{aligned}
cnot(|0\rangle|0\rangle) &= |0\rangle|0\rangle \\
cnot(|0\rangle|1\rangle) &= |0\rangle|1\rangle \\
cnot(|1\rangle|0\rangle) &= |1\rangle|1\rangle \\
cnot(|1\rangle|1\rangle) &= |1\rangle|0\rangle.
\end{aligned}
$$

We consider its algebraic, graphical and matrix representations below.

1. **Algebraic representation:** the circuit as a product of unitary matrices (quantum gates) on the 2-qubit tensor product space $\mathbb{C}^2 \otimes \mathbb{C}^2$. We have:

$$
cnot(H \otimes I) : \mathbb{C}^2 \otimes \mathbb{C}^2 \to \mathbb{C}^2 \otimes \mathbb{C}^2.
$$

H and *cnot* are linear on their respective domains, and I denotes the identity operator on $\mathbb{C}^2$. Hence, for example:

$$
\begin{aligned}
cnot(H \otimes I)(|0\rangle|0\rangle) &= cnot(\frac{1}{\sqrt{2}}(|0\rangle + |1\rangle) \otimes |0\rangle) \\
&= \frac{1}{\sqrt{2}}(cnot(|0\rangle \otimes |0\rangle) + cnot(|1\rangle \otimes |0\rangle)) \\
&= \frac{1}{\sqrt{2}}(|0\rangle|0\rangle + |1\rangle|1\rangle).
\end{aligned}
$$

We recognise the state $\frac{1}{\sqrt{2}}(|0\rangle|0\rangle + |1\rangle|1\rangle) \in \mathbb{C}^2 \otimes \mathbb{C}^2$ as the entangled EPR-Bell state.

In general, we have $H|x\rangle = \frac{1}{\sqrt{2}}(|0\rangle + (-1)^x|1\rangle)$ and:

$$
\begin{aligned}
cnot(H \otimes I)(|x\rangle|y\rangle) &= cnot(H(|x\rangle) \otimes |y\rangle) \\
&= cnot(\frac{1}{\sqrt{2}}(|0\rangle + (-1)^x|1\rangle) \otimes |y\rangle) \\
&= \frac{1}{\sqrt{2}}cnot(|0\rangle|y\rangle + (-1)^x|1\rangle|y\rangle) \\
&= \frac{1}{\sqrt{2}}(cnot|0\rangle|y\rangle + (-1)^x cnot|1\rangle|y\rangle) \\
&= \frac{1}{\sqrt{2}}(|0\rangle|y\rangle + (-1)^x|1\rangle|\overline{y}\rangle).
\end{aligned}
$$

From the above we obtain:

$$
\begin{aligned}
cnot(H \otimes I)(|0\rangle|0\rangle) &= \frac{1}{\sqrt{2}}(|0\rangle|0\rangle + |1\rangle|1\rangle) \\
cnot(H \otimes I)(|0\rangle|1\rangle) &= \frac{1}{\sqrt{2}}(|0\rangle|1\rangle + |1\rangle|0\rangle) \\
cnot(H \otimes I)(|1\rangle|0\rangle) &= \frac{1}{\sqrt{2}}(|0\rangle|0\rangle - |1\rangle|1\rangle) \\
cnot(H \otimes I)(|1\rangle|1\rangle) &= \frac{1}{\sqrt{2}}(|0\rangle|1\rangle - |1\rangle|0\rangle).
\end{aligned}
$$

2. **Graphical representation:** for $|x\rangle \in \mathbb{C}^2$ and $|y\rangle \in \mathbb{C}^2$ consider Figure 21.8 where, as before, for the orthonormal basis $\{|0\rangle, |1\rangle\}$ of $\mathbb{C}^2$, $H : \mathbb{C}^2 \to \mathbb{C}^2$ is the 1-qubit Hadamard operator

$$
H|0\rangle = \frac{1}{\sqrt{2}}(|0\rangle + |1\rangle) \qquad \text{and} \qquad H|1\rangle = \frac{1}{\sqrt{2}}(|0\rangle - |1\rangle),
$$

and $cnot : \mathbb{C}^2 \otimes \mathbb{C}^2 \to \mathbb{C}^2 \otimes \mathbb{C}^2$ is the 2-qubit operator:

$$
\begin{aligned}
cnot(|0\rangle|0\rangle) &= |0\rangle|0\rangle \\
cnot(|0\rangle|1\rangle) &= |0\rangle|1\rangle \\
cnot(|1\rangle|0\rangle) &= |1\rangle|1\rangle \\
cnot(|1\rangle|1\rangle) &= |1\rangle|0\rangle.
\end{aligned}
$$

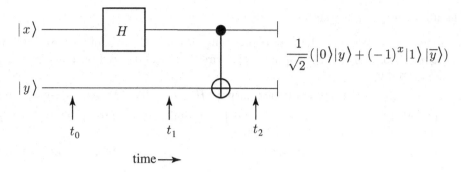

FIGURE 21.8
The Hadamard-cnot circuit for $x, y \in \mathbb{B}$.

For $x = y = 0$ we have:

$$|0\rangle|0\rangle \text{ at time } t_0, \quad \text{(unentangled input tensor)}$$

$$\frac{1}{\sqrt{2}}(|0\rangle + |1\rangle)|0\rangle \text{ at time } t_1, \quad \text{(unentangled tensor)}$$

$$\frac{1}{\sqrt{2}}(|0\rangle|0\rangle + |1\rangle|1\rangle) \text{ at time } t_2 \text{ (entangled output tensor)}.$$

Note that the output, in this case, is an entangled tensor. On the basis elements $\{|0\rangle|0\rangle, |0\rangle|1\rangle, |1\rangle|0\rangle, |1\rangle|1\rangle\}$ of $\mathbb{C}^2 \otimes \mathbb{C}^2$ the outputs of the gate determine an orthonormal entangled set of basis states of $\mathbb{C}^2 \otimes \mathbb{C}^2$ – known as the EPR or Bell states (see 1. Algebraic representation, above). These states are very significant in quantum mechanics and quantum computation. The vertical lines on the output 'wires' indicate that the outputs are entangled and the output tensor is placed between the two output 'wires' – as shown in Figure 21.8.

3. **Matrix representation:** the matrices of $H \otimes I$ and *cnot*, in the basis $\{|0\rangle|0\rangle, |0\rangle|1\rangle, |1\rangle|0\rangle, |1\rangle|1\rangle\}$ of $\mathbb{C}^2 \otimes \mathbb{C}^2$ are respectively:

$$\frac{1}{\sqrt{2}} \begin{pmatrix} 1 & 0 & 1 & 0 \\ 0 & 1 & 0 & 1 \\ 1 & 0 & -1 & 0 \\ 0 & 1 & 0 & -1 \end{pmatrix}$$

and

$$\begin{pmatrix} 1 & 0 & 0 & 0 \\ 0 & 1 & 0 & 0 \\ 0 & 0 & 0 & 1 \\ 0 & 0 & 1 & 0 \end{pmatrix}.$$

The matrix of $cnot(H \otimes I)$ is therefore the product:

$$\frac{1}{\sqrt{2}} \begin{pmatrix} 1 & 0 & 1 & 0 \\ 0 & 1 & 0 & 1 \\ 0 & 1 & 0 & -1 \\ 1 & 0 & -1 & 0 \end{pmatrix}.$$

From this we obtain:

$$cnot(H \otimes I)|0\rangle|0\rangle = \frac{1}{\sqrt{2}} \begin{pmatrix} 1 & 0 & 1 & 0 \\ 0 & 1 & 0 & 1 \\ 0 & 1 & 0 & -1 \\ 1 & 0 & -1 & 0 \end{pmatrix} \begin{pmatrix} 1 \\ 0 \\ 0 \\ 0 \end{pmatrix} = \begin{pmatrix} 1 \\ 0 \\ 0 \\ 1 \end{pmatrix} = \frac{1}{\sqrt{2}}(|0\rangle|0\rangle + |1\rangle|1\rangle)$$

etc.

The input/output possibilities for gates are:

1. unentangled input and unentangled output (e.g., the Toffoli, Peres gates on $|x\rangle|y\rangle|z\rangle$ for $x, y, z \in \mathbb{B}$). For such, the wired input and wired output representations are fine,

2. unentangled input and entangled output ($H \otimes I$ followed by $cnot_Q$, see above),

3. entangled input and entangled output (the unitary operator $U|x\rangle|y\rangle = |x\rangle|\overline{y}\rangle$ maps $|0\rangle|0\rangle + |1\rangle|1\rangle$ to $|0\rangle|1\rangle + |1\rangle|0\rangle$),

4. entangled input and unentangled output (the inverse of (2) i.e., $cnot_Q$ followed by $H \otimes I$).

21.7 The Hadamard gate on n-qubit registers

The Hadamard gate features in most known quantum algorithms, we therefore re-visit it in its most general form. The Hadamard gate $H : \mathbb{C}^2 \to \mathbb{C}^2$, on $\mathbb{C}^2$, may be expressed as

$$H|x\rangle = \frac{1}{\sqrt{2}} \sum_{y \in \{0,1\}} (-1)^{xy}|y\rangle$$

where $|x\rangle \in \mathbb{C}^2$, $|y\rangle \in \mathbb{C}^2$, and $xy = x \cdot y$. We note that for $x = 0$ we obtain

$$H|0\rangle = \frac{1}{\sqrt{2}}(|0\rangle + |1\rangle),$$

and for $x = 1$ we have

$$H|1\rangle = \frac{1}{\sqrt{2}}(|0\rangle - |1\rangle)$$

as required.

More generally, we can define $H^{\otimes n} : \bigotimes^n \mathbb{C}^2 \to \bigotimes^n \mathbb{C}^2$.

Lemma 21.2 *For* $|x\rangle = |x_1 x_2 \dots x_n\rangle \in \otimes^n \mathbb{C}^2$ *we have*

$$H^{\otimes n}|x\rangle = \bigotimes_{i=1}^{n} \left(\frac{1}{\sqrt{2}} \sum_{y_i=0}^{1} (-1)^{x_i y_i}|y_i\rangle \right).$$

Proof

$$
\begin{aligned}
H^{\otimes n}|x\rangle &= H|x_1\rangle H|x_2\rangle \cdots H|x_n\rangle, \quad \text{(by definition)} \\
&= \frac{1}{\sqrt{2}}(|0\rangle + (-1)^{x_1}|1\rangle)\frac{1}{\sqrt{2}}(|0\rangle + (-1)^{x_2}|1\rangle)\dots\frac{1}{\sqrt{2}}(|0\rangle + (-1)^{x_n}|1\rangle) \\
&= \bigotimes_{i=1}^{n} \frac{1}{\sqrt{2}}(|0\rangle + (-1)^{x_i}|1\rangle) \\
&= \bigotimes_{i=1}^{n} \left(\frac{1}{\sqrt{2}} \sum_{y_i=0}^{1} (-1)^{x_i y_i}|y_i\rangle \right).
\end{aligned}
$$

∎

We also have:

Lemma 21.3

$$H^{\otimes n}|x\rangle = \frac{1}{2^{\frac{n}{2}}} \sum_{y=0}^{2^n-1} (-1)^{x \cdot y}|y\rangle.$$

Proof

By induction: for $n = 1$ the right hand side of the above is

$$H|x\rangle = \frac{1}{\sqrt{2}} \sum_{y \in \{0,1\}} (-1)^{xy}|y\rangle$$

which is correct for $x \in \mathbb{B}$. Assuming the relation is true for n, we demonstrate that the truth for $n + 1$ is implied and the expression is therefore generally true. By definition we have

$$H^{\otimes(n+1)} = H^{\otimes n} \otimes H.$$

For $|x\rangle \in \otimes^n \mathbb{C}^2$ and $|z\rangle \in \mathbb{C}^2$ we have $|x\rangle|z\rangle \in \otimes^{n+1}\mathbb{C}^2$ and $H^{\otimes(n+1)}|x\rangle|z\rangle = H^{\otimes n}|x\rangle \otimes H|z\rangle$.

For $z = 0$ we obtain:

$$H^{\otimes(n+1)}|x\rangle|0\rangle = H^{\otimes n}|x\rangle \otimes H|0\rangle$$

$$= \frac{1}{2^{\frac{n}{2}}} \sum_{y=0}^{2^n-1} (-1)^{x \cdot y}|y\rangle \frac{1}{\sqrt{2}}(|0\rangle + |1\rangle)$$

$$= \frac{1}{2^{\frac{n+1}{2}}} \left(\sum_{y=0}^{2^n-1} (-1)^{x \cdot y}|y\rangle(|0\rangle + |1\rangle) \right)$$

$$= \frac{1}{2^{\frac{n+1}{2}}} \left(\sum_{y=0}^{2^n-1} (-1)^{x \cdot y}|y0\rangle + |y1\rangle \right)$$

$$= \frac{1}{2^{\frac{n+1}{2}}} \left(\sum_{q=0}^{2^{n+1}-1} (-1)^{x0 \cdot q}|q\rangle \right) \quad \text{where } q = yp \text{ for } p \in \mathbb{B}.$$

Similarly for $z = 1$, and the Lemma is true.

∎

Exercises

21.1 Show that the quantum *cnot* gate cannot be written as a tensor product of two 1-qubit gates.

21.2 Show that

$$H^{\otimes(n+1)}(\otimes^n|0\rangle\,|1\rangle) = \frac{1}{\sqrt{2^{n+1}}} \sum_{x \in \{0,1\}^n} |x\rangle(|0\rangle - |1\rangle).$$

22

Quantum algorithms 1

22.1 Objectives

The quantum algorithms covered in this chapter, i.e., the Deutsch, Deutsch-Jozsa and Bernstein-Vazirani algorithms, have been chosen for their relative simplicity and theoretical interest. They do not address problems of real practical value, their purpose being to demonstrate, using simple examples, the great potential power of quantum computation relative to the digital case. The most significant breakthrough in quantum computing is Shor's integer factorisation algorithm. Shor's algorithm is beyond the scope of this introductory text, but its significance lies in the fact that it demonstrates the potential to factor large integers efficiently. No efficient digital algorithms for factoring large integers exist, and many current encryption algorithms rely on this being the case. Shor's algorithm therefore threatens current encryption methods, and is responsible for much of the recent accelerated interest in quantum computation.

22.2 Preliminary lemmas

Lemma 22.1 *If $f : \mathbb{B} \to \mathbb{B}$, then in $\mathbb{C}^2$ we have:*

$$|f(x)\rangle - |1 \oplus f(x)\rangle = (-1)^{f(x)}(|0\rangle - |1\rangle).$$

Proof
As we have seen (Section 2.4) there are four functions of the type $f : \mathbb{B} \to \mathbb{B}$; two are balanced and two are constant.

(i) For f balanced.

(a) if $f(x) = x$ we have

$$
\begin{aligned}
|f(x)\rangle - |1 \oplus f(x)\rangle &= |x\rangle - |1 \oplus x\rangle \\
&= \begin{cases} |0\rangle - |1\rangle & \text{for } x = 0 = f(0) \\ |1\rangle - |0\rangle & \text{for } x = 1 = f(1) \end{cases} \\
&= (-1)^{f(x)}(|0\rangle - |1\rangle).
\end{aligned}
$$

DOI: 10.1201/9781003264569-22

(b) if $f(x) = \bar{x}$ we have

$$
\begin{aligned}
|f(x)\rangle - |1 \oplus f(x)\rangle &= |\bar{x}\rangle - |1 \oplus \bar{x}\rangle \\
&= |\bar{x}\rangle - |x\rangle \\
&= \begin{cases} |1\rangle - |0\rangle & \text{for } x = 0, f(0) = 1 \\ |0\rangle - |1\rangle & \text{for } x = 1, f(1) = 0 \end{cases} \\
&= (-1)^{f(x)}(|0\rangle - |1\rangle).
\end{aligned}
$$

(ii) For f constant.

(a) if $f(x) = 1$ we have

$$
\begin{aligned}
|f(x)\rangle - |1 \oplus f(x)\rangle &= |1\rangle - |0\rangle \\
&= (-1)^{f(x)}(|0\rangle - |1\rangle).
\end{aligned}
$$

(b) if $f(x) = 0$ we have

$$
\begin{aligned}
|f(x)\rangle - |1 \oplus f(x)\rangle &= |0\rangle - |1\rangle \\
&= (-1)^{f(x)}(|0\rangle - |1\rangle)
\end{aligned}
$$

hence the result.

■

Lemma 22.2 *If $p \in \mathbb{B}^n$ and $x \in \mathbb{B}^n$ then for $p \neq 0^n$ we have*

$$
\sum_{x=0}^{2^n-1} p \cdot x = 2^{n-1}.
$$

Proof
Clearly either $p \cdot x = 0$ or $p \cdot x = 1$. The Lemma states that for 2^{n-1} values of x we have $p \cdot x = 1$ and for the other 2^{n-1} values of x we have $p \cdot x = 0$.

Consider the case $p = 100 \cdots 0 \in \mathbb{B}^n$: then for all $x \in \mathbb{B}^n$ of the form $x = 0x'$, where $x' \in \mathbb{B}^{n-1}$ is arbitrary, we have $p \cdot x = 0$. There are 2^{n-1} strings of the form $0x'$ in $\mathbb{B}^n$. Strings x not of the form $x = 0x'$, are such that $x = 1x'$, with x' arbitrary. For these we have $p \cdot x = 1$. A similar argument holds for all other strings, $p \in \mathbb{B}^n$, of Hamming weight 1. Very similar arguments hold for strings $p \in \mathbb{B}^n$ of Hamming weights $2, 3, \ldots, n$. The Lemma is therefore true.

■

Corollary
For $p \in \mathbb{B}^n$ and $x \in \mathbb{B}^n$ we have

$$
\sum_{x=0}^{2^n-1} (-1)^{p \cdot x} = 2^n \quad \text{for } p = 0^n
$$

and

$$
\sum_{x=0}^{2^n-1} (-1)^{p \cdot x} = 0 \quad \text{for } p \neq 0^n.
$$

Proof

For $p = 0^n$ each term of the sum is 1. For $p \neq 0$, the Lemma shows that half the terms are 1 and half are -1 – the sum is therefore zero.

∎

22.3 Diagrammatic representation of the measurement process

In the graphical representations of quantum circuits and algorithms the measurement process, which we have in algebraic expressions denoted μ_Q, is usually represented by a meter – as shown in Figures 22.1 and 22.2. The real output channel producing the output $\lambda \in \mathbb{R}$ is indicated in Figure 22.2 by the pair of parallel lines.

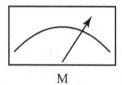

M

FIGURE 22.1
Meter representation of quantum measurement.

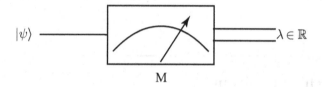

M

FIGURE 22.2
Measurement of an observable on the state $|\psi\rangle$ producing the output $\lambda \in \mathbb{R}$.

22.4 An introduction to computational complexity

Computational complexity (or complexity theory) is, with computability, one of the great achievements of the study of automated computation. Its roots date back to the Turing era, and it currently intertwines classical and quantum computation in a fascinating way. Complexity theory is a vast topic, and we consider only the minimum required to assess the potential performance improvements offered by the quantum algorithms we discuss later.

Complexity theory is concerned with estimating, or bounding, the resources required to compute solutions to particular problems. Both space (e.g., memory) and execution-time are resources that may be considered. Here we are only concerned with execution-time bounds – which are often a function of the problem 'size'. For example, consider the problem of sorting a set of n integers into ascending order. One would expect the time to sort a set of 1000 integers to be greater than that for sorting 10 integers. In this case the problem size is n, the number of integers to be sorted, and we denote the execution time by $t(n)$. That is, we have a function $t : \mathbb{N} \to \mathbb{R}$ which we refer to as the time-complexity function of the computation. In general t is an increasing function of n.

The simplest sorting algorithms are such that t is quadratic, i.e., t takes the form $t(n) = k_1 n^2$ for some constant k_1 and the execution time grows quadratically with respect to problem size n. The most efficient (known) digital sorting algorithms have execution times proportional to $n \log(n)$ i.e., $t(n) = k_2 n \log(n)$ for some constant k_2. The constants are, generally, not known and will depend on the implementation of the algorithm and on the hardware used. This being so it is usual to say that an algorithm is $O(n^2)$ if $t(n) = k_1 n^2$, and $O(n \log(n))$ if $t(n) = k_2 n \log(n)$ etc., when discussing the execution-time complexity. The notation being used here is referred to as big-O notation. When the execution time is independent of n, i.e., t is constant, we write $O(1)$ for its complexity class; $O(2^n)$ problems are said to be 'hard'.

22.5   Oracle-based complexity estimation

The algorithms we discuss in this chapter use oracles. We recall that oracles are black-boxes, or virtual devices, that respond immediately to queries – irrespective of the computation that may be necessary to 'answer' the query. Essentially, we assume that both quantum and classical oracles have this abstract property and that it is therefore acceptable to compare their respective efficiencies (i.e., complexities), by counting the number of oracle queries required, in both the classical and quantum algorithms, to solve a particular problem.

22.6   Deutsch's algorithm

Deutsch's algorithm determines whether a classical function $f : \mathbb{B} \to \mathbb{B}$ is constant or balanced. The possibilities for f are (i) $f(0) = f(1) = 0$, (ii) $f(0) = f(1) = 1$, i.e., f is constant, or (iii) $f(0) = 1, f(1) = 0$ or (iv) $f(0) = 0, f(1) = 1$ (i.e., f is balanced).

We note that if $f(0) \oplus f(1) = 0$ then f is constant and if $f(0) \oplus f(1) = 1$ then f is balanced (see Example 2.4.2).

We begin with the two-qubit state $|0\rangle \otimes |1\rangle$ and apply the tensor product Hadamard transform $H \otimes H$ to obtain:

$$\frac{1}{2}(|0\rangle + |1\rangle)(|0\rangle - |1\rangle).$$

We assume we have a quantisation of the function f; i.e., an implementation of the unitary operator $U_f |x\rangle |y\rangle = |x\rangle |y \oplus f(x)\rangle$ which delivers values for $f(x)$ when queried.

We apply U_f to $\frac{1}{2}(|0\rangle + |1\rangle)(|0\rangle - |1\rangle)$ to obtain:

$$U_f(\frac{1}{2}(|0\rangle + |1\rangle)(|0\rangle - |1\rangle)) = \frac{1}{2}\left[|0\rangle \left(|0 \oplus f(0)\rangle - |1 \oplus f(0)\rangle\right) + |1\rangle \left(|0 \oplus f(1)\rangle - |1 \oplus f(1)\rangle\right)\right]$$

$$= \frac{1}{2}\left[|0\rangle \left(|f(0)\rangle - |1 \oplus f(0)\rangle\right) + |1\rangle (|f(1)\rangle - |1 \oplus f(1)\rangle)\right].$$

Then, from Lemma 22.1, we have

$$U_f(\frac{1}{2}(|0\rangle + |1\rangle)(|0\rangle - |1\rangle)) = \frac{1}{2}[|0\rangle(-1)^{f(0)}(|0\rangle - |1\rangle) + |1\rangle(-1)^{f(1)}(|0\rangle - |1\rangle)]$$

$$= \frac{1}{2}[|0\rangle(-1)^{f(0)} + |1\rangle(-1)^{f(1)}] (|0\rangle - |1\rangle)$$

$$= (-1)^{f(0)} \frac{1}{2}[|0\rangle + (-1)^{f(0)\oplus f(1)}|1\rangle](|0\rangle - |1\rangle)$$

$$= (-1)^{f(0)} \frac{1}{\sqrt{2}}[|0\rangle + (-1)^{f(0)\oplus f(1)}|1\rangle] \frac{1}{\sqrt{2}}(|0\rangle - |1\rangle)$$

$$= \frac{1}{\sqrt{2}}[|0\rangle + (-1)^{f(0)\oplus f(1)}|1\rangle] \frac{1}{\sqrt{2}}(|0\rangle - |1\rangle).$$

(Note that global phase factors $((-1)^{f(0)}$ in this case) can be ignored in quantum mechanics).

Applying $H \otimes I$, where H is the Hadamard operator, transforms the state to:

$$\frac{1}{2}[|0\rangle + |1\rangle + (-1)^{f(0)\oplus f(1)}|0\rangle - (-1)^{f(0)\oplus f(1)}|1\rangle] \frac{1}{\sqrt{2}}(|0\rangle - |1\rangle)$$

which is equal to

$$\frac{1}{2}[(1 + (-1)^{f(0)\oplus f(1)})|0\rangle + (1 - (-1)^{f(0)\oplus f(1)})|1\rangle] \frac{1}{\sqrt{2}}(|0\rangle - |1\rangle).$$

The first qubit of the state above is not a superposition; because $f(0) \oplus f(1)$ is either 0 or 1 the state is therefore either $|0\rangle$ or $|1\rangle$ **before** measurement. In fact we have:

$$\frac{1}{2}[(1 + (-1)^{f(0)\oplus f(1)})|0\rangle + (1 - (-1)^{f(0)\oplus f(1)})|1\rangle] = |f(0) \oplus f(1)\rangle$$

and the state is, therefore, equal to:

$$|f(0) \oplus f(1)\rangle \frac{1}{\sqrt{2}}(|0\rangle - |1\rangle).$$

Hence, if 0 is measured for the observable:

$$A = \begin{pmatrix} 0 & 0 \\ 0 & 1 \end{pmatrix}$$

on the first qubit, it follows that $f(0) \oplus f(1) = 0$ and is f constant, and if 1 is measured then $f(0) \oplus f(1) = 1$ and f is balanced. We note that after the application of the operator U_f the second qubit, $\frac{1}{\sqrt{2}}(|0\rangle - |1\rangle)$, could have been ignored.

Mathematically the algorithm may be expressed as:

$$\mu_Q(H \otimes I)(U_f(H \otimes H))|0\rangle \otimes |1\rangle,$$

where μ_Q denotes measurement of the first qubit. A circuit for the algorithm is shown in Figure 22.3.

Complexity: digitally we compute:

$$if\ f(0) = f(1)\ then\ f\ \text{constant}\ else\ f\ \text{balanced}$$

i.e., two queries to an oracle for f are required. In the quantum case only one query is required – i.e., one application of the operator U_f is sufficient to determine the nature of f. If the evaluation of f is the primary 'overhead' then the quantum algorithm is 50% more 'efficient'. The next algorithm shows that a more significant improvement in efficiency, over a digital equivalent, is possible with a quantum algorithm.

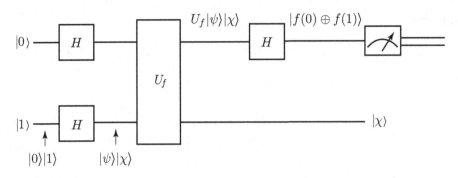

FIGURE 22.3
A circuit for Deutsch's algorithm, where $|\psi\rangle = \frac{1}{\sqrt{2}}(|0\rangle + |1\rangle)$ and $|\chi\rangle = \frac{1}{\sqrt{2}}(|0\rangle - |1\rangle)$.

22.7　The Deutsch-Jozsa algorithm

The Deutsch-Jozsa algorithm generalises Deutsch's algorithm to the functions $f : \mathbb{B}^n \to \mathbb{B}$. That is, we have an oracle for $f : \mathbb{B}^n \to \mathbb{B}$, which is guaranteed to be either constant or balanced; the purpose of the Deutsch-Jozsa algorithm is to decide which it is by querying the oracle for f.

The initial state for the algorithm is $(\otimes^n |0\rangle)|1\rangle$; i.e., the first n qubits are in state $|0\rangle$ and the last qubit is in state $|1\rangle$. The Hadamard gate is then applied to each qubit to obtain, using Lemma 21.3, the state

$$|\psi_1\rangle = \frac{1}{\sqrt{2^{n+1}}} \sum_{x \in \{0,1\}^n} |x\rangle(|0\rangle - |1\rangle).$$

The function f is available as a quantum oracle; it maps $|x\rangle|y\rangle$ to $|x\rangle|y \oplus f(x)\rangle$. Applying the oracle to $|\psi_1\rangle$ produces the state

$$|\psi_2\rangle = \frac{1}{\sqrt{2^{n+1}}} \sum_{x \in \{0,1\}^n} |x\rangle(|f(x)\rangle - |1 \oplus f(x)\rangle).$$

Given that for each x, $f(x)$ is either 0 or 1 the above may be written, using Lemma 22.1, as:

$$|\psi_2\rangle = \frac{1}{\sqrt{2^{n+1}}} \sum_{x \in \{0,1\}^n} (-1)^{f(x)}|x\rangle(|0\rangle - |1\rangle).$$

At this point the last qubit, i.e., $\frac{1}{\sqrt{2}}(|0\rangle - |1\rangle)$, can be ignored giving:

$$|\psi_2\rangle = \frac{1}{\sqrt{2^n}} \sum_{x \in \{0,1\}^n} (-1)^{f(x)}|x\rangle.$$

We now apply a Hadamard transform to each of the remaining qubits, using Lemma 21.3, to obtain

$$|\psi_3\rangle \;=\; \frac{1}{2^n} \sum_{x\in\{0,1\}^n} (-1)^{f(x)} \Big[\sum_{y\in\{0,1\}^n} (-1)^{x\cdot y}|y\rangle \Big]$$

$$=\; \frac{1}{2^n} \sum_{y\in\{0,1\}^n} \Big[\sum_{x\in\{0,1\}^n} (-1)^{f(x)}(-1)^{x\cdot y} \Big]|y\rangle \qquad\qquad (*)$$

where $x \cdot y = (x_0 \wedge y_0) \oplus (x_1 \wedge y_1) \oplus \cdots \oplus (x_{n-1} \wedge y_{n-1})$.

It is helpful to look at the case $n = 2$ in some detail; the general case is similar. For $n = 2$ we have:

$$|\psi_3\rangle \;=\; \frac{1}{4}[(-1)^{f(00)} + (-1)^{f(01)} + (-1)^{f(10)} + (-1)^{f(11)}]|00\rangle$$

$$+\frac{1}{4}[(-1)^{f(00)} + (-1)^{f(01)}(-1) + (-1)^{f(10)} + (-1)^{f(11)}(-1)]|01\rangle$$

$$+\frac{1}{4}[(-1)^{f(00)} + (-1)^{f(01)} + (-1)^{f(10)}(-1) + (-1)^{f(11)}(-1)]|10\rangle$$

$$+\frac{1}{4}[(-1)^{f(00)} + (-1)^{f(01)}(-1) + (-1)^{f(10)}(-1) + (-1)^{f(11)}]|11\rangle.$$

It should be noted that, in the case where f is either constant or balanced, this is not a superposition of quantum states; in fact, following a little arithmetic, we see that f is constant if and only if

$$|\psi_3\rangle = |00\rangle$$

and that f is balanced if and only if

$$|\psi_3\rangle = |01\rangle, \text{ or } |10\rangle, \text{ or } |11\rangle.$$

Hence, at least in the case $n = 2$, the algorithm generates states that are not in superposition. Such states can be measured deterministically using an observable which is easily determined to differentiate the outcomes by eigenvalue, and correct results are computed deterministically. In particular measuring $|\psi_3\rangle$ with the observable $B \otimes B$ where

$$B = \begin{pmatrix} 1 & 0 \\ 0 & 0 \end{pmatrix},$$

we note that if the outcome is 1 then f is constant, and if the outcome is 0 then f is balanced. These conclusions follow from the fact that $|00\rangle$ is an eigenvector of $B \otimes B$ with eigenvalue 1 and that $|01\rangle, |10\rangle$ and $|11\rangle$ are eigenvectors of $B \otimes B$ with eigenvalue 0.

For $n > 2$ the situation is similar. The coefficient of the vector $|0\rangle^{\otimes n}$ in the general case is:

$$\frac{1}{2^n} \sum_{x\in\{0,1\}^n} (-1)^{f(x)}$$

which is of magnitude 1 if f is constant and 0 if f is balanced. Hence measuring the state with the observable $\otimes^n B = B \otimes B \otimes \cdots \otimes B$ (which has eigenvalues 1 and 0; with $|00\cdots0\rangle$ as eigenstate for 1, and all other basis elements $|y\rangle$, for $y \in \{0,1\}^n$, corresponding to eigenvalue 0) we conclude that if 1 is measured then f is constant and if 0 is measured it is balanced. A circuit for this algorithm is shown in Figure 22.4.

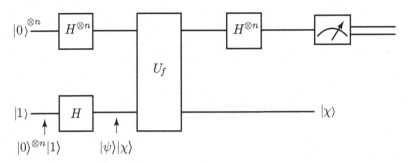

FIGURE 22.4
A circuit for the Deutsch-Jozsa algorithm, where $|\psi\rangle = \frac{1}{\sqrt{2^n}}\sum_{x\in\{0,1\}^n}|x\rangle$ and $|\chi\rangle = \frac{1}{\sqrt{2}}(|0\rangle - |1\rangle)$.

Complexity: the Deutsch-Jozsa algorithm may be thought to be $O(n)$ due to the application of the tensor product gate $H^{\otimes n}$. The question here is: does the tensor product take n times longer to execute than the operator H on a single qubit? This is a quantum mechanical issue beyond the scope of this book; however, it follows from the physics that the n applications of H may be executed simultaneously without time penalty and the complexity of the Deutsch-Jozsa algorithm is therefore $O(1)$. Classical digital computation would require $2^{n-1} + 1$ evaluations of f to determine its nature and the classical algorithm is therefore $O(2^{n-1})$.

22.8   The Bernstein-Vazirani algorithm

The Bernstein-Vazirani problem may be stated as follows: A function $f : \mathbb{B}^n \rightarrow \mathbb{B}$ is known to be of the form $f(x) = x \cdot a$ for some $a \in \mathbb{B}^n$; given an oracle for f, determine a. Here $x \cdot a$ is defined to be $x \cdot a = (x_1 \wedge a_1) \oplus (x_2 \wedge a_1) \oplus \cdots \oplus (x_n \wedge a_n)$.

A quantum algorithm follows the Deutsch-Jozsa algorithm computation as far as the relation marked (*) in the discussion of that algorithm. At this point we have:

$$|\psi_3\rangle = \frac{1}{2^n} \sum_{y\in\{0,1\}^n} \Big[\sum_{x\in\{0,1\}^n} (-1)^{f(x)}(-1)^{x\cdot y}\Big]|y\rangle \qquad (*)$$

which has been obtained with a single call to a quantum oracle for f. For the Bernstein-Vazirani algorithm, we have $f(x) = x \cdot a$ giving, in this case:

$$
\begin{aligned}
|\psi_3\rangle &= \frac{1}{2^n} \sum_{y\in\{0,1\}^n} \Big[\sum_{x\in\{0,1\}^n} (-1)^{x\cdot a}(-1)^{x\cdot y}\Big]|y\rangle \\
&= \frac{1}{2^n} \sum_{y=0}^{2^n-1} |y\rangle \sum_{x=0}^{2^n-1} (-1)^{x\cdot a}(-1)^{x\cdot y}.
\end{aligned}
$$

We note that, using the Corollary to Lemma 22.2,

$$\frac{1}{2^n} \sum_{x=0}^{2^n-1} (-1)^{x \cdot a} (-1)^{x \cdot y} = \frac{1}{2^n} \sum_{x=0}^{2^n-1} (-1)^{x \cdot (a+y)}$$

$$= \begin{cases} 1 & \text{if } a = y, \text{ i.e., } a + y = 0 \\ 0 & \text{if } a \neq y. \end{cases}$$

Hence $\dfrac{1}{2^n} \displaystyle\sum_{x=0}^{2^n-1} (-1)^{x \cdot a} (-1)^{x \cdot y} = \delta_{a,y}$ and, from the relation (*) above, we obtain:

$$\begin{aligned}
|\psi_3\rangle &= \frac{1}{2^n} \sum_{x=0}^{2^n-1} \sum_{y=0}^{2^n-1} (-1)^{x \cdot a} (-1)^{x \cdot y} |y\rangle \\
&= \sum_{y=0}^{2^n-1} \delta_{a,y} |y\rangle \\
&= |a\rangle.
\end{aligned}$$

In the above $\delta_{a,y}$ is the function defined by $\delta_{a,y} = 1$ if $y = a$ and 0 otherwise.

Measurement 1:

We can measure $|a\rangle \in \otimes^n \mathbb{C}^2$ qubit-by-qubit with the operators:

$$A \otimes I \otimes \cdots \otimes I \quad \text{measures} \quad |a_1\rangle \quad \text{with output} \quad a_1 \in \mathbb{B}$$
$$I \otimes A \otimes \cdots \otimes I \quad \text{measures} \quad |a_2\rangle \quad \text{with output} \quad a_2 \in \mathbb{B}$$
$$\vdots$$
$$I \otimes I \otimes \cdots \otimes A \quad \text{measures} \quad |a_n\rangle \quad \text{with output} \quad a_n \in \mathbb{B}$$

and the string $a = a_1 a_2 \cdots a_n$ is revealed. In the above A is represented by the matrix:

$$A = \begin{pmatrix} 0 & 0 \\ 0 & 1 \end{pmatrix}$$

which is self-adjoint on $\mathbb{C}^2$.

Measurement 2:

Alternatively, we can measure $|a\rangle \in \otimes^n \mathbb{C}^2$ with the observable,

$$\sum_{z=0}^{2^n-1} z |z\rangle \langle z| : \otimes^n \mathbb{C}^2 \to \otimes^n \mathbb{C}^2$$

the outcome of which is an eigenvalue $a \in \{0, \ldots, 2^n - 1\}$; the binary form of which is the required string $a \in \mathbb{B}^n$.

The matrix representation of the observable $\displaystyle\sum_{z=0}^{2^n-1} z |z\rangle \langle z| : \otimes^n \mathbb{C}^2 \to \otimes^n \mathbb{C}^2$ is diagonal, having the form:

$$\begin{pmatrix} 0 & 0 & 0 & 0 & \cdots & \\ 0 & 1 & 0 & 0 & & \\ 0 & 0 & 2 & 0 & & \\ 0 & 0 & 0 & 3 & & \\ \vdots & & & & \ddots & \\ 0 & \cdots & \cdots & & 0 & 2^n - 1 \end{pmatrix}$$

for which the eigenvalues are $0, 1, \ldots, 2^n - 1$ with eigenvectors $|0 \cdots 0\rangle, \ldots, |1 \cdots 1\rangle$ respectively.

The circuit for the Bernstein-Vazirani algorithm is, essentially, the same as that for the Deutsch-Jozsa algorithm, but with $f : \mathbb{B}^n \to \mathbb{B}$ defined by $f(x) = x \cdot a$, rather than $f : \mathbb{B}^n \to \mathbb{B}$ being either constant or balanced.

Complexity: Digitally, a single evaluation of a classical oracle for f can deliver at most one of the values $a_1, \ldots, a_n$. That is, we have

$$
\begin{aligned}
f(100 \cdots 0) &= (1 \wedge a_1) \oplus (0 \wedge a_2) \oplus \cdots \oplus (0 \wedge a_n) = a_1 \\
f(010 \cdots 0) &= (0 \wedge a_1) \oplus (1 \wedge a_2) \oplus \cdots \oplus (0 \wedge a_n) = a_2 \\
&\quad \vdots \\
f(00 \cdots 01) &= (0 \wedge a_1) \oplus (0 \wedge a_2) \oplus \cdots \oplus (1 \wedge a_n) = a_n,
\end{aligned}
$$

and the digital algorithm is therefore $O(n)$. The execution-time bound of the quantum algorithm is $O(1)$.

23

Quantum algorithms 2: Simon's algorithm

23.1 Objectives

Simon's algorithm relates to the computation of the period of a Boolean function; some aspects of periodicity are, therefore, covered in this chapter prior to consideration of the algorithm.

23.2 Periodicity, groups, subgroups and cosets

23.2.1 Real-valued functions

If $f : \mathbb{R} \to \mathbb{R}$ and, for some $P \in \mathbb{R} \setminus 0$ we have

$$f(x) = f(x + P)$$

for all x, then f is said to be periodic. The smallest P for which this is true is called the period of f.

Observation 23.1 *If $f : \mathbb{R} \to \mathbb{R}$ is periodic with period p then $f(x) = f(x \pm np)$ for all $n \in \mathbb{N}$.*

Proof
If $f(x + p) = f(x)$ for all $x \in \mathbb{R}$ then with $x^* = x + p$ we have

$$
\begin{aligned}
f(x + 2p) &= f(x^* + p) \\
&= f(x^*) \\
&= f(x + p) \\
&= f(x) \qquad \text{etc.}
\end{aligned}
$$

∎

The following are true:

1. If $f : \mathbb{R} \to \mathbb{R}$ is constant, i.e., $f(x) = c$, for all $x \in \mathbb{R}$, then f is periodic, but has no period.

2. If $f : \mathbb{R} \to \mathbb{R}$ is one-to-one then f is not periodic.

3. If $f : \mathbb{R} \to \mathbb{R}$ is such that $f(x+a) = f(x)$ for all x and for all a then f is constant.

We note that $\mathbb{R}$ is a commutative group under $+$. For $p \in \mathbb{R}, p > 0$ we define

$$
\begin{aligned}
\mathcal{Z}_p &= \{0, \pm np : n \in \mathbb{N}\} \\
&= \{0, \pm p, \pm 2p, \pm 3p, \ldots\}.
\end{aligned}
$$

DOI: 10.1201/9781003264569-23

For any given p, $\mathcal{Z}_p$ is a subgroup of $\mathbb{R}$; each $\mathcal{Z}_p$ isomorphic to $\mathbb{Z}$. Hence each $\mathcal{Z}_p$ is a discrete subgroup of $\mathbb{R}$ with infinitely many elements. We note that the group $\mathcal{Z}_p$ should not be confused with finite discrete groups $\mathbb{Z}_k$ defined by addition mod k, for $k \in \{2, 3, 4, \ldots\}$.

$\mathbb{R}/\mathcal{Z}_p$ denotes the coset space of $\mathbb{R}$ with respect to the subgroup $\mathcal{Z}_p$; the coset containing $x \in \mathbb{R}$ is by definition: (see Sections 8.5 and 8.6)

$$x + \mathcal{Z}_p = \{x, x \pm p, x \pm 2p, x \pm 3p, \ldots\}$$

which we denote by $[x]$. We have:

$$\bigcup_{0 \le x < p} [x] = \mathbb{R}$$

and for $x \ne y$, for $x, y \in [0, p)$,

$$[x] \cap [y] = \emptyset$$

where $\emptyset$ denotes the empty set.

Lemma 23.1 $f : \mathbb{R} \to \mathbb{R}$ *has period* $0 < p \in \mathbb{R}$ *if and only if it is constant on the cosets* $\mathbb{R}/\mathcal{Z}_p$; f *may, however, take the same value on distinct cosets. See Example 23.2.1.*

Proof
This follows directly from the observation above and Example 23.2.1 below.

■

Example 23.2.1

The function $f : \mathbb{R} \to \mathbb{R}$, $f(x) = \sin x$, shown in Figure 23.1 is continuous and has period $p = 2\pi$. Hence in this example we are concerned with the cosets of the subgroup $\mathcal{Z}_{2\pi}$ in $\mathbb{R}$. We have:

$$\mathcal{Z}_{2\pi} = \{0, \pm 2\pi, \pm 4\pi, \ldots\}$$

and the coset, determined by $x \in \mathbb{R}$, is by definition:

$$x + \mathcal{Z}_{2\pi} = \{x, \; x \pm 2\pi, \; x \pm 4\pi, \; \ldots\}$$

which is also denoted $[x]$. It follows that if $y \ne x \pm 2\pi n$, for some $n \in \mathbb{N}$, then $[y] \cap [x] = \emptyset$. In Figure 23.1 the x-values identified by the filled circles are elements of $[x]$ and those identified by the hollow circles are elements of $[y] = [\pi - x]$. We have $[x] \cap [y] = \emptyset$, but we note that the two distinct cosets are mapped to the same value by the sine function.

Example 23.2.2

The discontinuous function $f_d : \mathbb{R} \to \mathbb{R}$, shown in Figure 23.2 and defined by:

$$f_d(x) = \begin{cases} x, & 0 \le x < 1 \\ x - 1, & 1 \le x < 2 \\ x - 2, & 2 \le x < 3 \\ \vdots & \vdots \end{cases}$$

has period $p = 1$. Each of the cosets $\mathbb{R}/\mathcal{Z}_1$ maps to a distinct value in the range $[0, 1)$. Specifically the coset $x + \mathcal{Z}_1$ maps to x.

Conjecture: If f is continuous and periodic, with period p, then there will exist distinct cosets in $\mathbb{R}/\mathcal{Z}_p$ on which the value of f is the same.

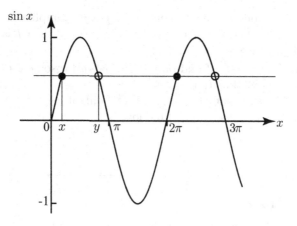

FIGURE 23.1
Graph (not to scale) of the sine function, with $y = \pi - x$.

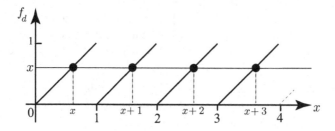

FIGURE 23.2
Graph of the periodic function f_d and the coset $x + \mathcal{Z}_1$.

23.2.2   A summary of the real-valued case

Below we summarise, in various alternative forms, the key aspects of periodicity for functions $f : \mathbb{R} \to \mathbb{R}$. This summary should be helpful in understanding the Boolean case discussed below.

f one-to-one

   f is one-to-one if $f(x) = f(y)$ implies $y = x$.
   f is one-to-one if $f(x) = f(y)$ implies $y - x = 0$.

f periodic

   f is periodic if there exists $0 \neq p \in \mathbb{R}$ such that $f(x) = f(x + p)$ for all $x \in \mathbb{R}$.
   f is periodic if there exists $0 \neq p \in \mathbb{R}$ such that $f(x) = f(y)$ implies $y = x + p$ for all $x \in \mathbb{R}$.
   f is periodic if there exists $0 \neq p \in \mathbb{R}$ such that $f(x) = f(y)$ implies $y - x = p$ for all $x \in \mathbb{R}$.
   $f : \mathbb{R} \to \mathbb{R}$ is periodic, with $0 < p \in \mathbb{R}$, if and only if it is constant on the cosets $\mathbb{R}/\mathcal{Z}_p$; f may, however, take the same value on distinct cosets.

   In the above p may not be unique – see the sine example, which is periodic with $p \in \{2\pi, 4\pi, 6\pi, \ldots\}$, i.e., it is periodic with respect to distinct translations involving multiples of 2π. We show later that Boolean functions may also be periodic with respect to distinct translations.

f combined statements 1

f is periodic, or one-to-one, if $f(x) = f(y)$ implies $y - x = p$ for some $p \neq 0$ (periodic) or $x - y = 0$ (one-to-one).

Equivalently: f is periodic, or one-to-one, if $f(x) = f(y)$ implies $y - x \in \{0, p\}$ for some $p \neq 0$; where if $y - x = p$ then f is periodic, and if $x - y = 0$ then f is one-to-one.

f combined statements 2

If f is such that $f(x) = f(y)$ if and only if $y = x + a$ for some $a \in \mathbb{R}, a \neq 0$ then f has period a and is a two-to-one function.

Equivalently: If f is such that $f(x) = f(y)$ if and only if $y - x = a$ for some $a \in \mathbb{R}, a \neq 0$ then f has period a and is a two-to-one function.

23.3   Boolean functions

We note that $(\mathbb{B}^n, \oplus)$ is a group and, following the case of real functions, we say that the Boolean function $f : \mathbb{B}^n \to \mathbb{B}^n$ is periodic if there exists $s \in \mathbb{B}^n$, $s \neq 0^n$, such that $f(x) = f(x \oplus s)$ for all $x \in \mathbb{B}^n$.

For $s \in \mathbb{B}^n$, $s \neq 0^n$, the sets $K_s = \{0, s\}$ are 2-element subgroups of $\mathbb{B}^n$. The cosets of K_s in $\mathbb{B}^n$ are the two element sets $\{x, x \oplus s\}$ for $x \in \mathbb{B}^n$ (see Section 8.5).

As in the real case the periodicity of a function implies it is constant on the cosets – but not necessarily taking different values on distinct cosets. That is, the following lemma is true.

Lemma 23.2 $f : \mathbb{B}^n \to \mathbb{B}^n$ *has period $s \in \mathbb{B}^n$ if and only if it is constant on the cosets* $\mathbb{B}^n/K_s$. *As in the real case, f may take the same value on distinct cosets – see Example 23.3.1.*

Proof

(i) If f is periodic then there is an $s \in \mathbb{B}^n$ such that $f(x) = f(x \oplus s)$ for all $x \in \mathbb{B}$; equivalently f is constant on the sets $\{x, x \oplus s\}$ for $x \in \mathbb{B}^n$ which are the cosets of K_s in $\mathbb{B}^n$.

(ii) If f is constant on the cosets $\mathbb{B}^n/K_s$ then, clearly, $f(x) = f(x \oplus s)$ for all $x \in \mathbb{B}^n$. ∎

Example 23.3.1

The function

$$
\begin{aligned}
000 &\rightarrow 000 \\
001 &\rightarrow 000 \\
010 &\rightarrow 000 \\
011 &\rightarrow 000 \\
100 &\rightarrow 101 \\
101 &\rightarrow 101 \\
110 &\rightarrow 111 \\
111 &\rightarrow 111
\end{aligned}
$$

has period 001 but takes the same value, i.e., 000, on the cosets $\{000, 001\}$ and $\{010, 011\}$ of $\mathbb{B}^3/K_{001}$.

Corollary

If $f : \mathbb{B}^n \to \mathbb{B}^n$ has period $s \in \mathbb{B}^n$ and f takes distinct values on the cosets $\mathbb{B}^n/K_s$ then f is a two-to-one function.

Observation

As in the real case, a non-constant, Boolean function may be periodic with respect to distinct translations. We show this below for functions $f : \mathbb{B}^3 \to \mathbb{B}^3$ with translations $s = 011$ and $s^* = 101$.

Proof

Assume $f : \mathbb{B}^n \to \mathbb{B}^n$ is such that:

$$
\begin{aligned}
f(x) &= f(x \oplus s) \text{ for all } x \\
&= f(x \oplus s^*) \text{ for all } x
\end{aligned}
$$

then with $x^* = x \oplus s$ we obtain the further relation:

$$
\begin{aligned}
f(x) = f(x^*) &= f(x^* \oplus s^*) \\
&= f(x \oplus s \oplus s^*).
\end{aligned}
$$

Hence for $x = 0^n$ we have:

$$
\begin{aligned}
f(0^n) &= f(s) \\
f(0^n) &= f(s^*) \\
f(0^n) &= f(s \oplus s^*)
\end{aligned}
$$

and for $x = 1^n$:

$$
\begin{aligned}
f(1^n) &= f(\bar{s}) \\
f(1^n) &= f(\bar{s^*}) \\
f(1^n) &= f(\overline{s \oplus s^*}).
\end{aligned}
$$

Consider the case $n = 3$ choosing $s = 011$ and $s^* = 101$ we obtain the relations:

$$
\begin{aligned}
f(000) &= f(011) \\
f(000) &= f(101) \\
f(000) &= f(110) \\
f(111) &= f(100) \\
f(111) &= f(010) \\
f(111) &= f(001).
\end{aligned}
$$

Choosing $A, B \in \mathbb{B}^3$, $A \neq B$ for the values of $f(000)$ and $f(111)$ respectively, we obtain a function $f : \mathbb{B}^3 \to \mathbb{B}^3$ defined by:

$$
\begin{array}{llll}
& f(x) & & \\
f(000) &= A & & \\
f(001) &= B & (x = \overline{s \oplus s^*}) \\
f(010) &= B & (x = \overline{s^*}) \\
f(011) &= A & (x = s) \\
f(100) &= B & (x = \bar{s}) \\
f(101) &= A & (x = s^*) \\
f(110) &= A & (x = s \oplus s^*) \\
f(111) &= B & &
\end{array}
$$

which is periodic in both $s = 011$ and $s = 101$. As A and B may be selected arbitrarily from $\mathbb{B}^3$, provided $A \neq B$, a total of ${}^8C_2 = 28$ functions, periodic in both $s = 011$ and $s^* = 101$, are defined by the above.

■

23.3.1 The subgroups of group $(\mathbb{B}^3, \oplus)$

- Subgroups with one element: $\{000 \in \mathbb{B}\}$

- Subgroups with two elements: $K_s = \{000, s\}$, where $s \in \mathbb{B}^3$ and $s \neq 0$

- Subgroups with four elements include:

  1. $\{000, 001, 010, 011\}$,
  2. $\{000, 001, 100, 101\}$,
  3. $\{000, 010, 100, 110\}$,
  4. $\{000, 010, 101, 111\}$,
  5. $\{000, 011, 100, 111\}$,
  6. $\{000, 011, 101, 110\}$,

  $$\vdots$$

23.3.2 The cosets $\mathbb{B}^3/K_s$

For $x \in \mathbb{B}^3$ the coset xK_s is $xK_s = \{x, x \oplus s\}$.

23.4 The hidden subgroup problem

Many known quantum algorithms are special cases of the hidden subgroup problem which we now state in a general form.

Let $f : G \to X$ be a function from the finite group G to a discrete set X that is constant on the cosets of a subgroup K, of G, but takes a different value on each coset. (i.e., the 'induced' mapping $f_{G/K} : G/K \to X$ is one-to one). Assuming we have an oracle for the unitary operator $U_f|g\rangle|h \oplus f(g)\rangle$, for $g \in G$ and $h \in X$, and $\oplus$ an appropriate binary operation on X, find K.

23.5 Simon's problem

For this problem we have an oracle for a function $f : \mathbb{B}^n \to \mathbb{B}^n$, where f is guaranteed to be two-to-one and periodic; equivalently for some unknown $s \in \mathbb{B}^n$, with $s \neq 0^n$, we have, for $y \neq x, f(x) = f(y)$ if and only if $y = x \oplus s$. The problem is to determine s by oracle query.

Simon's problem can be expressed as a particular case of the hidden subgroup problem, with $G = (\mathbb{B}^n, \oplus)$ and $K = \{0^n, s\}$, for some $s \in \mathbb{B}^n$. The periodic two-to-one function f is said to 'hide' the subgroup $\{0^n, s\}$ of $(\mathbb{B}^n, \oplus)$. f is constant on the cosets, $\mathbb{B}^n/\{0^n, s\}$, of $\mathbb{B}^n$ and takes a distinct value on each coset. Simon's problem may therefore be expressed as: given an oracle for f, determine the subgroup $K = \{0^n, s\}$ of $(\mathbb{B}^n, \oplus)$.

23.6   The complexity of digital solutions

Using 'brute-force' we query the oracle and test pairs $(f(0^n), f(b))$, for $b \in \mathbb{B}^n \setminus \{0^n\}$, for equality. Clearly up to $2^n - 1$ comparisons are required to determine the period of f and this approach is, therefore, $O(2^n)$ in execution time.

An alternative view is to investigate the probability of success when a particular number of distinct pairs $(x, y) \in \mathbb{B}^n \times \mathbb{B}^n$ are tested for $f(x) = f(y)$ from which the period, s, is then $s = x \oplus y$.

The 'solution' pairs (x, y), i.e., those for which $f(x) = f(y)$, with $y \neq x$ are sparse, in the complete set of distinct pairs $(x, y) \in \mathbb{B}^n \times \mathbb{B}^n$. Specifically, as f is two-to-one, there will be precisely 2^{n-1} solution pairs, amongst the total number of distinct pairs, which is:

$$2^n C_2 = \frac{2^n!}{2 \times (2^n - 2)!} = \frac{2^n(2^n - 1)}{2}.$$

Hence, the probability of a randomly selected pair (x, y) providing a solution is

$$\frac{2 \times 2^{n-1}}{2^n \times (2^n - 1)} = \frac{1}{2^n - 1}.$$

It follows that exponentially many pairs need to be checked for the probability of success to be high. Given the nature of Simon's problem there seem to be no 'smart' ways to determine s and that, digitally, the problem is 'hard'; i.e., no polynomially-bounded digital solutions are known.

23.7   Simon's quantum algorithm

We note that the algorithm may be expressed as shown in Figure 23.3. The input state is $|\psi_0\rangle = |0^n\rangle \otimes |0^n\rangle$, equivalently $|0\rangle^{\otimes n} \otimes |0\rangle^{\otimes n}$. The algorithm may be expressed as follows: defining

$$A = \mu_Q\{U_f[H^{\otimes n} \otimes I(|0\rangle^n \otimes |0\rangle^n)]\}$$

we follow this with

$$\mu_Q(H^{\otimes n} \otimes I(A))$$

to give

$$\mu_Q(H^{\otimes n} \otimes I(\mu_Q\{U_f[H^{\otimes n} \otimes I(|0\rangle^n \otimes |0\rangle^n)]\}))$$

as shown in Figure 23.3.

Following the first Hadamard transform the state is:

$$\begin{aligned} |\psi_1\rangle &= (H^{\otimes n}|0^n\rangle) \otimes |0^n\rangle \\ &= \frac{1}{2^{\frac{n}{2}}} \sum_{x \in \{0,1\}^n} |x\rangle \otimes |0^n\rangle \end{aligned}$$

and following U_f we have

$$\begin{aligned} |\psi_2\rangle &= U_f|\psi_1\rangle \\ &= \frac{1}{2^{\frac{n}{2}}} \sum_{x \in \{0,1\}^n} |x\rangle \otimes |f(x)\rangle. \end{aligned}$$

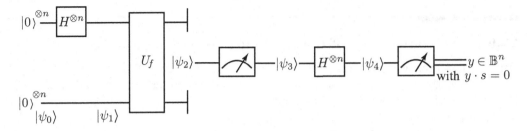

FIGURE 23.3
A circuit for a single pass of Simon's quantum computation.

At this point we follow Preskill's approach ('for pedagogical clarity' – see his on-line material) and measure the second register of the state $|\psi_2\rangle$. Specifically, we measure the self-adjoint operator

$$\mathcal{O} = I \otimes \sum_{z=0}^{2^n-1} z|z\rangle\langle z|$$

on the second register, $|f(x)\rangle$, of the state $|\psi_2\rangle$. The outcome is a value, $f(x^*)$, for some $x^* \in \mathbb{B}^n$, of the observable $\mathcal{O}$ and we can conclude that the normalised state, following the measurement of $\mathcal{O}$ on $|\psi_2\rangle$ is:

$$|\psi_3\rangle = \frac{1}{\sqrt{2}}(|x^*\rangle + |x^* \oplus s\rangle) \otimes |f(x^*)\rangle$$

where s is the period of f.

The reasons this conclusion may be drawn are:

1. the assumption that f is both periodic and two-to-one; hence the measured value $f(x^*)$ occurs also at $x^* \oplus s$; i.e., we have $f(x^* \oplus s) = f(x^*)$, where s is the period of f,

    and

2. the von Neumann-Lüders measurement postulate, which states that the post-measurement state is the normalised projection of $|\psi_2\rangle$ onto the subset of states compatible with the measured value, $f(x^*)$, of $\mathcal{O}$ on $|\psi_2\rangle$.

Measuring the first register of the state $\frac{1}{\sqrt{2}}(|x^*\rangle + |x^* \oplus s\rangle) \otimes |f(x^*)\rangle$ would not help to determine s; with probability $\frac{1}{2}$ we would obtain either x^* or $x^* \oplus s$ but neither of these outcomes reveals anything about s and the state collapses to either $|x^*\rangle \otimes |f(x^*)\rangle$ or $|x^* \oplus s\rangle \otimes |f(x^*)\rangle$.

Instead we apply the Hadamard transform, $H^{\otimes n}$, to the first register of $|\psi_3\rangle$. At this point the second register may be dropped and we compute:

$$H^{\otimes n}(\frac{1}{\sqrt{2}}(|x^*\rangle + |x^* \oplus s\rangle)) \quad = \quad \frac{1}{2^{(n+1)/2}} \sum_{y=0}^{2^n-1} [(-1)^{x^* \cdot y} + (-1)^{(x^* \oplus s) \cdot y}]|y\rangle. \tag{1}$$

We now note that

$$
\begin{aligned}
(-1)^{x^* \cdot y} + (-1)^{(x^* \oplus s) \cdot y} &= (-1)^{x^* \cdot y} + (-1)^{x^* \cdot y}(-1)^{s \cdot y} \\
&= (-1)^{x^* \cdot y}(1 + (-1)^{s \cdot y}) \\
&= \begin{cases} 0 & \text{if } s \cdot y = 1 \\ 2(-1)^{x^* \cdot y} & \text{if } s \cdot y = 0. \end{cases}
\end{aligned}
$$

Hence the only non-zero terms in the sum on the right-hand side of relation (1) are those for which $s \cdot y = 0$ and we therefore have:

$$
\begin{aligned}
|\psi_4\rangle &\equiv H^{\otimes n}(\frac{1}{\sqrt{2}}(|x^*\rangle + |x^* \oplus s\rangle)) \\
&= \frac{1}{2^{(n-1)/2}} \sum_{y \text{ with } y \cdot s = 0} (-1)^{x^* \cdot y}|y\rangle \\
&= \sum_{y \text{ with } y \cdot s = 0} \alpha_y|y\rangle
\end{aligned}
$$

where $\alpha_y = \dfrac{(-1)^{x^* \cdot y}}{2^{(n-1)/2}}$ and $|\alpha_y|^2 = \dfrac{1}{2^{n-1}}$.

Measurement of the state $|\psi_4\rangle$ with the observable $\sum_{z=0}^{2^n-1} z|z\rangle\langle z|$ therefore produces a $y \in \mathbb{B}^n$ with the property $y \cdot s = 0$. Specific y values each occur with probability $\frac{1}{2^{n-1}}$.

A single y satisfying $y \cdot s = 0$ is insufficient to determine s, $(n-1)$ linearly independent (in the Boolean sense) y values being required to enable the $(n-1) \times n$ homogeneous system of linear equations:

$$
\begin{aligned}
y_1 \cdot s &= 0 \\
y_2 \cdot s &= 0 \\
&\vdots \\
y_{n-1} \cdot s &= 0
\end{aligned}
$$

to be solved for s which identifies the hidden subgroup $K = \{0^n, s\}$. This step of Simon's algorithm is solved classically. Hence the quantum procedure is, first, repeated until a set $y_1, \ldots, y_{n-1}$ of linearly independent y's are obtained. Given that the probability of obtaining a specific y, using the quantum procedure, is low (i.e., $\frac{1}{2^{n-1}}$) it follows that with $O(n)$ repetitions, the probability of not having $(n-1)$ linearly independent y values is low. Hence the query complexity of the classical solution is exponential, i.e., $O(2^n)$, whilst that of the quantum algorithm is linear, i.e., $O(n)$. A wider view of complexity estimation might include the final step of Simon's algorithm (which is done classically) – i.e., solving the linear equations to determine s. As this is no worse than $O(n^3)$, Simon's solution remains significantly more efficient.

Simon's problem is of little practical interest; its purpose is to demonstrate the great potential power of quantum computation.

Writing $s = s_1 \cdots s_n$ and $y_i = y_{i1} \cdots y_{in}$ then the equation system above may be written as

$$
\begin{pmatrix}
y_{11} & y_{12} & \cdots & y_{1n} \\
y_{21} & y_{22} & \cdots & y_{2n} \\
\vdots & \vdots & \ddots & \vdots \\
y_{(n-1)1} & y_{(n-1)2} & \cdots & y_{(n-1)n}
\end{pmatrix}
\begin{pmatrix}
s_1 \\
s_2 \\
\vdots \\
s_n
\end{pmatrix}
=
\begin{pmatrix}
0 \\
0 \\
\vdots \\
0
\end{pmatrix}
$$

where the 'arithmetic' is Boolean, i.e.,

$$y_i \cdot s = (y_{i1} \wedge s_1) \oplus (y_{i2} \wedge s_2) \oplus \cdots \oplus (y_{in} \wedge s_n).$$

That is, $\wedge$ replaces multiplication, in the fields $\mathbb{R}$ and $\mathbb{C}$, and $\oplus$ replaces addition.

The simplest examples to illustrate the post-quantum determination of s are for $n = 2$. In this case we just have one equation, $y \cdot s = 0$, to solve for s, the period of f.

Example 23.7.1

Suppose $y = 01$ is measured, then the equation for $s = s_1 s_2$ is:

$$01 \cdot s_1 s_2 = 0.$$

We have

$$
\begin{aligned}
01 \cdot s_1 s_2 &= (0 \wedge s_1) \oplus (1 \wedge s_2) \\
&= 0 \oplus s_2 \\
&= 0 \qquad \text{if and only if } s_2 = 0.
\end{aligned}
$$

Given that $s \neq 00$, otherwise f would not be periodic, the solution is $s = 10$.

Example 23.7.2

Suppose $y = 11$ is measured, then the equation for $s = s_1 s_2$ is:

$$11 \cdot s_1 s_2 = 0.$$

We have

$$
\begin{aligned}
11 \cdot s_1 s_2 &= (1 \wedge s_1) \oplus (1 \wedge s_2) \\
&= s_1 \oplus s_2 \\
&= 0 \qquad \text{if and only if } s_1 = 1 \text{ and } s_2 = 1,
\end{aligned}
$$

$s = 00$ again, not being allowed. Hence the solution is $s = 11$.

A

Probability

A.1 Definition of probability

Understanding quantum computation relies on an understanding of probability. If an **event**, E say, (or **outcome**) is certain to occur, we say that the probability of it occurring, $P(E)$, is 1. If it is impossible for an event to occur it has probability 0, that is $P(E) = 0$. We see that in some circumstances the outcome of an event can be calculated exactly, that is, it is **deterministic**. All other probabilities lie between 0 and 1. Events with probabilities nearer to 1 are more likely to happen than not.

Definition A.1 Probability
For any event E, the probability of its occurrence, $P(E)$, is such that

$$0 \leq P(E) \leq 1.$$

The result of performing an experiment is often referred to as an event. Consider the experiment of tossing a coin and hoping to obtain a head, H. We may not be able to predict precisely whether a head will occur because the outcome is **stochastic** or **random**. If the coin is fair then the **theoretical probability** of obtaining a head is $\frac{1}{2}$. This means that on average one in every two tosses should result in a head. So $P(H) = 0.5$. Likewise the probability of obtaining a tail, T, is also $\frac{1}{2}$, that is $P(T) = 0.5$. No other outcomes are possible in such an experiment. Observe that $P(H) + P(T) = 1$, representing total probability: we know that obtaining a head or a tail is certain to happen. Note that to calculate theoretical probabilities, a priori knowledge about the possible events is essential. In this case we assumed that the coin was fair. In some circumstances we don't have sufficient information to calculate a theoretical probability. In such a case we can calculate an **experimental** probability by performing an experiment a large number of times, N say. Then we can observe the number of times, M say, that an event E occurs. The experimental probability of event E is then given by

$$P(E) = \frac{M}{N}.$$

For example, if we toss a biased coin 1000 times, and obtain a head on 650 occasions, the experimental probability of obtaining a head is

$$P(H) = \frac{650}{1000} = 0.65.$$

Example A.1.1

Suppose 6000 four-digit binary numbers are generated, e.g. 1101, 1111, 0011, 0101, ... Suppose that we observe the number 0000 on 423 occasions. Calculate

(a) the experimental probability of obtaining the event E: 0000,

DOI: 10.1201/9781003264569-A

(b) the theoretical probability of obtaining $E : 0000$ assuming that all outcomes are equally likely.

Solution

(a) The experimental probability $P(E) = \frac{423}{6000} = 0.0705$.

(b) There are $2^4 = 16$ possible outcomes when generating four-digit binary numbers. If each outcome is equally likely then $P(E) = \frac{1}{16} = 0.0625$. It would seem that in our experiment the number 0000 has been generated rather more times than we would have expected if the outcome was truly random.

Example A.1.2

When a quantum bit is measured the measured value is either 0 or 1. The probability of obtaining 0, $P(0)$, is sometimes labelled $|\alpha_0|^2$ and similarly, the probability of obtaining 1, $P(1)$, is sometimes labelled $|\alpha_1|^2$. These are the only possible outcomes and so

$$P(0) + P(1) = |\alpha_0|^2 + |\alpha_1|^2 = 1.$$

This rather simple fact has profound implications for how quantum states are described mathematically.

A.2   Discrete random variables and probability distributions

There are usually several possible outcomes of any given experiment. For example, if we measure some property of a quantum system, we find that the measured value can take on one of several possible values. Which value we actually achieve is not known a priori but what we do know is the probability of achieving that value. Thus the measured value is a **random variable**, X say.

When a random variable can assume *any* value in a given interval it is said to be **continuous**. On the other hand, when the variable must assume a value from a set of individually specified values it is said to be **discrete**.

Given a discrete random variable X, which in a single experiment can assume one of the n values $x_1, x_2, \ldots, x_n$, a **probability distribution** tells us how the total probability is distributed amongst the various possible values. For example, in a single throw of a fair die the probability of getting any one of the values $1, \ldots, 6$ is $\frac{1}{6}$. Table A.1 shows this probability distribution. Note that we write $P(X = x)$ to indicate the probability that the random variable X assumes the specific value x. Note also that the sum of the probabilities

TABLE A.1
A discrete probability distribution

x	1	2	3	4	5	6
$P(X = x)$	$\frac{1}{6}$	$\frac{1}{6}$	$\frac{1}{6}$	$\frac{1}{6}$	$\frac{1}{6}$	$\frac{1}{6}$

of all possible outcomes is 1 representing total probability.

The **expected value** or **expectation**, $E(X)$, of a discrete random variable X is found by multiplying each possible value by its probability and then summing the results. So, if the value x_i occurs with probability p_i then $E(X) = \sum_{i=1}^{n} p_i x_i$.

Definition A.2 Expected value of the discrete random variable X:

$$Expected\ value = E(X) = \sum_{i=1}^{n} p_i x_i.$$

If an experiment is repeated a large number of times, then the mean or average of the resulting measurements is approximately equal to the expected value. Again, if an experiment is repeated a large number of times, there will be a spread of values of resulting measurements. The **standard deviation** and the **variance** are both measures of this spread. If the value x_i occurs with probability p_i, then the variance, written σ^2, is found from

$$\sigma^2 = \sum_{i=1}^{n} p_i(x_i - E(X))^2.$$

The standard deviation, σ, is then the square root of the variance.

Definition A.3 Variance and standard deviation of the discrete random variable X:

$$variance = \sigma^2 = \sum_{i=1}^{n} p_i(x_i - E(X))^2.$$

The standard deviation is the square root of the variance.

Example A.2.1

A discrete random variable X can take two possible values 0 and 1 each with probability $\frac{1}{2}$ (Table A.2). Calculate the expected value and the variance.

TABLE A.2
A discrete
probability
distribution

x	0	1
$P(X = x)$	$\frac{1}{2}$	$\frac{1}{2}$

Solution

The expected value is given by

$$E(X) = \sum_{i=1}^{2} p_i x_i = \frac{1}{2} \times 0 + \frac{1}{2} \times 1 = \frac{1}{2}.$$

The variance is given by

$$\sigma^2 = \sum_{i=1}^{2} p_i(x_i - \frac{1}{2})^2 = \frac{1}{2}(0 - \frac{1}{2})^2 + \frac{1}{2}(1 - \frac{1}{2})^2 = \frac{1}{4}$$

It follows that the standard deviation is $\sigma = \frac{1}{2}$.

B

Trigonometric ratios and identities

The three common **trigonometric ratios** sine (sin), cosine (cos) and tangent (tan) are defined with reference to a right-angled triangle ABC (Figure B.1).

$$\sin\theta = \frac{\text{side opposite to }\theta}{\text{hypotenuse}} = \frac{BC}{AC}, \quad \cos\theta = \frac{\text{side adjacent to }\theta}{\text{hypotenuse}} = \frac{AB}{AC}$$

$$\tan\theta = \frac{\text{side opposite to }\theta}{\text{side adjacent to }\theta} = \frac{BC}{AB}$$

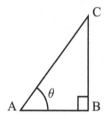

FIGURE B.1
A right-angled triangle.

The trigonometric ratios of $\frac{\pi}{6} = 30°$, $\frac{\pi}{4} = 45°$ and $\frac{\pi}{3} = 60°$ occur frequently in calculations. They can be calculated exactly by considering suitable right-angled triangles:

$$\sin\frac{\pi}{6} = \frac{1}{2}, \quad \cos\frac{\pi}{6} = \frac{\sqrt{3}}{2}, \quad \tan\frac{\pi}{6} = \frac{1}{\sqrt{3}}$$

$$\sin\frac{\pi}{4} = \frac{1}{\sqrt{2}}, \quad \cos\frac{\pi}{4} = \frac{1}{\sqrt{2}}, \quad \tan\frac{\pi}{4} = 1$$

$$\sin\frac{\pi}{3} = \frac{\sqrt{3}}{2}, \quad \cos\frac{\pi}{3} = \frac{1}{2}, \quad \tan\frac{\pi}{3} = \sqrt{3}$$

Whilst these ratios are defined with respect to a right-angled triangle, and so $0 < \theta < 90°$, the definition can be readily extended to angles of any size. Use of a calculator makes calculation of trigonometric ratios straightforward. The **trigonometric functions** $f : \theta \to \sin\theta$, $f : \theta \to \cos\theta$, $f : \theta \to \tan\theta$ then follow immediately. The graphs of $\sin\theta$ and $\cos\theta$ have been given in Chapter 2. Inspection of the graphs reveals other important properties of the trigonometric functions, particularly

$$\cos(-\theta) = \cos\theta \quad \text{and} \quad \sin(-\theta) = -\sin\theta$$

so that $\cos\theta$ is a so-called **even function** and $\sin\theta$ is a so-called **odd function**.

   Trigonometric identities provide a means to write a given expression involving trigonometric functions in terms of other trigonometric functions. Some common identities are given in Table B.1. (Note that $(\cos A)^2$ is usually written $\cos^2 A$. Likewise, $\sin^2 A$ means $(\sin A)^2$.)

DOI: 10.1201/9781003264569-B

TABLE B.1
Common trigonometric identities

$$\frac{\sin A}{\cos A} = \tan A$$
$$\sin^2 A + \cos^2 A = 1$$
$$\sin(A + B) = \sin A \cos B + \sin B \cos A$$
$$\sin(A - B) = \sin A \cos B - \sin B \cos A$$
$$\sin 2A = 2 \sin A \cos A$$
$$\cos(A + B) = \cos A \cos B - \sin A \sin B$$
$$\cos(A - B) = \cos A \cos B + \sin A \sin B$$
$$\cos 2A = \cos^2 A - \sin^2 A$$
$$\sin A + \sin B = 2 \sin \left(\frac{A+B}{2}\right) \cos \left(\frac{A-B}{2}\right)$$
$$\sin A - \sin B = 2 \sin \left(\frac{A-B}{2}\right) \cos \left(\frac{A+B}{2}\right)$$
$$\cos A + \cos B = 2 \cos \left(\frac{A+B}{2}\right) \cos \left(\frac{A-B}{2}\right)$$
$$\cos A - \cos B = -2 \sin \left(\frac{A+B}{2}\right) \sin \left(\frac{A-B}{2}\right)$$

Exercises

B.1   Show that $\frac{1}{2} - \frac{1}{2} \cos \theta = \sin^2 \frac{\theta}{2}$.

B.2   Show that $\frac{1}{2} + \frac{1}{2} \cos \theta = \cos^2 \frac{\theta}{2}$.

C

Coordinate systems

The coordinates of a point describe its position. When working in two spatial dimensions the **Cartesian coordinate system** is the most common. The system comprises two axes – usually referred to as the x axis and the y axis – intersecting at right angles at a point called the **origin** O. Referring to Figure C.1 the horizontal distance of point P from the y axis is the x coordinate of P, with positive values to the right of O and negative values to the left. Likewise, the vertical distance is the y coordinate. We state that the coordinates of the point P are (x, y), written $P(x, y)$. An alternative description is to give the distance of

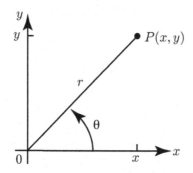

FIGURE C.1
P has Cartesian coordinates (x, y) and polar coordinates (r, θ).

point P from the origin, $r \geq 0$ say, and the angle, θ, that the arm OP makes with a reference axis, usually the positive x axis, and usually measured anticlockwise so that $-\pi < \theta \leq \pi$ (Figure C.1). Then, the coordinates (r, θ) are the **polar coordinates** of P. Cartesian and polar coordinates are related through the formulae:

$$x = r \cos \theta, \qquad y = r \sin \theta$$

$$r = \sqrt{x^2 + y^2}, \qquad \tan \theta = \frac{y}{x}.$$

When working in three spatial dimensions, the position of a point can be described by three Cartesian coordinates, so the coordinates of P are (x, y, z) (Figure C.2). In situations involving spherical symmetry, it is often more useful to work in **spherical polar coordinates** (r, θ, φ) which are defined with reference to Figure C.3. Cartesian and spherical polar coordinates are related through the formulae:

$$x = r \cos \varphi \sin \theta, \quad y = r \sin \varphi \sin \theta, \quad z = r \cos \theta$$

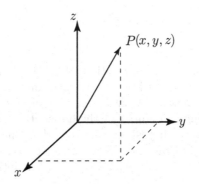

FIGURE C.2
The Cartesian coordinates of P are (x, y, z).

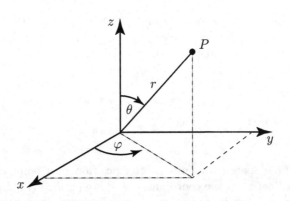

FIGURE C.3
The spherical polar coordinates of P are (r, θ, φ).

D

Field axioms

Fields are fundamental mathematical structures to which we make frequent reference in this book. Two particularly relevant fields are the set of real numbers $\mathbb{R}$ and the set of complex numbers $\mathbb{C}$. To be referred to as a field a set must possess two **binary operations**, that is a way of combining two elements from the set, which satisfy properties known as **field axioms**. These are given below.

A **field** $\mathbb{F}$ is a set with two binary operations, that we can conveniently call addition $+$ and multiplication $\cdot$, which satisfy the following[1]:

1. The set $\mathbb{F}$ must be closed under $+$ and $\cdot$. That is, for any $a, b \in \mathbb{F}$, $a + b \in \mathbb{F}$ and $a \cdot b \in \mathbb{F}$.

2. Addition is associative: for $a, b, c \in \mathbb{F}$, $a + (b + c) = (a + b) + c$.

3. There exists an identity element for addition, often denoted $0 \in \mathbb{F}$, such that $0 + a = a + 0 = a$ for any $a \in \mathbb{F}$.

4. For every element $a \in \mathbb{F}$ there is an additive inverse labelled $-a$ such that

$$a + (-a) = 0.$$

5. Addition is commutative: for $a, b \in \mathbb{F}$, $a + b = b + a$.

6. Multiplication is associative: for $a, b, c \in \mathbb{F}$, $a \cdot (b \cdot c) = (a \cdot b) \cdot c$.

7. Multiplication is commutative: for $a, b \in \mathbb{F}$, $a \cdot b = b \cdot a$.

8. There exists a multiplicative identity, often denoted by $1 \in \mathbb{F}$, such that $1 \cdot a = a$ for any $a \in \mathbb{F}$.

9. For every element $a \in \mathbb{F}$, with the exception of possibly 0, there is a multiplicative inverse labelled a^{-1} such that $a \cdot (a^{-1}) = 1$.

10. Multiplication is distributive over addition: for $a, b, c \in \mathbb{F}$, $a \cdot (b + c) = (a \cdot b) + (a \cdot c)$.

Notes

1. Axioms 1-5 mean that $(\mathbb{F}, +)$ is an Abelian group.

2. Axioms 1, 6-9 mean that $(\mathbb{F}, \cdot)$ is an Abelian group.

---

[1] other binary operations can be substituted for $+$ and $\cdot$, provided the field axioms are satisfied.

DOI: 10.1201/9781003264569-D

E

Solutions to selected exercises

Chapter 1.

1.1 $\emptyset, \{0\}, \{1\}, \{0,1\} = \mathbb{B}$.

1.2 The power set of $\mathbb{B} = \{\emptyset, \{0\}, \{1\}, \{0,1\}\}$. The cardinality is 4.

1.3 Let the set be $\{a, b, c\}$. The power set has $2^3 = 8$ elements:

$$\{\emptyset, \{a\}, \{b\}, \{c\}, \{a,b\}, \{a,c\}, \{b,c\}, \{a,b,c\}\}.$$

1.4 2^n.

1.5 Every even integer is clearly an integer. The set of odd integers is a subset of $\mathbb{Z}$.

1.7 The set of odd integers does not include the additive identity 0.

1.8 The elements which are self-inverse: $1, P_{12}, P_{13}, P_{23}$.

1.9 4.

1.10 2^n.

1.11 $s_1 = s_2 = 0$, or $s_1 = s_2 = 1$.

1.13 The identity element is 0.

1.14 The identity element is 0.

+	0	1	2
0	0	1	2
1	1	2	0
2	2	0	1

Chapter 1. End-of-chapter exercises.

1. $2^{(2^n)}$.

3. a) A, b) $A \cup \overline{B}$.

4. $[0] = \{\ldots, 0, 6, 12, \ldots\}$, $[1] = \{\ldots, 1, 7, 13, \ldots\}$, $[2] = \{\ldots, 2, 8, 14, \ldots\}$,
   $[3] = \{\ldots, 3, 9, 15, \ldots\}$, $[4] = \{\ldots, 4, 10, 16, \ldots\}$, $[5] = \{\ldots, 5, 11, 17, \ldots\}$.

6. If $y - x = y^* - x^*$ then $y^* - y = x^* - x$ and then $\frac{y^* - y}{x^* - x} = 1$. The slope of the line joining the two points has gradient 1. The set of equivalence classes is the set of straight lines with gradient 1.

7. a)

+	0	1
0	0	1
1	1	0

b)

×	0	1
0	0	0
1	0	1

c) 0. d) 1.

10. 5

DOI: 10.1201/9781003264569-E

Chapter 2.

2.1 (12345).

2.5 Yes

2.12 $f_0(x) = 0$ and $f_3(x) = 1$ cannot be inverted. They are not one-to-one.

2.16 $f_0(x) = 0$, $f_1(x) = x$, $f_2(x) = \overline{x}$, $f_3(x) = 1$, for all $x \in \mathbb{B}$.

2.18

(x, y)	$f(x, y) = (x, y \oplus 0)$
$(0, 0)$	$(0, 0)$
$(0, 1)$	$(0, 1)$
$(1, 0)$	$(1, 0)$
$(1, 1)$	$(1, 1)$

2.19

(x, y)	$f(x, y) = (x, y \oplus 1)$
$(0, 0)$	$(0, 1)$
$(0, 1)$	$(0, 0)$
$(1, 0)$	$(1, 1)$
$(1, 1)$	$(1, 0)$

Chapter 2. End-of-chapter exercises.

4. Bijective.

(x, y)	$f(x, y) = (y, \overline{x \oplus y})$
$(0, 0)$	$(0, 1)$
$(0, 1)$	$(1, 0)$
$(1, 0)$	$(0, 0)$
$(1, 1)$	$(1, 1)$

5. It is neither one-to-one nor two-to-one.

(x, y)	$f(x, y) = (x \wedge y, \overline{x \wedge y})$
$(0, 0)$	$(0, 1)$
$(0, 1)$	$(0, 1)$
$(1, 0)$	$(0, 1)$
$(1, 1)$	$(1, 0)$

6. From the table $(x \wedge \overline{y}) \vee (\overline{x} \wedge y)$ is equivalent to $x \oplus y$.

x	y	$x \oplus y$	$x \wedge \overline{y}$	$\overline{x} \wedge y$	$(x \wedge \overline{y}) \vee (\overline{x} \wedge y)$
0	0	0	0	0	0
0	1	1	0	1	1
1	0	1	1	0	1
1	1	0	0	0	0

7. It is not invertible.

(x, y)	$f(x, y) = (x \oplus y, x \wedge y)$
$(0, 0)$	$(0, 0)$
$(0, 1)$	$(1, 0)$
$(1, 0)$	$(1, 0)$
$(1, 1)$	$(0, 1)$

8. The function is bijective.

x	y	z	$x \oplus y$	$x \wedge y$	$z \oplus (z \wedge y)$
0	0	0	0	0	0
0	0	1	0	0	1
0	1	0	1	0	0
0	1	1	1	0	1
1	0	0	1	0	0
1	0	1	1	0	1
1	1	0	0	1	1
1	1	1	0	1	0

Thus

$$000 \rightarrow 000$$
$$001 \rightarrow 001$$
$$010 \rightarrow 010$$
$$011 \rightarrow 011$$
$$100 \rightarrow 110$$
$$101 \rightarrow 111$$
$$110 \rightarrow 101$$
$$111 \rightarrow 100$$

11. The range is $\{000, 001, 010, 011\}$. The range is a proper subset of the co-domain $\mathbb{B}^3$.

14. (a) Given b in the co-domain, choose $a = \frac{b-5}{3}$, then $f(a) = b$.

(b) Given b in the co-domain, choose $a = 2^b$, then $f(a) = b$.

Chapter 3.

3.1 $-\frac{1}{2} \pm i\frac{\sqrt{3}}{2}$.

3.2 (a) $-i$, (b) 1, (c) i.

3.3 $(x-2)(3x^2 - 5x + 6)$, $\frac{5}{6} \pm i\frac{\sqrt{47}}{6}$.

3.4 $x^2 - 10x + 29 = 0$. There are others.

3.5 a) $-\frac{1}{2}$, b) $\frac{1}{2}$.

3.6 a) $6i$, b) $6 + 10i$, c) $3 + 26i$, d) $-55 + 48i$, e) $7 - 30i$.

3.7 $\text{Re}(z) = 16$, $\text{Im}(z) = 11$.

3.10 $\frac{16-4i}{17}$, $1 + \frac{i}{4}$.

3.12 a) $-\frac{62}{85}$, $-\frac{61}{85}$; b) 0, -2; c) $-\frac{1}{2}$, $-\frac{1}{2}$.

3.13 a) 5, $\tan\theta = \frac{4}{3}$, $\theta = 53.13°$. b) $\sqrt{2}$, $-\frac{3\pi}{4} = -135°$.

c) 3, 0. d) 2, $\frac{\pi}{2}$. e) $\sqrt{2}\pi$, $\frac{\pi}{4}$.

3.14 2, $\frac{5\pi}{6}$; $4\sqrt{2}$, $\frac{\pi}{4}$; $z_1 z_2 = 8\sqrt{2} \angle \frac{13\pi}{12}$; $\frac{z_1}{z_2} = \frac{1}{2\sqrt{2}} \angle \frac{7\pi}{12}$.

3.17 (a) $|z_1||z_2| = |z_1 z_2| = \sqrt{533}$.

(b) $|\frac{z_1}{z_2}| = \frac{|z_1|}{|z_2|} = \frac{\sqrt{533}}{41}$.

3.19 (a) 1. (b) -1. (c) $-i$. (d) $-i$.

3.20 (a) $-\frac{1}{2} + \frac{\sqrt{3}}{2}i$. (b) $-\frac{1}{2} - \frac{\sqrt{3}}{2}i$. (c) $-\frac{1}{2} - \frac{\sqrt{3}}{2}i$. (d) $-\frac{1}{2} + \frac{\sqrt{3}}{2}i$.

3.22 ei.

3.23. $\cos\theta = \frac{e^{i\theta} + e^{-i\theta}}{2}$, $\sin\theta = \frac{e^{i\theta} - e^{-i\theta}}{2i}$.

3.25 $a = 1 + \cos 2\omega t$, $b = \sin 2\omega t$.

3.26 $e^{i\pi/6}$.

3.28 $f(0) = 2$, $f(1) = 4$, $f(2) = -10$, $f(3) = 6$.

3.29 $\hat{f}(0) = \frac{1}{\sqrt{3}}(f(0) + f(1) + f(2))$, $\hat{f}(1) = \frac{1}{\sqrt{3}}(f(0) + f(1)e^{-2\pi i/3} + f(2)e^{-4\pi i/3})$,

$\hat{f}(2) = \frac{1}{\sqrt{3}}(f(0) + f(1)e^{-4\pi i/3} + f(2)e^{-8\pi i/3})$.

Chapter 3. End-of-chapter exercises.

1. (a) $\frac{2}{13} - \frac{3}{13}i$. (b) $\frac{x}{x^2+y^2} + \frac{y}{x^2+y^2}i$

10. $z = 1\angle 2k\pi/3$, $k = 0, 1, 2$.

Chapter 4.

4.1 $\langle\psi| = \frac{1}{\sqrt{3}}(-i, 1-i)$.

4.2 $|\phi\rangle = \begin{pmatrix} 1/\sqrt{2} \\ 1/\sqrt{2} \end{pmatrix}$.

4.6 $\sqrt{69}$.

4.7 $\sqrt{3}$.

4.10 $\frac{1}{\sqrt{2}}\begin{pmatrix} 1 \\ 1 \end{pmatrix} = \frac{1}{\sqrt{2}}(|0\rangle + |1\rangle)$. $\qquad \frac{1}{\sqrt{2}}\begin{pmatrix} -1 \\ 1 \end{pmatrix} = \frac{1}{\sqrt{2}}(-|0\rangle + |1\rangle)$

4.11 $\begin{pmatrix} 1 \\ 0 \end{pmatrix} = \frac{\sqrt{2}}{2}(|\psi\rangle + |\phi\rangle)$. $\qquad \begin{pmatrix} 0 \\ 1 \end{pmatrix} = \frac{\sqrt{2}}{2}(|\psi\rangle - |\phi\rangle)$.

4.13 $u \cdot v = 18$, $\cos\theta = 0.9487$, $\theta = 18.4°$.

4.16 $z_1 w_1^* + z_2 w_2^*$; $z_1^* w_1 + z_2^* w_2$.

4.18 zw^* (definition 1).

4.19 a) 0, b) 0, c) 0, d) 0.

Chapter 4. End-of-chapter exercises.

2. $\begin{pmatrix} \sqrt{2/3} \\ \sqrt{1/3} \end{pmatrix}$; 1.

3. $\begin{pmatrix} 38 + 17i \\ 1 - 23i \end{pmatrix}$.

4. Definition 1: $\langle u, v \rangle = 11 - 4i$, $\langle v, u \rangle = 11 + 4i$, $\langle u, u \rangle = 15$, $\langle v, v \rangle = 34$.
Definition 2: $\langle u, v \rangle = 11 + 4i$, $\langle v, u \rangle = 11 - 4i$, $\langle u, u \rangle = 15$, $\langle v, v \rangle = 34$.

5. $\begin{pmatrix} 1 \\ 0 \end{pmatrix} = \frac{1}{2}\begin{pmatrix} 1 \\ 1 \end{pmatrix} + \frac{1}{2}\begin{pmatrix} 1 \\ -1 \end{pmatrix}$; $\begin{pmatrix} 0 \\ 1 \end{pmatrix} = \frac{1}{2}\begin{pmatrix} 1 \\ 1 \end{pmatrix} - \frac{1}{2}\begin{pmatrix} 1 \\ -1 \end{pmatrix}$.

12.

x	y	$x \cdot y$
$x_1 x_2 x_3$	000	0
$x_1 x_2 x_3$	001	x_3
$x_1 x_2 x_3$	010	x_2
$x_1 x_2 x_3$	011	$x_2 \oplus x_3$
$x_1 x_2 x_3$	100	x_1
$x_1 x_2 x_3$	101	$x_1 \oplus x_3$
$x_1 x_2 x_3$	110	$x_1 \oplus x_2$
$x_1 x_2 x_3$	111	$x_1 \oplus x_2 \oplus x_3$

13. a) 1, b) 0.

Chapter 5.

5.3 $\text{Tr}(A) = 5$. $\text{Tr}(A^T) = 5$. $\text{Tr}(I_{n \times n}) = n$.

5.6 $A + A^T = \begin{pmatrix} 18 & 7 \\ 7 & 4 \end{pmatrix}$ which is symmetric.

5.7 $A - A^T = \begin{pmatrix} 0 & 1 \\ -1 & 0 \end{pmatrix}$ which is skew-symmetric.

5.13 a) $\begin{pmatrix} 1 & 0 \\ 0 & 1 \end{pmatrix}$, b) $\begin{pmatrix} 0 & 1 \\ 1 & 0 \end{pmatrix}$.

5.18 (a) $\begin{pmatrix} 20 & 28 \\ -10 & -14 \end{pmatrix}$, (b) $\begin{pmatrix} 312 \\ -156 \end{pmatrix}$, (c) 78, (d) $\begin{pmatrix} 312 \\ -156 \end{pmatrix}$.

5.20 $|A| = |A^T| = 4$.

5.21 1.

5.23 $\begin{pmatrix} -1 & 1 \\ -2 & 3/2 \end{pmatrix}$.

5.24 $\frac{1}{15} \begin{pmatrix} -2 & 8 & 1 \\ 7 & 17 & -11 \\ 4 & -1 & -2 \end{pmatrix}$.

5.25 $\lambda = 4, -1$.

5.26 a) $R^T = \begin{pmatrix} \cos\theta & \sin\theta \\ -\sin\theta & \cos\theta \end{pmatrix}$.

5.29 $x_1 = -4\mu$, $x_2 = \mu$.

5.30 $x = 2\mu - \lambda$, $y = 1 + 2\lambda - 2\mu$, $z = \lambda$, $w = \mu$.

Chapter 5. End-of-chapter exercises.

1. $\lambda = 5, -1$.

2. $x = \frac{\mu-1}{2}$, $y = \mu$, $z = 3$, for $\mu \in \mathbb{R}$.

3. $AB = \begin{pmatrix} 11 & 0 & 20 \\ 7 & -20 & 15 \\ 5 & 21 & 25 \end{pmatrix}$, $(AB)^T = \begin{pmatrix} 11 & 7 & 5 \\ 0 & -20 & 21 \\ 20 & 15 & 25 \end{pmatrix}$

7. a) $H^\dagger = \begin{pmatrix} 3 & 1+i \\ 1-i & 2 \end{pmatrix}$, c) $HH^\dagger = \begin{pmatrix} 11 & 5+5i \\ 5-5i & 6 \end{pmatrix}$.

8. $[\sigma_2, \sigma_3] = \begin{pmatrix} 0 & 2i \\ 2i & 0 \end{pmatrix} = 2i\sigma_1$.

9. $HP = \begin{pmatrix} e^{i\theta_2} & e^{i\theta_1} \\ e^{i\theta_2} & -e^{i\theta_1} \end{pmatrix}$. $HPH = \begin{pmatrix} e^{i\theta_2} + e^{i\theta_1} & e^{i\theta_2} - e^{i\theta_1} \\ e^{i\theta_2} - e^{i\theta_1} & e^{i\theta_2} + e^{i\theta_1} \end{pmatrix}$.

Further

$$\begin{pmatrix} e^{i\theta_2} + e^{i\theta_1} & e^{i\theta_2} - e^{i\theta_1} \\ e^{i\theta_2} - e^{i\theta_1} & e^{i\theta_2} + e^{i\theta_1} \end{pmatrix} = \begin{pmatrix} e^{i\frac{\theta_2}{2}}e^{i\frac{\theta_2}{2}} + e^{i\frac{\theta_1}{2}}e^{i\frac{\theta_1}{2}} & e^{i\frac{\theta_2}{2}}e^{i\frac{\theta_2}{2}} - e^{i\frac{\theta_1}{2}}e^{i\frac{\theta_1}{2}} \\ e^{i\frac{\theta_2}{2}}e^{i\frac{\theta_2}{2}} - e^{i\frac{\theta_1}{2}}e^{i\frac{\theta_1}{2}} & e^{i\frac{\theta_2}{2}}e^{i\frac{\theta_2}{2}} + e^{i\frac{\theta_1}{2}}e^{i\frac{\theta_1}{2}} \end{pmatrix}$$

$$= e^{i\frac{\theta_2+\theta_1}{2}} \begin{pmatrix} e^{i\frac{\theta_2-\theta_1}{2}} + e^{-i\frac{\theta_2-\theta_1}{2}} & e^{i\frac{\theta_2-\theta_1}{2}} - e^{-i\frac{\theta_2-\theta_1}{2}} \\ e^{i\frac{\theta_2-\theta_1}{2}} - e^{-i\frac{\theta_2-\theta_1}{2}} & e^{i\frac{\theta_2-\theta_1}{2}} + e^{-i\frac{\theta_2-\theta_1}{2}} \end{pmatrix}$$

$$= e^{i\frac{\theta_2+\theta_1}{2}} \begin{pmatrix} e^{-i\frac{\theta_1-\theta_2}{2}} + e^{i\frac{\theta_1-\theta_2}{2}} & e^{-i\frac{\theta_1-\theta_2}{2}} - e^{i\frac{\theta_1-\theta_2}{2}} \\ e^{-i\frac{\theta_1-\theta_2}{2}} - e^{i\frac{\theta_1-\theta_2}{2}} & e^{-i\frac{\theta_1-\theta_2}{2}} + e^{i\frac{\theta_1-\theta_2}{2}} \end{pmatrix}$$

$$= e^{i\frac{\theta_2+\theta_1}{2}} \begin{pmatrix} e^{i\frac{\theta}{2}} + e^{-i\frac{\theta}{2}} & -(e^{i\frac{\theta}{2}} - e^{-i\frac{\theta}{2}}) \\ -(e^{i\frac{\theta}{2}} - e^{-i\frac{\theta}{2}}) & e^{i\frac{\theta}{2}} + e^{-i\frac{\theta}{2}} \end{pmatrix} \quad \text{where } \theta = \theta_1 - \theta_2$$

Note $2\cos\frac{\theta}{2} = e^{i\frac{\theta}{2}} + e^{-i\frac{\theta}{2}}$ and $2i\sin\frac{\theta}{2} = e^{i\frac{\theta}{2}} - e^{-i\frac{\theta}{2}}$. Therefore

$$HPH = e^{i\frac{\theta_2+\theta_1}{2}} \begin{pmatrix} 2\cos\frac{\theta}{2} & -2i\sin\frac{\theta}{2} \\ -2i\sin\frac{\theta}{2} & 2\cos\frac{\theta}{2} \end{pmatrix}$$

so that

$$\frac{1}{2}HPH = e^{i\frac{\theta_2+\theta_1}{2}} \begin{pmatrix} \cos\frac{\theta}{2} & -i\sin\frac{\theta}{2} \\ -i\sin\frac{\theta}{2} & \cos\frac{\theta}{2} \end{pmatrix}$$

where $\theta = \theta_1 - \theta_2$, as required.

12. e.g. $\begin{pmatrix} \frac{1}{\sqrt{2}} & \frac{1}{\sqrt{2}} \\ \frac{1}{\sqrt{2}} & -\frac{1}{\sqrt{2}} \end{pmatrix}$ is orthogonal but is not a permutation matrix.

Chapter 6.

6.1 The zero element is the zero polynomial $p(t) = 0$. The inverses are $-p_i(t)$, $i = 1, \ldots, 4$.
6.5 The zero element of $\mathbb{R}^2$ is $(0,0)$. This element is not in U and hence U cannot be a subspace.

6.11 The null space is $\mu \begin{pmatrix} -1 \\ 1 \end{pmatrix}$, $\mu \in \mathbb{R}$.

6.14 Linearly dependent: the third is the sum of the others.
6.15 Linearly independent.
6.18 For example $\frac{1}{\sqrt{2}}(1,1,0)$, $\frac{1}{\sqrt{6}}(-1,1,2)$, $\frac{1}{\sqrt{3}}(1,-1,1)$.

Chapter 6. End-of-chapter exercises.

3. d) $p = 7\bar{e}_0 - 2\bar{e}_1$.

5. $\begin{pmatrix} 8 \\ 11 \\ 1 \\ 5 \end{pmatrix} = 3 \begin{pmatrix} 1 \\ 2 \\ 0 \\ 1 \end{pmatrix} + 2 \begin{pmatrix} 2 \\ 2 \\ 0 \\ 1 \end{pmatrix} + \begin{pmatrix} 1 \\ 1 \\ 1 \\ 0 \end{pmatrix}$ so r is in the space spanned by u, v and w.

8. $(z_1, z_2) = -iz_1(i, 0) - iz_2(0, i)$.
9. $(3 + 4i, -7 - 2i) = 3(1,0) - 7(0,1) + 4(i,0) - 2(0,i)$.

Chapter 7.

7.1 Spectrum $= \{1, 0\}$. The algebraic multiplicity of each eigenvalue is 1.

7.5 $\lambda = \sqrt{13}$, $v = \begin{pmatrix} \frac{1+\sqrt{13}}{4} \\ 1 \end{pmatrix}$; $\lambda = -\sqrt{13}$, $v = \begin{pmatrix} \frac{1-\sqrt{13}}{4} \\ 1 \end{pmatrix}$.

7.6 $\lambda = 3, 4, 2$, $v = \begin{pmatrix} 1 \\ 2 \\ 1 \end{pmatrix}$, $\begin{pmatrix} 0 \\ 1 \\ 1 \end{pmatrix}$, $\begin{pmatrix} -1 \\ -1 \\ 1 \end{pmatrix}$ resp.

Chapter 7. End-of-chapter exercises.

2. $\lambda = 1, -1$, $v = \begin{pmatrix} 1 \\ 0 \end{pmatrix}$, $\begin{pmatrix} 0 \\ 1 \end{pmatrix}$ resp.

5. $\lambda = 1, 0$, $v = \begin{pmatrix} 0 \\ 1 \end{pmatrix}$, $\begin{pmatrix} 1 \\ 0 \end{pmatrix}$ resp.

7. a) 0,0,1,1; $v = \begin{pmatrix} 0 \\ 1 \\ 0 \\ 0 \end{pmatrix}$, $\begin{pmatrix} 1 \\ 0 \\ 0 \\ 0 \end{pmatrix}$, $\begin{pmatrix} 0 \\ 0 \\ 0 \\ 1 \end{pmatrix}$, $\begin{pmatrix} 0 \\ 0 \\ 1 \\ 0 \end{pmatrix}$ resp.

b) 0,0,1,1; $v = \begin{pmatrix} 0 \\ 0 \\ 1 \\ 0 \end{pmatrix}$, $\begin{pmatrix} 1 \\ 0 \\ 0 \\ 0 \end{pmatrix}$, $\begin{pmatrix} 0 \\ 0 \\ 0 \\ 1 \end{pmatrix}$, $\begin{pmatrix} 0 \\ 1 \\ 0 \\ 0 \end{pmatrix}$ resp.

8. a) 0,0,1,1; $v = \begin{pmatrix} -1 \\ 0 \\ 0 \\ 1 \end{pmatrix}$, $\begin{pmatrix} 0 \\ -1 \\ 1 \\ 0 \end{pmatrix}$, $\begin{pmatrix} 1 \\ 0 \\ 0 \\ 1 \end{pmatrix}$, $\begin{pmatrix} 0 \\ 1 \\ 1 \\ 0 \end{pmatrix}$ resp.

b) 0,0,1,1; $v = \begin{pmatrix} 1 \\ 0 \\ 0 \\ 1 \end{pmatrix}$, $\begin{pmatrix} 0 \\ -1 \\ 1 \\ 0 \end{pmatrix}$, $\begin{pmatrix} -1 \\ 0 \\ 0 \\ 1 \end{pmatrix}$, $\begin{pmatrix} 0 \\ 1 \\ 1 \\ 0 \end{pmatrix}$ resp.

c) 1,1,0,0; $v = \begin{pmatrix} 0 \\ 0 \\ 0 \\ 1 \end{pmatrix}, \begin{pmatrix} 1 \\ 0 \\ 0 \\ 0 \end{pmatrix}, \begin{pmatrix} 0 \\ 0 \\ 1 \\ 0 \end{pmatrix}, \begin{pmatrix} 0 \\ 1 \\ 0 \\ 0 \end{pmatrix}$ resp.

9. 0,0,0,1; $v = \begin{pmatrix} 0 \\ 0 \\ 1 \\ 0 \end{pmatrix}, \begin{pmatrix} 0 \\ 1 \\ 0 \\ 0 \end{pmatrix}, \begin{pmatrix} 0 \\ 0 \\ 0 \\ 1 \end{pmatrix}, \begin{pmatrix} 1 \\ 0 \\ 0 \\ 0 \end{pmatrix}$ resp.

10. $\lambda = 1$: $\begin{pmatrix} 1 \\ 0 \end{pmatrix}$. $\lambda = e^{i\phi}$: $\begin{pmatrix} 0 \\ 1 \end{pmatrix}$.

Chapter 8.

8.1 $(\mathbb{R}^+, +)$ does not include the identity element 0.
8.2 $(\mathbb{R}\backslash 0, +)$ does not include the identity element 0.
8.3 The element $0 \in \mathbb{R}$ does not possess a multiplicative inverse.
8.4 3.
8.5 7.
8.6

	I	L_1	L_2	R
I	I	L_1	L_2	R
L_1	L_1	I	R	L_2
L_2	L_2	R	I	L_1
R	R	L_2	L_1	I

8.7 For example, choose $\Phi(00) = I$, $\Phi(01) = L_1$, $\Phi(10) = L_2$, $\Phi(11) = R$.

Chapter 8. End-of-chapter exercises.

5. b) The identity element of the group is the identity matrix $I = \begin{pmatrix} 1 & 0 \\ 0 & 1 \end{pmatrix}$.

c) The group has an infinite number of elements.
9.
a)

+	0	1	2	3	4
0	0	1	2	3	4
1	1	2	3	4	0
2	2	3	4	0	1
3	3	4	0	1	2
4	4	0	1	2	3

d)

·	1	−1	i	−i
1	1	−1	i	−i
−1	−1	1	−i	i
i	i	−i	−1	1
−i	−i	i	1	−1

The group can be generated by both i and −i.

Chapter 9.

9.2 f is not a linear functional.
9.3 f is not a linear functional.

Chapter 9. End-of-chapter exercises.

1. $P = \begin{pmatrix} 1 & 0 & 0 \\ 0 & 1 & 0 \\ 0 & 0 & 0 \end{pmatrix}$.

2. $P = \begin{pmatrix} \frac{1}{2} & \frac{1}{2} \\ \frac{1}{2} & \frac{1}{2} \end{pmatrix}$.

7. a) The kernel contains only the zero vector. The dimension of the kernel is zero.
b) Any three linearly independent vectors, e.g. $(2,1,0)$, $(0,-3,2)$, $(0,0,-4)$.
c) $\dim \ker(f) + \dim \operatorname{Im}(f) = 0 + 3 = 3 = \dim \mathbb{R}^3$.

9. The kernel is spanned by $\begin{pmatrix} 1 \\ 0 \end{pmatrix}$.

Chapter 10.

10.5 a) $\frac{1}{2}(|00\rangle + |10\rangle + |01\rangle + |11\rangle)$.
b) $\frac{1}{2}(|00\rangle - |10\rangle - |01\rangle + |11\rangle)$.
10.6 The computational basis $B_{\otimes^3 \mathbb{C}^2}$ is

$$|000\rangle, |001\rangle, |010\rangle, \ldots, |111\rangle$$

which is equal to

$$\begin{pmatrix} 1 \\ 0 \\ 0 \\ 0 \\ 0 \\ 0 \\ 0 \\ 0 \end{pmatrix}, \begin{pmatrix} 0 \\ 1 \\ 0 \\ 0 \\ 0 \\ 0 \\ 0 \\ 0 \end{pmatrix}, \begin{pmatrix} 0 \\ 0 \\ 1 \\ 0 \\ 0 \\ 0 \\ 0 \\ 0 \end{pmatrix}, \ldots, \begin{pmatrix} 0 \\ 0 \\ 0 \\ 0 \\ 0 \\ 0 \\ 0 \\ 1 \end{pmatrix}.$$

Chapter 10. End-of-chapter exercises.

1. a) dimension: $2 \times 2 = 4$.

$$\begin{pmatrix} 1 \\ 0 \\ 0 \\ 0 \end{pmatrix}, \begin{pmatrix} 0 \\ 1 \\ 0 \\ 0 \end{pmatrix}, \begin{pmatrix} 0 \\ 0 \\ 1 \\ 0 \end{pmatrix}, \begin{pmatrix} 0 \\ 0 \\ 0 \\ 1 \end{pmatrix}.$$

b) dimension: $3 \times 2 = 6$.

$$\begin{pmatrix} 1 \\ 0 \\ 0 \\ 0 \\ 0 \\ 0 \end{pmatrix}, \begin{pmatrix} 0 \\ 1 \\ 0 \\ 0 \\ 0 \\ 0 \end{pmatrix}, \begin{pmatrix} 0 \\ 0 \\ 1 \\ 0 \\ 0 \\ 0 \end{pmatrix}, \begin{pmatrix} 0 \\ 0 \\ 0 \\ 1 \\ 0 \\ 0 \end{pmatrix}, \begin{pmatrix} 0 \\ 0 \\ 0 \\ 0 \\ 1 \\ 0 \end{pmatrix}, \begin{pmatrix} 0 \\ 0 \\ 0 \\ 0 \\ 0 \\ 1 \end{pmatrix}.$$

c) dimension: $2 \times 4 = 8$.

$$
\begin{pmatrix} 1 \\ 0 \\ 0 \\ 0 \\ 0 \\ 0 \\ 0 \\ 0 \end{pmatrix},
\begin{pmatrix} 0 \\ 1 \\ 0 \\ 0 \\ 0 \\ 0 \\ 0 \\ 0 \end{pmatrix},
\begin{pmatrix} 0 \\ 0 \\ 1 \\ 0 \\ 0 \\ 0 \\ 0 \\ 0 \end{pmatrix},
\begin{pmatrix} 0 \\ 0 \\ 0 \\ 1 \\ 0 \\ 0 \\ 0 \\ 0 \end{pmatrix},
\begin{pmatrix} 0 \\ 0 \\ 0 \\ 0 \\ 1 \\ 0 \\ 0 \\ 0 \end{pmatrix},
\begin{pmatrix} 0 \\ 0 \\ 0 \\ 0 \\ 0 \\ 1 \\ 0 \\ 0 \end{pmatrix},
\begin{pmatrix} 0 \\ 0 \\ 0 \\ 0 \\ 0 \\ 0 \\ 1 \\ 0 \end{pmatrix},
\begin{pmatrix} 0 \\ 0 \\ 0 \\ 0 \\ 0 \\ 0 \\ 0 \\ 1 \end{pmatrix}.
$$

2. Not entangled: $(|0\rangle + |1\rangle) \otimes (|0\rangle - |1\rangle)$.

3. 10.

4. $-100 - 60i$.

5. a) $\frac{2}{\sqrt{2}}(|0\rangle \otimes |0\rangle) = \sqrt{2}|00\rangle$.

b) $\frac{2}{\sqrt{2}}(|0\rangle \otimes |1\rangle) = \sqrt{2}|01\rangle$.

c) $\frac{2}{\sqrt{2}}(|1\rangle \otimes |1\rangle) = \sqrt{2}|11\rangle$.

d) $\frac{2}{\sqrt{2}}(|1\rangle \otimes |0\rangle) = \sqrt{2}|10\rangle$

8. $\begin{pmatrix} 0 & 1 & 0 & 0 \\ 1 & 0 & 0 & 0 \\ 0 & 0 & 0 & 1 \\ 0 & 0 & 1 & 0 \end{pmatrix}$.

Chapter 11.

11.7 $(\mathcal{A} + \mathcal{B})(z_1, z_2) = (2z_1, 2z_1)$.

$\quad (k\mathcal{A})(z_1, z_2) = k(z_1 - z_2, z_1 + z_2) = (k(z_1 - z_2), k(z_1 + z_2))$.

11.8 a) $(\mathcal{A}\mathcal{B})(z_1, z_2) = (2z_2, 2z_1)$. b) $(\mathcal{B}\mathcal{A})(z_1, z_2) = (2z_1, -2z_2)$.

11.9 $\begin{pmatrix} 0 & 1 \\ 1 & 0 \end{pmatrix}$; $\begin{pmatrix} 1 & 0 \\ 0 & -1 \end{pmatrix}$.

11.10 a) $(0, \frac{1}{\sqrt{2}})$, b) $(0, -\frac{1}{\sqrt{2}})$, c) $\frac{1}{2}$, d) $-\frac{1}{2}$, e) $-\frac{1}{2}$, f) $\frac{1}{2}$. $\begin{pmatrix} \frac{1}{2} & -\frac{1}{2} \\ -\frac{1}{2} & \frac{1}{2} \end{pmatrix}$.

11.11 b) $\lambda = 1, v = \begin{pmatrix} 1 \\ 0 \end{pmatrix}$; $\lambda = -1, v = \begin{pmatrix} 0 \\ 1 \end{pmatrix}$.

11.12 $\begin{pmatrix} 0 & 1 \\ -1 & 0 \end{pmatrix}$. $\lambda = i, \begin{pmatrix} 1 \\ i \end{pmatrix}$; $\lambda = -i, \begin{pmatrix} 1 \\ -i \end{pmatrix}$.

11.14 $\begin{pmatrix} 0 & -i \\ i & 0 \end{pmatrix}$.

11.16 $(H \otimes H)|01\rangle = \frac{1}{2}\begin{pmatrix} 1 \\ -1 \\ 1 \\ -1 \end{pmatrix}$, $(H \otimes H)|10\rangle = \frac{1}{2}\begin{pmatrix} 1 \\ 1 \\ -1 \\ -1 \end{pmatrix}$, $(H \otimes H)|11\rangle = \frac{1}{2}\begin{pmatrix} 1 \\ -1 \\ -1 \\ 1 \end{pmatrix}$.

11.19 a) $\frac{1}{\sqrt{2}}(|00\rangle + |01\rangle)$, b) $\frac{1}{\sqrt{2}}(|00\rangle - |01\rangle)$, c) $-\frac{1}{\sqrt{2}}(|10\rangle + |11\rangle)$, d) $-\frac{1}{\sqrt{2}}(|10\rangle - |11\rangle)$

Chapter 11. End-of-chapter exercises.

7. $|11\rangle = \begin{pmatrix} 0 \\ 0 \\ 0 \\ 1 \end{pmatrix}$.

9. a) $(\frac{1}{\sqrt{2}}, 0)$, b) $(\frac{1}{\sqrt{2}}, 0)$, c) $\frac{1}{2}$, d) $\frac{1}{2}$, e) $\frac{1}{2}$, f) $\frac{1}{2}$. $\begin{pmatrix} \frac{1}{2} & \frac{1}{2} \\ \frac{1}{2} & \frac{1}{2} \end{pmatrix}$.

10. a) $(\frac{1}{\sqrt{2}}, \frac{1}{\sqrt{2}})$, b) $(-\frac{1}{\sqrt{2}}, \frac{1}{\sqrt{2}})$, c) 1, d) 0, e) 0, f) -1. $\begin{pmatrix} 1 & 0 \\ 0 & -1 \end{pmatrix}$.

Chapter 13.

13.1 As the only eigenvalue of the identity operator is 1 all measurements will produce this value. We also note that all vectors of $\otimes^n \mathbb{C}^2$ are eigenvectors. Considering the case $\mathbb{C}^2 \otimes \mathbb{C}^2$ and choosing any orthonormal basis $|\phi_1\rangle, |\phi_2\rangle, |\phi_3\rangle, |\phi_4\rangle$, we have, for any $|\psi\rangle \in C^2 \otimes \mathbb{C}^2$, that

$$|\psi\rangle = \alpha_1|\phi_1\rangle + \alpha_2|\phi_2\rangle + \alpha_3|\phi_3\rangle + \alpha_4|\phi_4\rangle$$

for some $\alpha_i \in \mathbb{C}$. Measuring I on $|\psi\rangle$ will produce the eigenvalue 1 with probability 1; and the post-measurement state will be $|\psi\rangle$, because each of the states $|\phi_1\rangle, |\phi_2\rangle, |\phi_3\rangle, |\phi_4\rangle$ are eigenvectors of I with eigenvalue 1. The extension of the solution to $\otimes^n \mathbb{C}^2$ is clear.

13.2 Again we consider the solution in $\otimes^2 \mathbb{C}^2$, the extension to $\otimes^n \mathbb{C}^2$ being clear. Assume $|\phi_1\rangle, |\phi_2\rangle, |\phi_3\rangle, |\phi_4\rangle$ are eigenvectors of both A and B and assume the corresponding eigenvalues of A and B are, respectively, $\lambda_1, \ldots, \lambda_4$ and $\mu_1, \ldots, \mu_4$.

Measuring A first we will measure λ_k with probability $|\alpha_k|^2$ with post-measurement state $|\phi_k\rangle$. Measuring B on $|\phi_k\rangle$ produces μ_k with probability 1. The result is the pair of measurements (λ_k, μ_k) occurring with probability $|\alpha_k|^2$.

Measuring B first we will measure μ_q with probability $|\alpha_q|^2$ with post-measurement state $|\phi_q\rangle$. Measuring A on $|\phi_q\rangle$ produces λ_q with probability 1. The result is the pair of measurements (λ_q, μ_q) occurring with probability $|\alpha_q|^2$.

The possible outcomes in each case are therefore $(\lambda_1, \mu_1), \ldots, (\lambda_4, \mu_4)$, with (λ_i, μ_i) occurring with probability $|\alpha_i|^2$.

Chapter 17.

17.3
$$\hat{f}(x_1, x_2, y) = (x_1, x_2, y \oplus f(x_1, x_2)) = (x_1, x_2, y \oplus (x_1 \wedge x_2)).$$

This is the Toffoli gate.

Chapter 18.

18.1 By evaluating $T_D^{\vee} \circ T_D^{\vee}$, it follows that $T_D^{\vee}$ is self inverse.

18.2 A suitable circuit, using the digital $T_D, T_D^{\vee}$ and *swap* gates, is shown in Figure E.1.

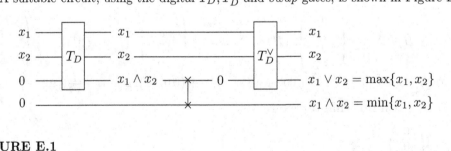

FIGURE E.1
A circuit, using invertible digital gates, to compute $\max\{x_1, x_2\}$ and $\min\{x_1, x_2\}$.

Chapter 21.

21.1 Assuming that the quantum *cnot* gate may be expressed as the tensor product of two 1-qubit gates implies that we have:

$$
\begin{pmatrix} 1 & 0 & 0 & 0 \\ 0 & 1 & 0 & 0 \\ 0 & 0 & 0 & 1 \\ 0 & 0 & 1 & 0 \end{pmatrix} = \begin{pmatrix} a & b \\ c & d \end{pmatrix} \otimes \begin{pmatrix} A & B \\ C & D \end{pmatrix}.
$$

for some a, b, c, d and A, B, C, D. Comparing the upper left 2×2 matrices of each side gives $B = C = 0$, and this is incompatible with the equivalence of the lower 2×2 matrices of each side. Hence, no such representation of the quantum *cnot* gate is possible.

References

The material presented in this book is based largely on the work of others. The authors have benefitted from the following formally published material on mathematics, quantum mechanics and computation.

In addition to the references cited below, the authors have benefitted from lecture notes, and other materials, kindly made available by teachers and researchers in quantum computing including: Coecke (Oxford University), Hannabuss (Oxford University), Jozsa (Cambridge University), Watrous (University of Waterloo), de Wolf (University of Amsterdam), Childs (University of Maryland), Preskill (California institute of Technology) and Rieffel and Polak (MIT).

[1] J.M. Jauch. *Foundations of Quantum Mechanics*. Addison Wesley, 1968.

[2] G.W. Mackey. *Mathematical Foundations of Quantum Mechanics*. Benjamin, 1963.

[3] M.A. Nielsen and I.L. Chuang. *Quantum Computation and Quantum Information, 10th Anniversary Edition*. Cambridge University Press, 2010.

Index

Printed in the United States
by Baker & Taylor Publisher Services